AF556701

ENGINEERING MATHEMATICS

SEMESTER - II

ENGINEERING MATHEMATICS

SEMESTER - II

By

Dr. Anand Swaroop Sharma

DISCOVERY PUBLISHING HOUSE PVT. LTD.
NEW DELHI-110 002

First Published-2009

ISBN 978-81-8356-405-2

Published by:

DISCOVERY PUBLISHING HOUSE PVT. LTD.

4831/24, Ansari Road, Prahlad Street,
Darya Ganj, New Delhi-110002 (India)
Phone: 23279245 • Fax: 91-11-23253475
E-mail: dphbooks@rediffmail.com
dphtemp@indiatimes.com
Website: www.discoverypublishinghouse.com

Printed at:
Sachin Printers, Delhi

Preface

The aim of this book is to present elements of Mathematics as applied to Scientific and Engineering students whose main interest in the subject lie in finding the particular solution so rather than the general theory the book has been designed to source as the textbook of formal courses in Engineering Mathematics of B.Ed. and B.Tech. students of all Indian Universities.

The subject matter has been discussed in a systematic way starting from basic concepts, keeping in mind the actual difficulties of students. Considerable more number of worked examples has been included in the text against each topic in all the chapters to make it more flexible.

I shall be highly obliged and grateful to receive suggestions for improvement of text.

Author

CONTENTS

1

Asymptotes

INTRODUCTION

It is interesting to enquire what happens to the tangent when the point at which the tangent is drawn moves further and further away from the origin. There are three possibilities corresponding to the three possibilities in the lose of a series *viz.* That a series may be divergent oscillatory or convergent. Thus, it is possible that the tangent may go further and further away from the origin or it may keep oscillating or it may the straight line to which the tangent tends is called the asymptote.

Definitions

(a) *An asymptote is a straight line which cuts a curve in two points at an infinite distance from the origin and yet is not itself wholly at infinity.*

(b) *An asymptote is a straight line, at a finite distance from the origin, to which a tangent to a curve tends, as the distance from the origin of the point of contact tends to infinity.*

INTERSECTION OF A CURVE AND ITS ASYMPTOTES

To prove that any asymptote of an algebraic curve of the nth degree cuts the curve in (n – 2) points.

A straight line $y = mx + c$ ...(1)

cuts the curve of the nth degree

$$x^n\phi_n\left(\frac{y}{x}\right) + x^{n-1}\phi_{n-1}\left(\frac{y}{x}\right) + x^{n-2}\phi_{n-2}\left(\frac{y}{x}\right) + ... = 0 \quad ...(2)$$

in n points real or imaginary.

Eliminating y from (1) and (2), we get

$$x^n\phi_n\left(m+\frac{c}{x}\right)+x^{n-1}\phi_{n-1}\left(m+\frac{c}{x}\right)+x^{n-2}\phi_{n-2}\left(m+\frac{c}{x}\right)+...=0.$$

Expanding each term by Taylor's theorem and arranging in descending powers of x, we get

$$x^n\phi_n(m)|\left[c\phi_n'(m)+\phi_{n-1}(m)\right]x^{n-1}$$

$$+\left[\frac{c^2}{2!}\phi_n''(m)+\frac{c}{1!}\phi_{n-1}'(m)+\phi_{n-2}(m)\right]x^{n-2}+...=0. \quad ....(3)$$

The equation (3) gives the abscissae of the points of intersection of (1) and (2).

If y = mx + c is an asymptote of (2), then $\phi_n(m) = 0$

and $$c\phi_n'(m)+\phi_{n-1}(m)=0.$$

Consequently (3) reduces to an equation of (n – 1 degree in x and therefore the asymptote (1) cuts the curve (2) in (n – 2) points.

Cor. *The, n asymptotes of a curve of the nth degree cut it in n(n – 2) points. In general, a curve of the degree n – 2, or less, can be made to pass through these n (n – 2) points.*

CASE OF PARALLEL ASYMPTOTES

(i) Two Parallel Asymptotes: Suppose the equation $\phi_n(m) = 0$ gives us two equal values of m. This repeated value of m makes $\phi_n'(m)=0$. In case it does not make $\phi_{n-1}(m)$ equal to zero, the value of c determined by the equation $c=-\frac{\phi_{n-1}(m)}{\phi_n'(m)}$ comes out to be infinite, and hence the asymptotes corresponding to this value of m do not exist. Thus, for the existence of the asymptotes corresponding to this value of m, it is necessary that it must make $\phi_{n-1}(m)$ equal to zero. But then the equation $c\phi_n'(m)+\phi_{n-1}(m)=0$ from which c is usually determined reduces to the identity

$$0.c + 0 = 0,$$

and we cannot find the value of c in this way. To determine c in this case, we equate to zero the coefficient of x^{n-2} in the equation (3) of 3, and we get the equation

$$\frac{c^2}{2!}\phi_n''(m)+\frac{c}{1!}\phi_{n-1}'(m)+\phi_{n-2}(m)=0.$$

This equation is a quadratic in c. It gives us two values of c, say c_1 and c_2, corresponding to that repeated value of m. The corresponding asymptotes are $y = mx + c_1$ and $y = mx + c_2$, which are obviously parallel.

(ii) Three Parallel Asymptotes: If the equation $\phi_n(m) = 0$ gives us three equal values of m, then this repeated value of m makes $\phi'_n(m)$ and $\phi''_n(m)$ equal to zero. For the existence of the corresponding asymptotes it must $\phi_{n-1}(m)$ equal to zero. If it also makes $\phi'_{n-1}(m)$ and $\phi_{n-2}(m)$ equal to zero, then the equation to determine c reduces to the identity

$$0.c^2 + 0.c + 0 = 0,$$

and we cannot find the values of c in this way. To determine c in this case, we equate to zero the coefficient of x^{n-3} in the equation (3) of 3, and we get the equation

$$\frac{c^3}{3!}\phi'''_n(m)+\frac{c^2}{2!}\phi''_{n-1}(m)+\frac{c}{1!}\phi'_{n-2}(m)+\phi_{n-3}(m)=0.$$

This equation gives us three values of c corresponding to that repeated value of m and accordingly we get three parallel asymptotes. In a similar way we can discuss the case of more than three parallel asymptotes.

Example 1:

Find the asymptotes of $x^3 + 3x^2y - 4y^3 - x + y + 3 = 0$.

Solution:

Putting y = m and x = 1 in the third degree and second degree terms separately, we get

$$\phi_3(m) = 1 + 3m - 4m^3 \text{ and } \phi_2(m) = 0.$$

(Not that there are no second degree terms).

$$\therefore \phi'_3(m) = 3 - 12m^2.$$

Now the slopes of the asymptotes are given by the equation

$$\phi_3(m) = 0 \textit{ i.e.}, 1 + 3m - 4m^3 = 0 \textit{ i.e.}, (1 - m)(1 + 2m)^2 = 0.$$

$$\therefore m = 1, -\frac{1}{2}, -\frac{1}{2}.$$

Again, c, is given by the equation

$$c = -\frac{\phi_2(m)}{\phi'_3(m)} = -\frac{0}{3-12m^2}. \quad ...(1)$$

When m = 1, we have c = 0 and the corresponding asymptote is
$y = x + 0$ *i.e.*, $y = x$.

When $m = -\frac{1}{2}$, the equation (1) fails to give c. In this case c is to be determined from the equation

$$\frac{c^2}{2!}\phi_3''(m) + \frac{c}{1!}\phi_2'(m) + \phi_1(m) = 0.$$

Putting y = m and x = 1 in the first degree terms of the equation of the curve, we get $\phi_1(m) = -1 + m$. Also $\phi_3''(m) = -24m$ and $\phi_2'(m) = 0$. Hence for $m = -\frac{1}{2}$, c is to be given by

$$\frac{1}{2}c^2(-24m) + 0 + m - 1 = 0.$$

Putting $m = -\frac{1}{2}$ in this equation, we get

$$6c^2 - \frac{3}{2} = 0 \text{ or } c^2 = \frac{1}{4} \text{ or } c = \pm\frac{1}{2}.$$

Hence $y = -\frac{1}{2}x + \frac{1}{2}$ and $y = -\frac{1}{2}x - \frac{1}{2}$ are two parallel asymptotes corresponding to the slope $m = -\frac{1}{2}$.

Therefore the required asymptotes are
$y = x$ and $x + 2y = \pm 1$.

DETERMINATION OF ASYMPTOTES

If y + mx + c is an asymptote to the curve y = f(x), to show that

$$m = \lim_{x\to\infty}\left(\frac{y}{x}\right).$$

Let y = mx + c be an asymptote of the curve y = f(x) or f(x, y) = 0 so that m and c are both finite.

The equation of the tangent at P(x, y) to the curve y = f(x) is

$$Y - y = \left(\frac{dy}{dx}\right)(X - x)$$

or
$$Y = \left(\frac{dy}{dx}\right)X + \left[y - x\left(\frac{dy}{dx}\right)\right]. \quad ...(1)$$

From (1), $x \to \infty$, $\frac{dy}{dx}$ and $y - x\frac{dy}{dx}$ must both tend to finite limits, say m and c, in order that an asymptote might exist. Thus the tangent (1) tends to the asymptote y = mx + c if

$$\lim_{x\to\infty}\frac{dy}{dx} = m \text{ and } \lim_{x\to\infty}\left(y - x\frac{dy}{dx}\right) = c.$$

Then we have $\lim_{x\to\infty}\frac{y - x\left(\frac{dy}{dx}\right)}{x} = 0$, [∵ c is finite]

or $\lim_{x\to\infty}\left(\frac{y}{x} - \frac{dy}{dx}\right) = 0$ or $\lim_{x\to\infty}\frac{y}{x} = \lim_{x\to\infty}\frac{dy}{dx} = m$.

Also $c = \lim_{x\to\infty}\left(y - x\frac{dy}{dx}\right) = \lim_{x\to\infty}(y - mx)$, $\left[\because \lim_{x\to\infty}\frac{dy}{dx} = m\right]$

Therefore, if y = mx + c is an asymptote to the curve y = f(x), then

$$m = \lim_{x\to\infty}\frac{y}{x} \text{ and } c = \lim_{x\to\infty}(y - mx).$$

THE ASYMPTOTES OF THE GENERAL RATIONAL ALGEBRAIC CURVE

Let f(x, y) = 0 be the equation of any rational algebraic curve of the nth degree.

Let the equation of the curve on being arranged in groups of homogeneous terms, be

$$a_0y^n + a_1y^{n-1} + a_2y^{n-2}x^2 + \ldots + a_{n-1}yx^{n-1} + a_nx^n$$
$$+ b_1y^{n-1} + b_2y^{n-2}x + \ldots + b_{n-1}yx^{n-2} + b_nx^{n-1}$$
$$+ c_2y^{n-2} + \ldots + \ldots = 0. \quad \ldots(1)$$

This equation can also be written as

$$x^n\phi_n\left(\frac{y}{x}\right) + x^{n-1}\phi_{n-1}\left(\frac{y}{x}\right) + \ldots = 0, \quad \ldots(2)$$

where $\phi_r\left(\frac{y}{x}\right)$ is a polynomial in $\frac{y}{x}$ of degree r.

Let y = mx + c be a line which is not parallel to the y-axis. It meets (2) in points whose abscissae are given by

$$x_n\phi_n\left(\frac{mx+c}{x}\right) + x^{n-1}\phi_{n-1}\left(\frac{mx+c}{x}\right) + \ldots = 0$$

$$\text{or } x_n\phi_n\left(m+\frac{c}{x}\right)+x^{n-1}\phi_{n-1}\left(m+\frac{c}{x}\right)+...=0.$$

Expanding each term of the type $\phi_r\left(m+\frac{c}{x}\right)$, by Taylor's theorem, we get

$$x^n\left[\phi_n(m)+\left(\frac{c}{x}\right)\phi_n'(m)+\left(\frac{c^2}{2x^2}\right)\phi_n''(m)+...\right]$$

$$+x^{n-1}\left[\phi_{n-1}(m)+\left(\frac{c}{x}\right)\phi_{n-1}(m)+...\right]$$

$$+x^{n-2}\left[\phi_{n-2}(m)+\left(\frac{c}{x}\right)\phi_{n-2}'(m)+...\right]+...=0.$$

Arranging the terms according to the descending powers of x, we have

$$x^n\phi_n(m) + x^{n-1}[\phi_{n-1}(m) + c\phi_n'(m)]$$

$$+ x^{n-2}\left[\phi_{n-2}(m)+c\phi_{n-1}'(m)+\frac{1}{2}c^2\phi_n''(m)\right] + ... = 0. \quad ...(3)$$

If $y = mx + c$ is an asymptote, this equation must have two infinite roots and consequently

$$\phi_n(m) = 0, \quad ...(4)$$

and $$\phi_{n-1}(m) + c\phi_n'(m) = 0, \quad ...(5)$$

as can be seen on dividing (3) by x^n.

Solving equations (4) and (5) simultaneously, we get the values for m and c, and hence the asymptotes are determined by substituting the corresponding values of m and c in the equation $y = mx + c$.

Note: All the imaginary values of m will be rejected.

ASYMPTOTES WORKING RULE FOR FINDING

(i) Substitute $mx + c$ for y in the equation of the curve and arrange it in descending powers of x.

(ii) By equating the coefficients of the two highest powers of x to zero find m and c.

(iii) Substitute these values of m and c in $y = mx + c$.

(iv) If a value of m, makes the coefficient of x^{n-1} identically zero, find c from the equation obtained by equating to zero the coefficient of x^{n-2}; and so on.

A Short Cut Method

(i) Put x = 1 any y = m in the highest *i.e.* nth degree terms and get $\phi_n(m)$. Put $\phi_n(m) = 0$ and solve the resulting equation for m. Let

$$m = m_1 m_2, \ldots, m_n \text{ be its roots.}$$

(ii) Form $\phi_{n-1}(m)$ by putting x = 1 and y = m in the terms of degree n – 1. Then the values of c, say $c_1, c_2, \ldots, c_n$ are found by substituting $m = m_1, m_2 \ldots, m_n$, in turn, in the formula

$$c = -\frac{\phi_{n-1}(m)}{\phi_n'(m)}.$$

The asymptotes then are

$$y = m_1 x + c_1;\ y = m_2 x + c_2;\ \ldots$$

CIRCULAR ASYMPTOTES

Definition: *Let the equation of a curve be* $r = f(\theta)$.

If $\lim_{\theta \to \infty} f(\theta) = l$, *then the circle* $r = l$ *is called the circular asymptote of the curve* $r = f(\theta)$.

Example:

Find the circular asymptote of the curve

$$r = a.\frac{\theta}{\theta - 1}.$$

Solution:

The circular asymptote is given by

$$r = a \lim_{\theta \to \infty} \frac{\theta}{\theta - 1} = a.$$ Thus, r = a is the circular asymptote.

ASYMPTOTES BY EXPANSION

To show that $y = mx + c$, *is an asymptote of the curve*

$$y = mx + c + \left(\frac{A}{x}\right) + \left(\frac{B}{x^2}\right) + \left(\frac{C}{x^3}\right) + \ldots, \qquad \ldots(1)$$

where the series $\left(\frac{A}{x}\right) + \left(\frac{B}{x^2}\right) + \left(\frac{C}{x^3}\right) + \cdots$ *is convergent for sufficiently large values of x.*

Differentiating (1), we get

$$\frac{dy}{dx} = m - \frac{A}{x^2} - \frac{2B}{x^3} - \ldots$$

$\therefore$ the equation of the tangent to (1) at (x, y) is

$$Y - y = \left(m - \frac{A}{x^2} - \frac{2B}{x^3} - \ldots\right)(X - x)$$

$$\text{or } Y = \left(m - \frac{A}{x^2} - \frac{2B}{x^3} - \ldots\right)X + y - \left(m - \frac{A}{x^2} - \frac{2B}{x^3} - \ldots\right)x$$

$$\text{or } Y = \left(m - \frac{A}{x^2} - \frac{2B}{x^3} - \ldots\right)X + c + \frac{2A}{x} + \frac{3B}{x^2} + \ldots \qquad \ldots(2)$$

substituting the value of y from (1).

Now when $x \to \infty$, equation (2) tends to the equation $Y = mX + c$.

Hence $y = mx + c$ is the asymptote of the curve

$$y = mx + c + \frac{A}{x} + \frac{B}{x^2} + \frac{C}{x^3} + \ldots$$

ALTERNATIVE METHODS OF FINDING ASYMPTOTES OF ALGEBRAIC CURVES

The methods of finding asymptotes of algebraic curves already discussed are sufficient to determine the asymptotes. But the following methods of finding asymptotes are sometimes found to be more convenient than the previous methods.

Let the equation of the curve be of degree n in x and y. Obviously the asymptotes of the curve are parallel to the lines obtained by equating to zero the linear factors of the highest degree terms in its equation.

Case I: *Let $(y - m_1x)$ be a non-repeated linear factor of the nth degree terms in the given equation of the curve.*

Then the equation of the curve can be put into the form

$$(y - m_1x)\, F_{n-1} + P_{n-1} = 0, \qquad \ldots(1)$$

where F_{n-1} contains only terms of degree $(n - 1)$, and P_{n-1} contains terms of various degrees, not higher than $(n - 1)$. Obviously, one of the values of m (slope of an asymptote) in this case is m_1 and so there is an asymptote $y = m_1x = c_1$, provided we can find a finite value of c_1.

Also $c_1 = \lim\limits_{x\to\infty,\, y/x\to m_1} (y - m_1x)$, where (x, y) lies on (1).

But when (x, y) lies on (1),

$$y - m_1x = -\frac{P_{n-1}}{F_{n-1}}. \quad \therefore\ c_1 = \lim_{x\to\infty,\, y/x\to m_1}\left(-\frac{P_{n-1}}{F_{n-1}}\right).$$

Hence $y - m_1 x + \lim_{x \to \infty,\, y/x \to m_1} \left(-\frac{P_{n-1}}{F_{n-1}} \right) = 0$, is an asymptote of the curve (1).

Thus, dividing (1) by F_{n-1} and taking limit as $x \to \infty$, $\frac{y}{x} \to m_1$, we get an asymptote of (1). Similarly, we can find asymptotes corresponding to the other non-repeated linear factors.

Case II: *Let $(y - m_1x)^2$ be a factor of the nth degree terms of the equation of the curve and $(y - m_1x)$ be not a factor of the $(n - 1)^{th}$ degree terms.*

Then, proceeding as in Case I, we find that

$$\lim_{x \to \infty,\, y/x \to m_1} (y - m_1 x)^2 \text{ is either } +\infty \text{ or } -\infty.$$

Hence there is no asymptote corresponding to the factor $(y - m_1x)^2$ in this case.

Case III: *Let the equation to the curve be of the form $(y - m_1x)^2 F_{n-2} + (y - m_1x) G_{n-2} + P_{n-2} = 0$, where F_{n-2} and G_{n-2} contain only terms of $(n - 2)^{th}$ degree and P_{n-2} contains terms none of which is of a degree higher than $(n - 2)$.*

Dividing the equation of the curve by F_{n-2}, we have

$$(y - m_1x)^2 + (y - m_1x).\frac{G_{n-2}}{F_{n-2}} + \frac{P_{n-2}}{F_{n-2}} = 0.$$

Taking limits as $x \to \infty$ and $\frac{y}{x} \to m_1$, we get an equation of the form $(y - m_1x)^2 + \lambda(y - m_1x) + \mu = 0$, which on being solved for $(y - m_1x)$ gives us two parallel asymptotes of the form $y = m_1x + c_1$ and $y = m_1x + c_2$.

Case IV: *Let the nth degree terms contain $(y - m_1x)^3$ or a higher power of $(y - m_1x)$, as a factor.*

In this case we can proceed exactly in the same way as in case III, to find the asymptotes.

Remark: If the equation to the curve is of the form,

$$(ax + by + c)\, F_{n-1} + P_{n-1} = 0,$$

where F_{n-1} and P_{n-1} contain terms none of which is of degree higher than $n - 1$, and F_{n-1} contains at least one term of degree $n - 1$, a little consideration will show that the asymptote corresponding to the factor $ax + by + c$ is

$$(ax + by + c) + \lim_{x \to \infty,\, y/x \to -a/b} \left(\frac{P_{n-1}}{F_{n-1}} \right) = 0.$$

A similar modification can be made in cases III and IV.

Working Rule: In order to find the asymptotes (nor parallel to y-axis) of any given curve.

1. Factorise the highest degree terms or (the terms including all the highest degree terms) into real linear factors.
2. Proceed as in Case I or Cases II, III and IV according as a factor $(y - m_1x)$ or $(ax + by + c)$ is a non-repeated one or a repeated one, and get asymptotes.

NON-EXISTENCE OF ASYMPTOTES

If $\phi_n(m) = 0$, gives one or more values of m, such that they make $\phi'_n(m) = 0$, whereas $\phi_{n-1}(m) \neq 0$, then from the equation $c = -\dfrac{\phi_{n-1}(m)}{\phi'_n(m)}$, we get $c = +\infty$ or $-\infty$ and this corresponds to the case when the tangent goes farther and farther away from the origin as $x \to \infty$. Therefore, for such values of m, we shall get no asymptotes.

Example 1:

Find the asymptotes of the curve $y^3 = x^2 + 3x$.

Solution:

Putting $x = 1$, $y = m$ in the third degree and second degree terms separately, we have

$$\phi_3(m) = m^3 \text{ and } \phi_2(m) = -1.$$

$$\therefore \phi'_3(m) = 3m^2.$$

Now $\phi_3(m) = 0$ *i.e.*, $m^3 = 0$ gives $m = 0, 0, 0$.

Also $c = -\dfrac{\phi_2(m)}{\phi'_3(m)} = -\dfrac{-1}{3m^2}$ which is ∞ for $m = 0$.

Hence $y^3 = x^2 + 3x$ has no asymptotes.

Example 2:

Find the asymptotes of $y^3 + x^2y + 2xy^2 - y + 1 = 0$.

Solution:

Putting $y = m$ and $x = 1$ in the third degree and second degree terms separately, we get

$$\phi_3(m) = m^3 + m + 2m^2, \text{ and } \phi_2(m) = 0.$$

$\therefore \qquad \phi_3'(m) = 3m^2 + 4m + 1.$

Now the slopes of the asymptotes are given by the equation $\phi_3(m) = 0$

i.e., $\qquad m^3 + 2m^2 + m = 0$ *i.e.,* $m(m^2 + 2m + 1) = 0$

i.e., $\qquad m(m + 1)^2 = 0.$

$\therefore \qquad m = 0, -1, -1.$

Again, c, is given by the equation

$$c = -\frac{\phi_2(m)}{\phi_3'(m)} = -\frac{0}{3m^2 + 4m + 1} \qquad \text{...(1)}$$

When $m = 0$, we have $c = -\frac{0}{1} = 0$, and the corresponding asymptote is $y = 0.x + 0$ *i.e.,* $y = 0$.

When $m = -1$, the equation (1) gives $c = -\frac{0}{0}$ which is indeterminate form. So the equation (1) fails to give c when $m = -1$. In this case c is to be determined from the equation

$$\frac{c^2}{2!}\phi_3''(m) = \frac{c}{1!}\phi_2'(m) + \phi_1(m) = 0.$$

Putting $y = m$ and $x = 1$ in the first degree terms of the equation of the curve, we get $\phi_1(m) = -m$. Also $\phi_3''(m) = 6m + 4$, $\phi_2'(m) = 0$.

Therefore for $m = -1$, c is given by the equation

$$\frac{1}{2}c^2(6m + 4) + c.0 - m = 0 \text{ i.e., } (3m + 2)c^2 - m = 0.$$

Putting $m = -1$ in this equation, we get

$$-c^2 + 1 = 0 \text{ or } c^2 = 1 \text{ or } c = \pm 1.$$

Hence $y = -x + 1$ and $y = -x - 1$ are two parallel asymptotes corresponding to the slope $m = -1$.

$\therefore$ all the three asymptotes of the curve are

$$y = 0,\ y + x - 1 = 0 \text{ and } y + x + 1 = 0.$$

Example 3:

Find all the asymptotes of the curve

$$(x + y)^2 (x + 2y + 2) = x + 9y + 2.$$

Solution:

The equation of the given curve can be written as

$$(x + y)^2 (x + 2y) + 2(x + y)^2 - (x + 9y) - 2 = 0.$$

Putting y = m and x = 1 in the third degree and second degree terms separately, we have

$$\phi_3(m) = (1 + m)^2 (1 + 2m) \text{ and } \phi_2(m) = 2(1 + m)^2.$$

$$\therefore \phi_3'(m) = 2(1 + m)(1 + 2m) + 2(1 + m)^2.$$

The slopes of the asymptotes are given by the equation

$\phi_3(m) = 0$ *i.e.*, $(1 + m)^2 (1 + 2m) = 0$.

The determine c, we have the equation

$$c = -\frac{\phi_2(m)}{\phi_3'(m)} = -\frac{2(1+m)^2}{2(1+m)(1+2m)+2(1+m)^2} \quad ...(1)$$

When $m = -\frac{1}{2}$, we have

$$c = -\frac{2\left(1-\frac{1}{2}\right)^2}{2\left(1-\frac{1}{2}\right)(1-1)+2\left(1-\frac{1}{2}\right)^2} = -1,$$

and the corresponding asymptote is $y = -\frac{1}{2}x - 1$ *i.e.*, $2y + x + 2 = 0$.

When m = – 1, the equation (1) fails to give c, because then the value of c takes the indeterminate form $\frac{0}{0}$. In this case c is to be determined from the equation

$$\frac{c^2}{2!}\phi_3''(m) + \frac{c}{1!}\phi_2'(m) + \phi_1(m) = 0.$$

Now $\phi_3''(m) = 2(1 + 2m) + 4(1 + m) + 4(1 + m)$,

$\phi_2'(m) = 4(1 + m)$. Also putting y = m and x = 1 in the first degree terms of the equation of the curve, we get $\phi_1(m) = -(1 - 9m)$. Hence for m = – 1, c is given by

$$\frac{1}{2}c^2\{2(1 + 2m) + 8(1 + m)\} + c\{4(1 + m)\} - (1 + 9m) = 0.$$

Putting m = – 1 in this equation, we get

$$-c^2 + 8 = 0 \text{ or } c = \pm 2\sqrt{2}.$$

Hence $y = -x + 2\sqrt{2}$ and $y = -x - 2\sqrt{2}$ are two parallel asymptotes corresponding to the slope m = – 1.

$\therefore$ the required asymptotes are

$$2y + x + 2 = 0 \text{ and } y + x = \pm\, 2\sqrt{2}.$$

ASYMPTOTES BY INSPECTION

If the equation of a curve is of the form $F_n + F_{n-2} = 0$, where F_n is of degree n (i.e., contains terms of degree n and may also contain terms of lower degree) and F_{n-2} is of degree (n – 2) at the most, and if $F_n = 0$ can be broken up into n linear factors so as to represent n straight lines no two of which are parallel or coincident, then $F_n = 0$ gives all the asymptotes of the curve.

Example 1:

Find the asymptotes of the curve

$$xy(x^2 - y^2)\,(x^2 - 4y^2) + 3xy(x^2 - y^2) + x^2 + y^2 - 7 = 0.$$

Solution:

The given equation can be put in the form

$$[xy(x^2 - y^2)\,(x^2 - 4y^2)] + [3xy\,(x^2 - y^2) + x^2 + y^2 - 7] = 0.$$

This equation is of the form $F_n + F_{n-2} = 0$, where F_n can be broken up into n linear factors so as to represent n straight lines no two of which are parallel or coincident.

$\therefore$ all the asymptotes are given by $F_n = 0$

or $\qquad xy(x - y)\,(x + y)\,(x - 2y)\,(x + 2y) = 0.$

Then by inspection, the asymptotes are

$$x = 0,\; y = 0,\; x - y = 0,\; x + y = 0,\; x - 2y = 0$$

and $\qquad x + 2y = 0.$

ASYMPTOTES PARALLEL TO THE CO-ORDINATE AXES

(i) **Asymptotes Parallel to y-axis of a Rational Algebraic Curve:** Asymptotes parallel to y-axis cannot be determined by the method given in 3 because the equation to a straight line parallel to y-axis cannot be put in the form y = mx + c. Here we shall discuss the method of finding asymptotes parallel to y-axis.

Let the equation of the curve, when arranged in descending powers of y, be

$$y^m\phi(x) + y^{m-1}\phi_1(x) + y^{m-2}\phi_2(x) + \ldots = 0, \qquad \ldots(1)$$

where $\phi(x)$, $\phi_1(x)$, $\phi_2(x)$ etc., are polynomials in x.

Dividing the equation (1) by y^m, we get

$$\phi(y)+\frac{1}{y}\phi_1(x)+\frac{1}{y^2}\phi_2(x)+...=0 \qquad ...(2)$$

If x = k is an asymptote parallel to y-axis of (1), then obviously $k = \lim_{y \to \infty} x$, where (x, y) lies on the curve (1). Therefore taking limit of (2) as $x \to \infty$ and remembering that $x \to k$ as $y \to \infty$, we get $\phi(k) = 0$.

Therefore, k is a root of the equation $\phi(x) = 0$.

If k_1, k_2 etc., be the roots of $\phi(x) = 0$, then the asymptotes of (1) parallel to y-axis are $x = k_1$, $x = k_2$, etc.

From algebra, we know that if k_1 is a root of $\phi(x) = 0$, then $x - k_1$ must be a factor of $\phi(x)$. Also $\phi(x)$ is the coefficient of the highest power of y *i.e.*, y^m in the equation of the curve. Hence we have the following simple rule:

The *asymptotes parallel to the axis of y are obtained by equating to zero the coefficient of the highest power of y in the equation of the curve.* In case the coefficient of highest power of y, is a constant or if its linear factors are all imaginary, there will be no asymptotes parallel to y-axis.

(ii) Asymptotes Parallel to x-axis: Proceeding as above, we have the following rule for finding the asymptotes parallel to x-axis of a rational algebraic curve:

The asymptotes parallel to the axis of x are obtained be equating to zero the coefficient of the highest power of x, in the equation of the curve. In case the coefficient of the highest power of x, is a constant or its linear factors are all imaginary, there will be no asymptotes parallel to x-axis.

TOTAL NUMBER OF ASYMPTOTES OF A CURVE

The number of asymptotes, real or imaginary, of an algebraic curve of the nth degree cannot exceed n.

The slopes of the asymptotes which are not parallel to y-axis are given as the roots of the equation $\phi_n(m) = 0$ which is of degree n at the most. If the equation of the curve possesses some asymptotes parallel to y-axis, then we can easily see that the degree of $\phi_n(m) = 0$ will be smaller than n by atleast the same number. In general, one value of m gives only one value of c. In case the equation for determining c is a quadratic, the equation $\phi_n(m) = 0$ has two equal roots. Similarly if the equation for determining c is a cubic, the equation $\phi_n(m) = 0$ has three equal roots.

Hence *a curve of degree n cannot have more than n asymptotes.* But the number of real asymptotes can be less than n. Some roots of the equation $\phi_n(m) = 0$ may come out to be imaginary or even corresponding to a real value of m the value of c may come out to be infinite.

COMPLETE WORKING RULE FOR FINDING THE ASYMPTOTES OF RATIONAL ALGEBRAIC CURVES

(i) A curve of degree n cannot have more than n asymptotes real or imaginary.

(ii) Equating to zero the coefficient of the highest power of y in the equation of the curve, we get asymptotes parallel to y-axis. Similarly equation to zero the coefficient of the highest power of x in the equation of the curve, we get asymptotes parallel to x-axis.

If $y = mx = c$ is an asymptote not parallel to y-axis, then the values of m and c are found as follows:

(iii) Putting $y = m$ and $x = 1$ in the highest *i.e.*, nth degree terms in the equation of the curve, we get $\phi_n(m)$. Solving the equation $\phi_n(m) = 0$, we get the slopes of the asymptotes. If some values of m are imaginary, we reject them.

(iv) Corresponding to a value of m, the value of c is given by the equation

$$c\phi'_n(m) + \phi_{n-1}(m) = 0,$$

where $\phi_{m-1}(m)$ is obtained by putting $y = m$ and $x = 1$ in the $(n - 1)$th degree terms in the equation of the curve. The asymptotes corresponding to $m = 0$ are already found in (ii). So we need not find the value of c corresponding to $m = 0$.

(v) If corresponding to two equal values of m, the equation for determining c, given in (iv) reduces to the identity $0.c + 0 = 0$, then the values of c are given by

$$\frac{c^2}{2!}\phi''_n(m) + \frac{c}{1!}\phi'_{n-1}(m) + \phi_{n-2}(m) = 0.$$

(vi) Similarly, if three values of m are equal and the equation for determining c, given in (v), reduces to the identity

$$0.c^2 + 0.c + 0 = 0,$$

then the corresponding values of c are given by

$$\frac{c^3}{3!}\phi'''_n(m) + \frac{c^2}{2!}\phi''_{n-1}(m) + \frac{c}{1!}\phi'_{n-2}(m) + \phi_{n-3}(m) = 0.$$

SOLVED EXAMPLES

Example 1:

Show that the asymptotes of the curve

$$x^2y^2 - a^2(x^2 + y^2) - a^3(x + y) + a^4 = 0$$

form a square through two of whose angular points the curve passes.

Solution:

Since the curve is of degree 4, therefore it cannot have more than four asymptotes.

Equating to zero the coefficient of the highest power of x (*i.e.*, of x^2), we get the asymptotes parallel to x-axis as $y^2 - a^2 = 0$ *i.e.*, $x = \pm a$.

Also equating to zero the coefficient of the highest power of x (*i.e.*, of x^2), we get the asymptotes parallel to x-axis as $y^2 - a^2 = 0$ *i.e.*, $y = \pm a$.

Thus all the four asymptotes of the curve have been found and they are $x = \pm a$, $y = \pm a$. These lines form a square whose sides are parallel to the axes each of length 2a.

The angular points of the square are (a, a), (a, – a), (– a, a) and (– a, – a). It is evident that the curve passes through two angular points (a, – a) and (– a, a).

Example 2:

Find all the asymptotes of the curve

$$x^3 - x^2y - xy^2 + y^3 + 2x^2 - 4y^2 + 2xy + x + y + 1 = 0.$$

Solution:

Putting y = m and x = 1 in the third and second degree terms separately, we get

$$\phi_3(m) = 1 - m - m^2 + m^3 \text{ and } \phi_2(m) = 2 - 4m^2 + 2m.$$

$$\therefore \phi_3'(m) = -1 - 2m + 3m^2.$$

Now the slopes of the asymptotes are given by the equation

$$\phi_3(m) = 0 \text{ } i.e., \text{ } 1 - m - m^2 + m^3 = 0$$

i.e., $(1 - m)^2 (1 + m) = 0.$

$\therefore m = 1, 1, -1.$

Again c is given by the equation

$$c = -\frac{\phi_2(m)}{\phi_3'(m)} = -\frac{2-4m^2+2m}{-1-2m+3m^2} = \frac{2-4m^2+2m}{1+2m-3m^2}. \quad ...(1)$$

When m = – 1, we have from (1), c = 1. The corresponding asymptotes is y = – x *i.e.*, x + y = 1.

When m = 1, the equation (1) gives $c = \frac{0}{0}$ and thus the equation (1) fails to give c for m = 1.

In this case c is to be determined from the equation

$$\left(\frac{c^2}{2!}\right)\phi_3''(m) = \left(\frac{c}{1!}\right)\phi_2'(m) + \phi_1(m) = 0.$$

Putting y = m and x = 1 in the first degree terms of the equation of the curve, we get

$\phi_1(m) = 1 + m$. Also $\phi_3''(m) = -2 + 6m$, and $\phi_2'(m) = -8m + 2$.

Hence for m = 1, c is to be given by

$$\frac{1}{2}c^2(-2 + 6m) + c(-8m + 2) + 1 + m = 0.$$

Putting m = 1 in this equation, we get

$$2c^2 - 6c + 2 = 0 \text{ i.e., } c^2 - 3c + 1 = 0.$$

$$\therefore\ c = \frac{3 \pm \sqrt{(9-4)}}{2} = \frac{3 \pm \sqrt{5}}{2}.$$

Hence the two parallel asymptotes corresponding to the slope m = 1 are

$$y = x + \frac{1}{2}\left(3 \pm \sqrt{5}\right).$$

∴ the required asymptotes are

$$x + y = 1,\ 2x - 2y + 3 \pm \sqrt{5} = 0.$$

Example 3:

Find the asymptotes of the curve

$$y^2(x-2) = x^2(y-1).$$

Solution:

The given curve is

$$y^2(x-2) - x^2(y-1) = 0$$

or $\quad y^2x - x^2y - 2y^2 + x^2 = 0.$ ...(1)

Since the curve (1) is of degree 3, it cannot have more than three asymptotes.

Equating to zero the coefficient of the highest power of y(*i.e.*, of y^2) the asymptote parallel to y-axis is given by

$x - 2 = 0$ *i.e.*, $x = 2$.

Again equating to zero the coefficient of the highest power of x(*i.e.*, of x^2) the asymptote parallel to x-axis is given by

$y - 1 = 0$ *i.e.*, $y = 1$.

To find the remaining oblique asymptotes, we put y = m and x = 1 in the highest *i.e.*, third degree terms in the equation (1) and we get

$$\phi_3(m) = m^2 - m.$$

The slopes of the asymptotes are given by the equation

$$\phi_3(m) = 0 \text{ i.e., } m^2 - m = 0 \text{ i.e., } m(m-1) = 0.$$

$\therefore$ $m = 0, 1$.

Again putting y = m and x = 1 in the second degree terms, we get

$$\phi_2(m) = -2m^2 + 1.$$

Now c is given by the equation

$$c\phi_3'(m) + \phi_2(m) = 0$$

i.e., $c(2m - 1) - 2m^2 + 1 = 0$.

The asymptote corresponding to m = 0 is parallel to x-axis and has already been found. So no need of finding c for m = 0.

When m = 1, we have $c - 1 = 0$ *i.e.*, $c = 1$ and so the corresponding asymptote is $y = 1x + 1$ *i.e.*, $y = x + 1$. Hence all the three asymptotes of the given curve are $x = 2$, $y = 1$ and $y = x + 1$.

Example 4:

Find the asymptotes of the curve

$$\left(\frac{a^2}{x^2}\right) - \left(\frac{b^2}{y^2}\right) = 1.$$

Solution:

The equation of the given curve can be written as

$$a^2y^2 - b^2x^2 = x^2y^2 \text{ or } x^2y^2 - a^2y^2 + b^2x^2 = 0.$$

Since the curve is of degree 4, therefore it cannot have more than four asymptotes.

Equating to zero the coefficient of the highest power of y(*i.e.*, of y^2) the asymptotes parallel to y-axis are given by

$x^2 - a^2 = 0$ *i.e.*, $x = \pm a$.

Also equating to zero the coefficient of the highest power of x (*i.e.*, of x^2) the asymptotes parallel to x-axis are given by

$$y^2 + b^2 = 0,$$

which gives two imaginary asymptotes.

Thus, all the four possible asymptotes of the curve have been found and the only real asymptotes are $x = \pm a$.

Example 5:

Find the asymptotes of the curve

$$x^3 + x^2y - xy^2 - y^3 - 3x - y - 1 = 0.$$

Solution:

The given curve is of degree 3, so it cannot have more than three asymptotes.

There are no asymptotes parallel to y-axis because the coefficient of the highest power of y *i.e.*, of y^3 is merely a constant. Similarly, there are no asymptotes parallel to x-axis.

Now putting $y = m$ and $x = 1$ in the highest *i.e.*, third degree terms in the equation of the curve, we get

$$\phi_3(m) = 1 + m - m^2 - m^3.$$

The slopes of the asymptotes are given by the equation

$$\phi_3(m) = 0 \text{ i.e., } 1 + m - m^2 - m^3 = 0$$

or $\quad (1 + m) - m^2(1 + m) = 0$

or $\quad (1 + m)(1 - m^2) = 0$

or $\quad (1 + m)^2 (1 - m) = 0.$

$\therefore \quad m = 1, -1, -1.$

Again putting $y = m$ and $x = 1$ in the second degree terms in the equation of the curve, we get

$\phi_2(m) = 0$, since there are no second degree terms.

To determine c we have the equation

$$c\phi_3'(m) + \phi_2(m) = 0$$

i.e., $(1 - 2m - 3m^2) + 0 = 0.$ ...(1)

For $m = 1$, the equation (1) gives $-4c = 0$ or $c = 0$ and the corresponding asymptote is $y = 1.x + 0$ *i.e.*, $y = x$.

For $m = -1$, the equation (1) reduces to the identity $c. 0 = 0$ and thus it fails to give c. In this case c is to be determined by the equation

$$\frac{c^2}{2!}\phi_3''(m) + \frac{c}{1!}\phi_2'(m) + \phi_1(m) = 0.$$

Now putting y = m and x = 1 in the first degree terms in the equation of the curve, we get

$$\phi_1(m) = -3 - m.$$

So for m = – 1, c is to be given by the equation

$$\frac{1}{2}c^2.(-2-6m) + 0.c - 3 - m = 0$$

or $$c^2(1 + 3m) + 3 + m = 0.$$

Putting m = – 1 in this equation, we get

$$-2c^2 + 2 = 0 \text{ or } c^2 = 1 \text{ or } c = \pm 1.$$

∴ the asymptotes corresponding to m = – 1 are

$$y = -1x + 1 \text{ and } y = -1x - 1.$$

Hence all the three asymptotes of the given curve are

$$y = x,\ y = -x + 1 \text{ and } y = -x - 1.$$

Example 6:

Find the asymptotes of the curve

$$y^2(x^2 - a^2) = x.$$

Solution:

Since the given curve is of degree 4, therefore it cannot have more than four asymptotes.

Equating to zero the coefficient of the highest power of y(*i.e.*, of y^2) the asymptotes parallel to y-axis are given by

$$x^2 - a^2 = 0 \text{ i.e., } x = \pm a.$$

Again equating to zero the coefficient of the highest power of x(*i.e.*, of x^2) the asymptotes parallel to x-axis are given by

$$y^2 = 0 \text{ i.e., } y = 0,\ y = 0 \qquad \text{(two coincident asymptotes).}$$

Thus all the four asymptotes have been found.

Example 7:

Find the asymptotes of the curve

$$x^2y^2 = a^2(x^2 + y^2).$$

Solution:

The given curve is of degree 4, so it cannot have more than four asymptotes.

Equating to zero the coefficient of the highest power of y(*i.e.*, of y^2) the asymptotes parallel to y-axis are given by

$$x^2 - a^2 = 0 \text{ i.e., } x = \pm a.$$

Again equating to zero the coefficient of the highest power of x (*i.e.*, of x^2) the asymptotes parallel to x-axis are given by

$$y^2 - a^2 = 0 \text{ i.e., } y = \pm a.$$

Thus all the four asymptotes have been found and they are

$$x = \pm a. y = \pm a.$$

Example 8(a):

Find the asymptotes of the curve

$$x^2(x^2 + y^2) = a^2(y^2 - x^2).$$

Solution:

The equation of the curve is $x^2 (x^2 + y^2) - a^2(y^2 - x^2) = 0$. Since the curve is of degree 4, therefore it cannot have more than four asymptotes.

Equating to zero the coefficient of the highest power of y(*i.e.*, of y^2) the asymptotes parallel to y-axis are given by

$$x^2 - a^2 = 0 \text{ i.e., } x = \pm a.$$

The coefficient of the highest power of x, *i.e.*, (of x^4) is merely a constant. Hence there is no asymptote parallel to x-axis.

To find the remaining oblique asymptotes, we put y = m and x = 1 in the highest *i.e.*, fourth degree terms of and we get

$$\phi_4(m) = 1 + m^2.$$

The roots of the equation $\phi_4(m) = 0$, *i.e.*, $1 + m^2 = 0$ are imaginary and consequently the corresponding asymptotes are imaginary.

Hence the only real asymptotes of the curve are $x = \pm a$

Example 8(b):

Find the asymptotes of the curve

$$x^2y^3 + x^3y^2 = x^3 + y^3.$$

Solution:

The given curve is

$$x^2y^3 + x^3y^2 - x^3 - y^3 = 0 \quad ...(1)$$

The curve (1) is of degree 5, so it cannot have more than five asymptotes.

The asymptotes parallel to y-axis are given by

$$x^2 - 1 = 0 \text{ i.e., } x = \pm 1.$$

To find the remaining oblique asymptotes, we put y = m and x = 1 in the highest *i.e.*, fifth degree terms in the equation (1) and we get

$$\phi_5(m) = m^3 + m^2.$$

The slopes of the asymptotes are given by the equation

$$\phi_5(m) = 0 \text{ i.e., } m^3 + m^2 = 0 \text{ i.e., } m^2 (m + 1) = 0.$$

$\therefore$ $m = 0, 9, -1.$

Again putting y = m and x = 1 in the next highest *i.e.*, fourth degree terms in the equation (1), we get

$\phi_4(m) = 0.$ [Note that there are no fourth degree terms in the equation (1) and so $f_4(m) = 0$.]

Nów c is given by the equation

$$c\phi_5'(m) + \phi_4(m) = 0$$

i.e., $c\,(3m^2 + 2m) + 0 = 0.$ ...(2)

The asymptotes corresponding to m = 0 are parallel to x-axis and have already been found and so no need of finding c for m = 0.

When m = – 1, we have from (2), c = 0 and so the corresponding asymptotes is $y = -1x + 0$ *i.e.*, $y = -x$.

Hence all the five asympstotes of the given curve are

$$x = \pm 1,\ y = \pm 1,\ y = -x.$$

Example 9:

Find the asymptotes of the following curves:

(a) $x^2y - xy^2 + 3x^2 - 2y^2 = 0;$

(b) $x^3 - xy^2 + 6y^2 = 0.$

Solution:

(a) The given curve is of degree 3. So it cannot have more than 3 asymptotes.

Equating to zero the coefficient of the highest power of y(*i.e.*, of y^2), we get the asymptote parallel to y-axis as $-x-2=0$ *i.e.*, $x = -2$. Also equating to zero the coefficient of the highest power of x(*i.e.*, of x^2) the asymptote parallel to x-axis is given by $y + 3 = 0$, *i.e.*, $y = -3$.

Now we proceed to find the remaining oblique asymptotes.

Putting y = m and x = 1 in the third degree and second degree terms separately, we get

$$\phi_3(m) = m - m^2 \text{ and } \phi_2(m) = 3 - 2m^2.$$

The slopes of the asymptotes are given by the equation

$$\phi_3(m) = 0 \text{ i.e., } m - m^2 = 0 \text{ i.e., } m(1 - m) = 0. \therefore m = 0, 1.$$

To determine c, we have the equation

$$c\phi_3'(m)+\phi_2(m)=0 \text{ i.e. } c(1-2m)+3-2m^2=0. \quad ...(1)$$

The asymptote corresponding to m = 0 (*i.e.*, parallel to x-axis) has already been found. So putting m = 1 in (1), we get

c(– 1) + 1 = 0 *i.e.*, c = 1 and the corresponding asymptote is y = x + 1.

∴ the required asymptotes are x = – 2.y = – 3, and y = x + 1.

(b) The given curve is of degree 3 and so it cannot have more than three asymptotes. The coefficients of the highest powers of y and x are (– x + 6) and 1 respectively. The asymptote parallel to y-axis is given by – x + 6 = 0, *i.e.*, x = 6. Also there is no asymptote parallel to x-axis.

For oblique asymptotes, we have

$$\phi_3(m) = 1 - m^2 \text{ and } \phi_2(m) = 6m^2.$$

The equation $\phi_3(m) = 0$ gives m = 1, – 1.

The determine c, we have the equation

$$c=-\frac{\phi_2(m)}{\phi_3'(m)}=-\frac{6m^2}{-2m}=\frac{3m^2}{m}=3m.$$

When m = 1, c = 3 and when m = – 1, c = – 3.

Hence the oblique asymptotes are y = x + 3 and y = – x – 3.

∴ the required asymptotes are x = 6, y = x + 3 and y = – x – 3.

Example 10:

Find all the asymptotes of the curve

$$y^3 - xy^2 - x^2y + x^3 + x^2 - y^2 - 1 = 0.$$

Solution:

Putting y = m and x = 1 in the highest *i.e.*, third degree terms of the equation of the curve, we get

$$\phi_3(m) = m^3 - m^2 - m + 1.$$

The slopes of the asymptotes are given by

$$\phi_3(m) = 0 \text{ i.e., } m^3 - m^2 - m + 1 = 0 \text{ i.e., } (m-1)^2(m+1) = 0.$$

∴ m = 1, 1, – 1.

Now putting y = m and x = 1 in the next highest *i.e.*, second degree terms, we get $\phi_2(m) = 1 - m^2$.

To determine c, we have $c\phi_3'(m)+\phi_2(m)=0$

i.e., $c(3m^2 - 2m - 1) + (1 - m^2) = 0.$ ...(1)

When $m = -1$, we have $c = 0$ and the corresponding asymptote is

$$y = -x + 0 \text{ i.e., } y + x = 0.$$

When $m = 1$, the equation (1) reduces to the identity

$$c.0 + 0 = 0$$

and we cannot determine c from it. In this case c is to be determined from the equation

$$\frac{c^2}{2!}\phi_3''(m) + \frac{c}{1!}\phi_2'(m) + \phi_1(m) = 0.$$

Putting $y = m$ and $x = 1$ in the first degree terms in the equation of the curve, we get $\phi_1(m) = 0$, since there are no first degree terms.

Hence for $m = 1$, c is to be given by

$$\left(\frac{c^2}{2}\right)(6m - 2) + c(-2m) = 0$$

i.e., $(3m - 1)c^2 - 2mc = 0.$

For $m = 1$, this becomes $2c^2 - 2c = 0$ *i.e.*, $c(c - 1) = 0.$

$\therefore$ $c = 0$ and 1.

Hence $y = x + 1$ and $y = x + 0$ are two parallel asymptotes corresponding to the slope $m = 1$.

$\therefore$ the required asymptotes are

$$y + x = 0,\ y - x = 0,\ y - x - 1 = 0.$$

Example 11:

Find the asymptotes of the curve

$$x^3 - 2x^2y + xy^2 + x^2 - xy + 2 = 0.$$

Solution:

The given curve is of degree 3. So it cannot have more than three asymptotes.

Equating to zero the coefficient of the highest power of y (*i.e.*, of y^2), we get the asymptote parallel to y-axis as $x = 0$. Also there is no asymptote parallel to x-axis because the coefficient of x^2 is merely a constant.

Now, we proceed to find the remaining oblique asymptotes.

Putting $y = m$ and $x = 1$ in the third degree and second degree terms separately, we get

$$\phi_3(m) = 1 - 2m + m^2 \text{ and } \phi_2(m) = 1 - m.$$

The slopes of the asymptotes are given by the equation

$\phi_3(m) = 0$ *i.e.*, $1 - 2m + m^2 = 0$ *i.e.*, $(1 - m)^2 = 0$. $\therefore$ m = 1, 1.

To determine c, we have the equation

$$c\phi_3'(m)+\phi_2(m)=0 \text{ i.e., } c(-2+2m)+(1-m)=0. \qquad ...(1)$$

For m = 1, the equation (1) reduces to the identity c.0 + 0 = 0 and thus it fails to give c. In this case c is to be determined by the equation

$$\frac{c^2}{2!}\phi_3''m+\frac{c}{1!}\phi_2'(m)+\phi_1(m)=0.$$

Now $\phi_3''(m)=2, \phi_2'(m)=-1$ and $\phi_1(m) = 0$ because there are no first degree terms in the equation of the curve. So for m = 1, c is to be given by

$$\frac{1}{2}c^2.(2) + c.(-1) + 0 = 0 \text{ i.e., } c^2 - c = 0 \text{ i.e., } c(c-1) = 0.$$

$\therefore$ c = 0, 1.

Hence y = x + 0 and y = x + 1 are two parallel asymptotes corresponding to the slope m = 1.

$\therefore$ the required asymptotes are x = 0, y = x and y = x + 1.

Example 12(a):

Find the asymptotes of the curve

$$x^2(x-y)^2 + a(x^2-y^2) - a^2xy = 0.$$

Solution:

Since the given curve is of degree 4, therefore it cannot have more than four asymptotes.

Equating to zero the coefficient of the highest power of y (*i.e.*, of y^2), we get the asymptotes parallel to y-axis as $x^2 - a^2 = 0$ *i.e.*, $x = \pm a$. But there is no asymptote parallel to x-axis because the coefficient of the highest power of x (*i.e.*, of x^4) is merely a constant.

To find the remaining oblique asymptotes, we put y = m and x = 1 in the highest *i.e.*, fourth degree terms of the equation of the curve, and we get $\phi_4(m) = (1 - m)^2$.

The slopes of the asymptotes are given by the equation

$\phi_4(m) = 0$ *i.e.*, $(1 - m)^2 = 0$. $\therefore$ m = 1, 1.

Now $\phi_3(m) = 0$ because there are no third degree terms in the equation of the curve. To determine c we have the equation

$$c\phi_3'(m)+\phi_3(m)=0$$

i.e., $c\{-2(1-m)\}+0=0.$ ...(1)

For m = 1, the equation (1) reduces to the identity c.0 + 0 = 0 and thus it fails to give c. In this case c is to be determined by the equation

$$\frac{c^2}{2!}\phi_4''(m)+\frac{c}{1!}\phi_3'(m)+\phi_2(m)=0. \qquad ...(2)$$

Now $\phi_4''(m)=2$, $\phi_3'(m)=0$ and $\phi_2(m)=-a^2m$.

Putting these values in (2), we get $c^2-a^2m=0$.

Putting m = 1 in this equation, we get $c^2-a^2=0$ *i.e.*, $c=\pm a$.

Hence y = x + a and y = x – a are two parallel asymptotes corresponding to the slope m = 1.

∴ the required asymptotes are

$x=\pm a$, $y=x+a$ and $y=x-a$.

Example 12(b):

Find the asymptotes of the hyperbola $\left(\frac{x^2}{a^2}\right)-\left(\frac{y^2}{b^2}\right)=1$.

Solution:

The given curve may be written as $y^2=\frac{b^2}{a^2}\left(x^2-a^2\right)$.

$$\therefore\ y=\pm\frac{b}{a}x\left(1-\frac{a^2}{x^2}\right)^{1/2}$$

$$=\pm\frac{b}{a}x\left[1-\frac{a^2}{2x^2}-\frac{a^4}{8x^4}\cdots\right], \qquad \text{(by binomial theorem)}$$

$$=\pm\frac{b}{a}\left[x-\frac{a^2}{2x}-\frac{a^4}{8x^3}-\cdots\right].$$

Hence by 9, the asymptotes of the curve are $y=\pm\left(\frac{b}{a}\right)x$.

Example 12(c):

Find the asymptotes of the curve, $y^3=x^2(x-a)$.

Solution:

The given curve may be written as

$$y=x\left(1-\frac{a}{x}\right)^{1/3}=x\left[1-\frac{a}{3x}-\frac{a^2}{9x^2}-\cdots\right], \qquad \text{(by binomial theorem)}$$

$$\text{or } y = x - \frac{a}{3} - \frac{a^2}{9x} - \ldots$$

Hence by 9, the asymptote of the curve is $y = x - \frac{1}{3}a$.

Example 13:

Find the asymptotes of the curve

$(x + y)^2 (x + 2y + 2) = x + 9y - 2.$

Solution:

The given curve is of degree 3. So it cannot have more than three asymptotes. Obviously the asymptotes of the curve are parallel to the lines $x + y = 0$ and $x + 2y + 2 = 0$.

Asymptotes parallel to the line $x + y = 0$ are given by

$$(x+y)^2 = \lim_{x\to\infty,\ y/x\to-1} \frac{x+9y-2}{x+2y+2}$$

$$= \lim_{x\to\infty,\ y/x\to-1} \frac{1+9\left(\frac{y}{x}\right)-\frac{2}{x}}{1+2\left(\frac{y}{x}\right)+\frac{2}{x}} = \frac{1-9}{1-2} = 8$$

Thus, $x + y = \pm\sqrt{8} = \pm 2\sqrt{2}$ are two asymptotes of the curve.

The asymptote parallel to the line $x + 2y + 2 = 0$ is given by

$$x + 2y + 2 = \lim_{x\to\infty,\ y/x\to-\frac{1}{2}} \frac{x+9y-2}{(x+y)^2}$$

$$= \lim_{x\to\infty,\ y/x\to-\frac{1}{2}} \frac{\frac{1}{x}+9\left(\frac{y}{x}\right)\left(\frac{1}{x}\right)-\frac{2}{x^2}}{\left(1+\frac{y}{x}\right)^2} = 0$$

Hence the required asymptotes are

$x + x + y = \pm\sqrt{2}$ and $x + 2y + 2 = 0$.

Example 14:

Find the all the asymptotes of $y^2(x^2 - a^2) = x^2(x^2 - 4a^2)$.

Solution:

Factorising the highest degree terms, the given equation can be re-written as

$x^2(y - x)(y + x) + 4a^2x^2 - y^2a^2 = 0.$

Equating the coefficient of the highest power of y (*i.e.*, of y^2) to zero, we get $x^2 - a^2 = 0$ or $x = \pm a$ as the asymptotes parallel to y-axis.

$$y - x = \lim_{x\to\infty,\ y/x\to 1}\left[\frac{a^2y^2 - 4a^2x^2}{x^2(y+x)}\right]$$

$$= \lim_{x\to\infty,\ y/x\to 1}\left[\frac{a^2\left(\frac{y^2}{x^2}\right)\left(\frac{1}{x}\right) - 4a^2\left(\frac{1}{x}\right)}{\left(\frac{y}{x+1}\right)}\right] = 0, \text{ (on taking limit).}$$

$\therefore$ $y - x = 0$ is an asymptote of the curve.

Again the asymptote corresponding to the factor $(y + x)$ is

$$y + x = \lim_{x\to\infty,\ y/x\to 1}\left[\frac{a^2y^2 - 4a^2x^2}{x^2(y-x)}\right] = 0, \qquad \text{(as above).}$$

Hence the required asymptotes are

$x = \pm a,\ y = \pm x.$

Example 15:

Find all the asymptotes of the curve

$(x^2 - y^2)(x + 2y + 1) + x + y + 1 = 0.$

Solution:

Factorising the highest degree terms into real linear factors, the given equation of the curve may be written as

$(x + y)(x - y)(x + 2y) + (x^2 - y^2) + (x + y + 1) = 0.$

The slope m of the asymptote corresponding to the factor $x + y$ is -1.

Hence the asymptote corresponding to the factor $(x + y)$ is

$$x + y = \lim_{x\to\infty,\ y/x\to -1}\frac{-(x^2 - y^2) - (x - y + 1)}{(x+2y)(x-y)}, \qquad [\because m = -1]$$

$$= \lim_{x\to\infty,\ y/x\to -1}\frac{(y^2 - x^2)}{(x-y)(x+2y)} - \lim_{x\to\infty,\ y/x\to -1}\frac{x+y+1}{(x-y)(x+2y)}$$

$$= \lim_{x\to\infty,\ y/x\to -1}\frac{\left\{\left(\frac{y^2}{x^2}\right) - 1\right\}}{\left\{1 - \left(\frac{y}{x}\right)\right\}\left\{1 + 2\left(\frac{y}{x}\right)\right\}}$$

$$-\lim_{x\to\infty,\ y/x\to-1}\frac{\frac{1}{x}+\frac{y}{x}.\frac{1}{x}+\frac{1}{x^2}}{\left\{1-\left(\frac{y}{x}\right)\right\}\left\{1+2\left(\frac{y}{x}\right)\right\}},$$

on dividing the numerator and denominator by x^2

$$=\frac{1-1}{(1+1)(1-2)}-0=0.$$

Thus, $x + y = 0$ is one asymptote of the curve. Again the asymptote corresponding to the factor $(x - y)$ is

$$x-y=\lim_{x\to\infty,\ y/x\to-1}\frac{\left(y^2-x^2\right)}{(x+y)(x+2y)}-\lim_{x\to\infty,\ y/x\to-1}\frac{x+y+1}{(x+y)(x+2y)},$$

[$\because$ m = 1]

$$=\lim_{x\to\infty,\ y/x\to-1}\frac{\left(\frac{y^2}{x^2}\right)-1}{\left\{1+\left(\frac{y}{x}\right)\right\}\left\{1+2\left(\frac{y}{x}\right)\right\}}$$

$$-\lim_{x\to\infty,\ y/x\to-1}\frac{\frac{1}{x}+\frac{y}{x}.\frac{1}{x}+\frac{1}{x^2}}{\left\{1+\left(\frac{y}{x}\right)\right\}\left\{1+2\left(\frac{y}{x}\right)\right\}}$$

$$=\frac{(1-1)}{(1+1)(1+2)}-0=0.$$

Thus, $x - y = 0$ is another asymptote of the curve. The third asymptote of the curve is

$$x+2y=\lim_{x\to\infty,\ y/x\to-1/2}\frac{\left(y^2-x^2\right)+\text{terms of degree lower than 2}}{(x-y)(x+y)}$$

$$=\lim_{x\to\infty,\ y/x\to-1/2}\frac{\left\{\left(\frac{y^2}{x^2}\right)-1\right\}+\text{terms which}\to 0}{\left\{1-\left(\frac{y}{x}\right)\right\}\left\{1+\left(\frac{y}{x}\right)\right\}}$$

$$=\frac{\left(\frac{1}{4}-1\right)+0}{\left(1+\frac{1}{2}\right)\left(1-\frac{1}{2}\right)}=\frac{-\frac{3}{4}}{\frac{3}{4}}=-1.$$

Therefore, $x + 2y + 1 = 0$ is the third asymptote of the curve.

Important Remark: It should be noted that while taking limit we should reject all the terms in the numerator whose degree is lower than the degree denominator. All such terms $\to 0$ as $x \to \infty$.

Example 16:

Find all the asymptotes of the curve

$$(a + x)^2 (b^2 + x^2) = x^2y^2.$$

Solution:

The given curve may be written as

$$x^2(y - x)(y + x) = 2ax^3 + (a^2 + b^2)x^2 + 2ab^2x + a^2b^2.$$

$\therefore$ the asymptotes are

$x^2 = 0$ *i.e.*, $x = 0$, $x = 0$ (the asymptotes parallel to y-axis)

$$\text{and } y - x = \lim_{x\to\infty,\ y/x\to-1}\left[\frac{2ax^3 + (a^2 + b^2)x^2 + 2ab^2x + a^2b^2}{x^2(y + x)}\right] = -a.$$

$\therefore$ the required asymptotes are $x = 0$, $y = x + a$, $y = -x - a$.

Example 17:

Find the asymptotes of the curve

$$y^2(x - 2) = x^2(y - 1).$$

Solution:

The given curve may be written as

$$xy(y - x) - 2y + x^2 = 0.$$

Here $y - 1 = 0$ is the asymptote parallel to x-axis,

and $x - 2 = 0$ is the asymptote parallel to y-axis.

The other asymptote is given by

$$y - x = \lim_{x\to\infty,\ y/x\to1}\left[\frac{2y^2 - x^2}{xy}\right] = 1,$$

[dividing the numerator and the denominator by x^2 and taking the limits].

$\therefore$ the required asymptotes are $x = 2$, $y = 1$ and $y = x + 1$.

Example 18:

Find the asymptotes of the curve

$$x^2y - xy^2 + xy + y^2 + x - y = 0.$$

Solution:

Factorising the highest degree terms, the given equation can be re-written as $xy(y - x) = xy + y^2 + x - y$.

Equating to zero the coefficient of the highest powers of x and y we get $y = 0$ and $x = 0$ as asymptotes parallel to the co-ordinate axes.

Now the asymptote corresponding to the factor $(y - x)$ is

$$y - x = \lim_{x\to\infty,\ y/x\to 1}\left[\frac{xy + y^2 + x - y}{xy}\right]$$

$$= \lim_{x\to\infty,\ y/x\to 1}\left[\frac{\left(\frac{y}{x}\right)+\left(\frac{y^2}{x^2}\right)+\left(\frac{1}{x}\right)-\left(\frac{y}{x}\right)\left(\frac{1}{x}\right)}{\left(\frac{y}{x}\right)}\right].$$

Taking limits, we have $y - x = 2$ or $y = x + 2$.

Hence the required asymptotes are $x = 0$, $y = 0$ and $y = x + 2$.

Example 19:

Find the asymptotes of the curve

$$(x^2 - y^2)(y^2 - 4x^2) - 6x^3 + 5x^2y + 3xy^2 - 2y^3 - x^2 + 3xy = 1.$$

Solution:

Factorising the highest degree terms, the given equation can be re-written as

$$(x + y)(x - y)(y + 2x)(y - 2x) = 6x^3 - 5x^2y - 3xy^2 + 2y^3 + x^2 - 3xy + 1.$$

$\therefore$ the asymptotes are parallel to the lines $x + y = 0$, $x - y = 0$, $y + 2x = 0$ and $y - 2x = 0$.

Asymptote parallel to $x + y = 0$ is

$$x + y = \lim_{x\to\infty,\ y/x\to -1}\left[\frac{6x^3 - 5x^2y - 3xy^2 + 2y^3 + x^2 - 3xy + 1}{(x - y)(y + 2x)(y - 2x)}\right]$$

$[\because m = -1]$

$$= \lim\left[\frac{6 - 5\left(\frac{y}{x}\right) - 3\left(\frac{y^2}{x^2}\right) + 2\left(\frac{y^3}{x^3}\right) + \left(\frac{1}{x}\right) - 3\left(\frac{y}{x^2}\right) + \left(\frac{1}{x^3}\right)}{\left(1 - \frac{y}{x}\right)\left(2 + \frac{y}{x}\right)\left(-2 + \frac{y}{x}\right)}\right],$$

on dividing the numerator and denominator by x^3

$$= \frac{6+5-3-2}{2.1.(-3)} = \frac{6}{-6} = -1.$$

Hence x + y + 1 = 0 is one asymptote of the curve.

The second asymptote parallel to x – y = 0 is

$$x - y = \lim_{x\to\infty,\ y/x\to 1}\left[\frac{6x^3 - 5x^2 - 3xy^2 + 2y^3 + x^2 - 3xy + 1}{(x+y)(y+2x)(y-2x)}\right].$$

Taking limits as above, we get

x – y = 0 as the second asymptote of the curve.

The third asymptote parallel to y + 2x = 0 is

$$x - y = \lim_{x\to\infty,\ y/x\to -2}\left[\frac{6x^3 + 5x^2y - 3xy^2 + 2y^3 + x^2 - 3xy + 1}{(x-y)(x+y)(y-2x)}\right].$$

Taking limits as above, we get

y + 2x = – 1, as the third asymptote of the curve.

The fourth asymptote parallel to (y – 2x) is

$$y - 2x = \lim_{x\to\infty,\ y/x\to 2}\left[\frac{6x^3 - 5x^2y - 2y^3 + x^2 - 3xy + 1}{(x-y)(x+y)(y+2x)}\right].$$

Taking limits as above, we get

y – 2x = 0, as the fourth asymptote of the curve.

Hence the required asymptotes are

x + y + 1 = 0, x – y = 0, 2x + y + 1 = 0 and 2x – y = 0.

Example 20:

Find the asymptotes of the curve

$(x - y)^2 (x^2 + y^2) - 10(x - y)x^2 + 12y^2 + 2x + y = 0.$

Solution:

The given curve is

$(x - y)^2 (x^2 + y^2) - 10(x - y)x^2 + 12y^2 + 2x + y = 0.$

The equation of the curve is of fourth degree and so there are at most four asymptotes. Dividing each term by the coefficient $x^2 + y^2$ of $(x - y)^2$ and taking limits, the asymptotes. Dividing each term by the coefficient $x^2 + y^2$ of $(x - y)^2$ and taking limits, the asymptotes corresponding to the factor $(x - y)^2$ are given by

$(x-y)^2(x^2+y^2)-10(x-y)x^2+12y^2+2x+y=0.$

$$(x-y)^2-10(x-y)\lim_{x\to\infty,\, y/x\to 1}\frac{x^2}{x^2+y^2}+\lim_{x\to\infty,\, y/x\to 1}\frac{12y^2+2x+y}{x^2+y^2}=0$$

or $$(x-y)^2-10(x-y)\lim_{x\to\infty,\, y/x\to 1}\frac{1}{1+\left(\frac{y}{x}\right)^2}$$

$$+\lim_{x\to\infty,\, y/x\to 1}\frac{12\left(\frac{y}{x}\right)^2+\left(\frac{2}{x}\right)+\left(\frac{y}{x}\right)\left(\frac{1}{x}\right)}{1+\left(\frac{y}{x}\right)^2}=0,$$

or $$(x-y)^2-10(x-y).\left(\frac{1}{2}\right)+\left(\frac{12}{2}\right)=0$$

or $$(x-y)^2-5(x-y)+6=0$$

i.e., $x-y=\dfrac{5\pm\sqrt{(25-24)}}{2}=\dfrac{5\pm 1}{2}=2$ or 3

i.e., $x-y=2$ and $x-y=3$.

Again $x^2+y^2=(x+iy)(x-iy)$ implies that the remaining linear factors of the fourth degree terms in the equation of the curve are imaginary and so the other two asymptotes are imaginary.

Example 21:

Find the asymptotes of the curve

$(x^2-y^2)(x+2y)=y^2-y+1.$

Solution:

Factorising the highest degree terms, the given equation can be re-written as $(x-y)(x+y)(x+2y)=y^2-y+1$.

$\therefore$ the asymptotes are parallel to the lines $x-y=0$, $x+y=0$ and $x+2y=0$.

The asymptote parallel to $x-y=0$ is

$$x-y=\lim_{x\to\infty,\, y/x\to 1}\left[\frac{y^2-y+1}{(x+y)(x+2y)}\right]$$

$$=\lim_{x\to\infty,\, y/x\to 1}\left[\frac{\left(\frac{y}{x}\right)^2-\left(\frac{y}{x}\right)\left(\frac{1}{x}\right)+\left(\frac{1}{x^2}\right)}{\left\{1+\left(\frac{y}{x}\right)\right\}\left\{1+2\left(\frac{y}{x}\right)\right\}}\right],$$

on dividing the numerator and the denominator by x^2

$$=\frac{1-0+0}{2.(1+2)}=\frac{1}{6}.$$

Hence $x-y=\frac{1}{6}$ is an asymptote of the curve.

The second asymptote parallel to x + y = 0 is

$$x+y=\lim_{x\to\infty,\ y/x\to-1}\left[\frac{y^2-y+1}{(x-y)(x+2y)}\right]$$

$$=\lim_{x\to\infty,\ y/x\to-1}\left[\frac{\left(\frac{y^2}{x^2}\right)-\left(\frac{y}{x}\right)\left(\frac{1}{x}\right)+\left(\frac{1}{x^2}\right)}{\left\{1-\left(\frac{y}{x}\right)\right\}\left\{1+2\left(\frac{y}{x}\right)\right\}}\right]=-\frac{1}{2}.$$

Thus, $x + y = -\frac{1}{2}$ is the second asymptote of the curve.

The third asymptote parallel to x + 2y = 0 is

$$x+2y=\lim_{x\to\infty,\ y/x\to-1/2}\left[\frac{y^2-y+1}{(x+y)(x-y)}\right]$$

$$=\lim_{x\to\infty,\ y/x\to-1/2}\left[\frac{\left(\frac{y}{x}\right)^2-\left(\frac{y}{x}\right)\left(\frac{1}{x}\right)+\left(\frac{1}{x}\right)^2}{\left\{1+\left(\frac{y}{x}\right)\right\}\left\{1-\left(\frac{y}{x}\right)\right\}}\right]$$

$$=\frac{\left(-\frac{1}{2}\right)^2-\left(-\frac{1}{2}\right).0+0}{\left\{1+\left(-\frac{1}{2}\right)\right\}\left\{1-\left(-\frac{1}{2}\right)\right\}}=\frac{\frac{1}{4}}{\frac{1}{2}.\frac{3}{2}}=\frac{1}{3}.$$

Thus, $x + 2y = \frac{1}{3}$ is the third asymptote of the curve.

Hence the required asymptotes are

$$x-y=\frac{1}{6},\ x+y=-\frac{1}{2},\ x+2y=\frac{1}{2}.$$

Example 22:

Find all the asymptotes of the curve $y^2(x - b) = x^3 + a^3$.

Solution:

The given curve is $x(y - x)(y + x) = a^3 + by^2$.

∴ the required asymptotes are

$$y - x = \lim_{x\to\infty,\, y/x\to 1} \left[\frac{a^3 + by^2}{x(y+x)}\right]$$

$$= \lim_{x\to\infty,\, y/x\to 1} \left[\frac{\left(\frac{a^3}{x^2}\right) + b\left(\frac{y^2}{x^2}\right)}{\left(\frac{y}{x}+1\right)}\right] = \frac{1}{2}b,$$

and $$y + x = \lim_{x\to\infty,\, y/x\to -1} \left[\frac{a^3 + by^2}{x(y-x)}\right] = -\frac{1}{2}b.$$

Also equating the coefficient of y^2 to zero, we get $x - b = 0$ as an asymptote parallel to y-axis.

$\therefore\ \pm y = x \pm \frac{1}{2}b$ and $x = b$ are the required asymptotes.

Example 23:

Find the asymptotes of the curve

$$(y - x)(y - 2x)^2 + (y + 3x)(y - 2x) + 2x + 2y - 1 = 0.$$

Solution:

The asymptotes corresponding to the factor $(y - 2x)^2$ are given by

$$(y-2x)^2 + (y-2x) \lim_{x\to\infty,\, y/x\to 2} \left[\frac{y+3x}{y-x}\right] + \lim_{x\to\infty,\, y/x\to 2} \left[\frac{2x+2y-1}{y-x}\right] = 0$$

or $$(y-2x)^2 + (y-2x) \lim_{x\to\infty,\, y/x\to 2} \left[\frac{\left(\frac{y}{x}\right)+3}{\left(\frac{y}{x}\right)-1}\right]$$

$$+ \lim_{x\to\infty,\, y/x\to 2} \left[\frac{2 + 2\left(\frac{y}{x}\right) - \left(\frac{1}{x}\right)}{\left(\frac{y}{x}\right)-1}\right] = 0$$

or $(y - 2x)^2 + 5(y - 2x) + 6 = 0$

i.e., $y - 2x = \frac{1}{2}\left[-5 \pm \sqrt{(25-24)}\right] = \frac{1}{2}(-5 \pm 1)$

i.e., $y - 2x = \frac{1}{2}(-5+1)$ and $y - 2x = \frac{1}{2}(-5-1)$.

$\therefore$ $y = 2x - 2$ and $y = 2x - 3$ are the asymptotes corresponding to the factor $(y - 2x)^2$.

Also the asymptote corresponding to the factor (y – x) is

$$y-x=\lim_{x\to\infty,\ y/x\to 1}\left[\frac{-(y+3x)(y-2x)-2x-2y+1}{(y-2x)^2}\right]$$

$$=\lim_{x\to\infty,\ y/x\to 1}\left[\frac{-\left\{\left(\frac{y}{x}\right)+3\right\}\left\{\left(\frac{y}{x}\right)-2\right\}-2\left(\frac{1}{x}\right)-2\left(\frac{y}{x^2}\right)+\left(\frac{1}{x^2}\right)}{\left\{\left(\frac{y}{x}\right)-2\right\}^2}\right]=4,$$

on taking limits.

Hence y = x + 4 is another asymptote of the given curve.

Example 24:

Find the asymptotes of the curve

$$(y-x)^2x-3y(y-x)+2x=0.$$

Solution:

The given equation is of third degree. So there are at most three asymptotes.

Dividing each term by x and taking limits, the asymptotes corresponding to the factor $(y-x)^2$ are given by

$$(y-x)^2-3(y-x)=\lim_{x\to\infty,\ y/x\to 1}\left(\frac{y}{x}\right)+\lim_{x\to\infty,\ y/x\to 1}\left(\frac{2x}{x}\right)=0$$

or $\quad (y-x)^2-3(y-x)+2=0.$

Hence $y-x=\frac{1}{2}\left[3\pm\sqrt{(9-8)}\right]$ *i.e.*, = 2, or 1.

Therefore, y – x = 2 and y – x = 1 are the two asymptotes.

Also equating to zero the coefficient of the highest power of y, we get the asymptote parallel to y-axis as x = 3.,

Hence the required asymptotes are x = 3, y – x = 2 and y – x = 1.

Example 25:

Find all the asymptotes of the curve

$$x^2(x+y)(x-y)^2+ax^2(x-y)-a^2y^3=0.$$

Solution:

The given equation is of fifth degree. So there are at most five asymptotes. Dividing each term by the coefficient of $(x-y)^2$ and taking the limits, the asymptotes corresponding to the factor $(x-y)^2$ are given by

$$(x-y)^2 + (x-y) \lim_{x\to\infty,\, y/x\to 1}\left[\frac{a}{x+y}\right] - \lim_{x\to\infty,\, y/x\to 1}\left[\frac{a^3y^3}{x^2(x+y)}\right] = 0$$

or $$(x-y)^2 + (x-y)\lim\left[\frac{\frac{a}{x}}{1+\frac{y}{x}}\right] - \lim\left[\frac{a^2\left(\frac{y^3}{x^2}\right)}{1+\left(\frac{y}{x}\right)}\right] = 0.$$

Taking limits, we get $(x-y)^2 + (x-y).0 - \frac{1}{2}a^2 = 0$

or $x - y = \pm\frac{a}{\sqrt{2}}$ as the two asymptotes.

Also the asymptote corresponding to the factor (x + y) is

$$x + y = \lim_{x\to\infty,\, y/x\to -1}\left[\frac{ax^2(y-x)+a^2y^3}{x^2(x-y)^2}\right]$$

$$= \lim_{x\to\infty,\, y/x\to -1}\left[\frac{a\left\{\left(\frac{y}{x}\right)-1\right\}\left(\frac{1}{x}\right)+a^2\left(\frac{y^3}{x^3}\right)\left(\frac{1}{x}\right)}{\left\{1-\left(\frac{y}{x}\right)\right\}^2}\right].$$

Taking limits we have x + y = 0, as an asymptote.

Also equating to zero the coefficient of the highest power of y, we get the asymptotes parallel to y-axis as $x^2 - a^2 = 0$ *i.e.*, $x = \pm a$.

Hence the required asymptotes are

$$x = \pm a,\; x - y = \frac{a}{\sqrt{2}} \text{ and } x + y = 0.$$

Example 26:

Find the asymptotes of the curve

$$y(x + y)^2 - y + 1 = 0.$$

Solution:

Here y = 0 is an asymptote parallel to x-axis.

The asymptotes corresponding to the factor $(y + x)^2$ are given by

$$(y + x)^2 = \lim_{x\to\infty,\, y/x\to -1}\left[\frac{y-1}{y}\right]$$

$$= \lim_{x\to\infty,\ y/x\to -1}\left[\frac{\left(\frac{y}{x}\right)-\left(\frac{1}{x}\right)}{\left(\frac{y}{x}\right)}\right] = 1$$

$\therefore$ the required asymptotes are $y + x = \pm 1$, and $y = 0$.

Example 27(a):

Find all the asymptotes of the curve

$$(y - a)^2 (x^2 - a^2) = x^4 + a^4.$$

Solution:

The given curve is

$$y^2(x^2 - a^2) - 2ay(x^2 - a^2) + a^2(x^2 - a^2) = x^4 + a^4$$

or $$x^2(y - x)(y + x) - 2ax^2y + a^2(x^2 - y^2) + 2a^3y - 2a^4 = 0.$$

Equating to zero the coefficient of the highest power of y (*i.e.*, of y^2), we get $x^2 - a^2 = 0$ or $x = \pm a$ as the asymptotes parallel to y-axis.

The other asymptotes are

$$y - x = \lim_{x\to\infty,\ y/x\to 1}\left[\frac{2ax^2y - a^2(x^2 - y^2) - 2a^3y + 2a^4}{x^2(y+x)}\right]$$

$$= \lim_{x\to\infty,\ y/x\to 1}\left[\frac{2a\left(\frac{y}{x}\right) - a^2\left\{\left(\frac{1}{x}\right)-\left(\frac{y^2}{x^3}\right)\right\} - 2a^3\left(\frac{y}{x^3}\right) + 2a^4\left(\frac{1}{x^3}\right)}{\left[1+\left(\frac{y}{x}\right)\right]}\right]$$

$= a,$

and $$= y + x = \lim_{x\to\infty,\ y/x\to -1}\left[\frac{2ax^2y - a^2(x^2 - y^2) - 2a^3y + 2a^4}{x^2(y-x)}\right]$$

$$= \lim_{x\to\infty,\ y/x\to -1}\left[\frac{2a\left(\frac{y}{x}\right) - a^2\left\{\left(\frac{1}{x}\right)-\left(\frac{y^2}{x^3}\right)\right\} - 2a^3\left(\frac{y}{x^3}\right) + 2a^4\left(\frac{1}{x^3}\right)}{\left(\frac{y}{x}\right)-1}\right]$$

$= a.$

$\therefore$ the required asymptotes are $x = \pm a$ and $y = \pm a$ and $y = \pm x + a$.

Example 27(b):

Find the asymptotes of the curve

$$x^2(x - y)^2 + 9(x^2 -- y^2) = 9xy.$$

Solution:

The given equation can be written can b re-written as

$$x^2(x - y)^2 + 9(x - y)(x + y) = 9xy.$$

Dividing each term by the coefficient of $(x - y)_2$ (*i.e.*, x^2) and taking limits, the asymptotes corresponding to the factor $(x - y)^2$ are given by

$$(x - y)^2 + 9(x - y) \lim_{x\to\infty,\ y/x\to 1}\left[\frac{(x+y)}{x^2}\right] - \lim_{x\to\infty,\ y/x\to 1}\left[\frac{9xy}{x^2}\right] = 0$$

or $\quad (x - y)^2 = 9$. Then $x - y = \pm 3$ are two asymptotes.

Also equating zero the coefficient of highest power of y, we get the asymptotes parallel to y-axis as $x^2 - 9 = 0$ *i.e.*, $x = \pm 3$.

Hence the required asymptotes are

$$x = \pm 3 \text{ and } x - y = \pm 3.$$

Example 27(c):

Find the asymptotes of the curve

$$x(y - 3)^3 = 4y(x - 1)^2.$$

Solution:

The equation of the given curve can be written as

$$xy(y - 2x)(y + 2x) = 9xy^2 - 12yx^2 - 15xy + 27x - 4y.$$

Here the coefficient of the highest power of $y = x$.

Therefore $x = 0$ is an asymptote parallel to y-axis.

Again the coefficient of the highest power of $x = 4y$.

Therefore $y = 0$ is an asymptote parallel to x-axis.

The other asymptotes are given by

$$y - 2x = \lim_{x\to\infty,\ y/x\to 2}\left[\frac{9xy^2 - 12yx^2 - 15xy + 27x - 4y}{xy(y+2x)}\right]$$

$$= \frac{3}{2};\ \textit{i.e.},\ 2y - 4x = 3;$$

$$\text{and } y + 2x = \lim_{x\to\infty,\ y/x\to -2}\left[\frac{9xy^2 - 12yx^2 - 15xy + 27x - 4y}{xy(y-2x)}\right]$$

$$= \frac{15}{2} \text{ i.e., } 2y + 4x = 15.$$

Example 28:

Find the asymptotes of the curve

$$x(y^2 - x^2) = 2ay^2 - a^3.$$

Solution:

The given curve is $(x - 2a)\, y^2 - x^3 + a^3 = 0.$ ...(1)

The asymptote parallel to y-axis is $x - 2a = 0$ or $x = 2a$.

The equation (1) can be written as

$$x(y - x)\,(y + x) = 2ay^2 - a^3.$$

Hence the other asymptotes are

$$y - x = \lim_{x\to\infty,\ y/x\to 1}\left[\frac{2ay^2 - a^3}{x(y+x)}\right] = \frac{2a}{2} = a,$$

and $$y + x = \lim_{x\to\infty,\ y/x\to -1}\left[\frac{2ay^2 - a^3}{x(y-x)}\right] = \frac{2a}{-2} = -a.$$

$\therefore$ the required asymptotes are $x = 2a$, $y = x + a$, $y = -x - a$.

Example 29:

Find the asymptotes of the curve

$$(y - x)\,(y - 2x)^2 + (y + 3x)\,(y - 2x) + 2x + 2y - 1 = 0.$$

Solution:

The given equation of third degree and so there are almost three asymptotes. Dividing each term by the coefficient of $(y - 2x)^2$ and taking limits, the asymptotes corresponding to the factor $(y - 2x)^2$ are given by

$$(y-2x)^2 + (y-2x)\lim_{x\to\infty,\ y/x\to 2}\left[\frac{(y+3x)}{(y-x)}\right] + \lim_{x\to\infty,\ y/x\to 2}\left[\frac{2x+2y-1}{y-x}\right] = 0$$

or $$(y - 2x)^2 + (y - 2x)\lim\left[\frac{\left(\frac{y}{x}\right)+3}{\left(\frac{y}{x}\right)-1}\right] + \lim\left[\frac{2+2\left(\frac{y}{x}\right)-\frac{1}{x}}{\left(\frac{y}{x}\right)-1}\right] = 0.$$

Taking limits, we have

$$(y - 2x)^2 + 5(y - 2x) + 6 = 0$$

giving $\quad y - 2x = \frac{1}{2}[-5 \pm 1] = -2, \text{ or } -3.$

Thus $\quad y - 2x + 2 = 0$ and $y - 2x + 3 = 0$ are the two asymptotes.

Also the asymptote corresponding to the factor $(y - x)$ is

$$y-x=\lim_{x\to\infty,\, y/x\to 1}\left[\frac{-(y+3x)(y-2x)-2x-2y+1}{(y-2x)^2}\right]$$

$$=\lim_{x\to\infty,\, y/x\to 1}\left[\frac{-\left(\frac{y}{x}+3\right)\left(\frac{y}{x}-2\right)-2\left(\frac{1}{x}\right)-2\left(\frac{y}{x^2}\right)+\left(\frac{1}{x^2}\right)}{\left\{\left(\frac{y}{x}\right)-2\right\}^2}\right].$$

Taking limits, we have $y - x = 4$ as another asymptote.

Hence the required asymptotes of the curve are

$$x - y + 4 = 0,\; y - 2x + 2 = 0 \text{ and } y - 2x + 3 = 0.$$

Example 30:

Find all the asymptotes of the curve

$$(x - y + 1)(x - y + 2)(x + y) = 8x - 1.$$

Solution:

The given curve may be written as

$$(x + y)(x - y)^2 - (x^2 - y^2) - 10x - 2y + 1 = 0.$$

The asymptote corresponding to the factor $(x + y)$ is

$$x+y=\lim_{x\to\infty,\, y/x\to -1}\left[\frac{8x-1}{(x-y+1)(x-y-2)}\right]$$

$= 0$, [dividing the numerator and the denominator by x^2 and taking the limits].

Also the asymptotes corresponding to the factor $(x - y)^2$ are given by

$$(x - y)^2 - (x - y) = \lim_{x\to\infty,\, y/x\to 1}\left[\frac{10x+2y-1}{x+y}\right]=6$$

or $$x-y=\frac{1}{3}\left[1\pm\sqrt{(1+24)}\right]=3 \text{ or } -2$$

i.e., $$y = x - 3 \text{ and } y = x + 2.$$

∴ the required asymptotes are

$$y + x = 0,\ y = x - 3 \text{ and } y = x + 2.$$

Example 31:

Find the asymptotes of the curve

$$x^2(x - y)^2 + a^2(x^2 - y^2) = a^2xy.$$

Solution:

The equation of the curve is

$$x^2(x - y)^2 + a^2(x - y)(x + y) = a^2xy.$$

The asymptotes corresponding to the factor $(x - y)^2$ are given by

$$(x - y)^2 + a^2(x - y) \lim_{x\to\infty,\ y/x\to 1}\left[\frac{x+y}{x^2}\right] = \lim_{x\to\infty,\ y/x\to 1} \frac{a^2xy}{x^2}$$

or $\qquad (x - y)^2 = a^2$ or $x - y = \pm a$.

Also $x = \pm a$ are asymptotes parallel to y-axis.

∴ the required asymptotes are $x = \pm a$ and $x - y = \pm a$.

Example 32:

Find all the asymptotes of the curve

$$y^2(x - 2) = x^2(y - 1) \text{ or } xy(y - x) = 2y^2 - x^2.$$

Solution:

Equating the coefficients of the highest powers of x and y to zero, we get

$y - 1 = 0$ as an asymptote parallel to x-axis

and $x - 2 = 0$ as an asymptote parallel to y-axis.

The other asymptote corresponding to the factor $(y - x)$ is

$$y - x = \lim_{x\to\infty,\ y/x\to 1}\left[\frac{2y^2 - x^2}{xy}\right] = 1.$$

∴ the required asymptotes are $y = 1$, $x = 2$ and $y - x = 1$.

Example 33(a):

Find the asymptotes of the curve

$$x^3 + 3x^2y - 4y^3 - x + y + 3 = 0.$$

Solution:

The equation of the curve may be written as

$$(x - y)(x + 2y)^2 = x - y - 3.$$

The asymptote corresponding to the factor (x – y) is

$$x - y = \lim_{x\to\infty,\ y/x\to 1}\left[\frac{x-y-3}{(x+2y)^2}\right] = 0.$$

Also the asymptotes corresponding to the factor $(x + 2y)^2$ are given by

$$(x+2y)^2 = \lim_{x\to\infty,\ y/x\to -1/2}\left[\frac{x-y-3}{x-y}\right] = 1$$

or $\quad x + 2y = \pm 1.$

∴ the required asymptotes are x – y = 0 and x + 2y = ± 1.

Example 33(b):

Find all the asymptotes of the curve

$$(x^2 - y^2)^2 = 4y^2 - y.$$

The asymptotes corresponding to the factors $(x - y)^2$ and $(x + y)^2$ are

$$(x - y)^2 = \lim_{x\to\infty,\ y/x\to 1}\left[\frac{4y^2-y}{(x+y)^2}\right] = 1, \text{ i.e., } x - y = \pm 1,$$

and

$$(x + y)^2 = \lim_{x\to\infty,\ y/x\to -1}\left[\frac{4y^2-y}{(x-y)^2}\right] = 1, \text{ i.e., } x + y = \pm 1.$$

∴ the required asymptotes are x – y = ± 1 and x + y = ± 1.

Example 33(c):

Find the asymptotes of the curve

$$x^3 + 2x^2y + xy^2 - x^2 - xy + 2 = 0.$$

Solution:

The given curve may be written as

$$x(x + y)^2 - x(x + y) + 2 = 0.$$

Here an asymptote parallel to y-axis is x = 0.

The asymptotes corresponding to the factor $(x + y)^2$ are given by

$$(x + y)^2 - (x + y) + \lim_{x\to\infty,\ y/x\to -1}\left[\frac{2}{x}\right] = 0$$

or $\quad (x + y)^2 - (x + y) + 0 = 0$

or $\quad (x + y)(x + y - 1) = 0.$

∴ the required asymptotes are

$$x = 0,\ x + y = 0 \text{ and } x + y - 1 = 0.$$

Example 34:

Find the asymptotes of the curve

$$x^3 - 4xy^2 + x^2y - 4y^3 - 2x^3 + 8y^2 + 3x + 4y - 5 = 0.$$

Solution:

The given equation can be put in the form

$$(x - 2y)(x + 2y)(x + y - 2) + (3x + 4y - 5) = 0.$$

By inspection, the asymptotes are

$$x - 2y = 0,\ x + 2y = 0,\ x + y - 2 = 0.$$

Example 35:

Find the equation of the cubic curve whose asymptotes are $x + a = 0$, $y - a = 0$ and $x + y + a = 0$ and which touches the axis of x at the origin and passes through the point $(-2a, -2a)$.

Solution:

The asymptotes are

$$x + a = 0,\ y - a = 0 \text{ and } x + y + a = 0.$$

∴ the combined equation of the asymptotes is

$$(x + a)(y - a)(x + y + a) = 0.$$

Now the equation of any curve, having these asymptotes, is of the form

$$(x + a)(y - a)(x + y + a) + F_1 = 0,$$

where F_1 is of 1st degree in x and y (say $F_1 = bx + cy + d$).

So let the equation of the required curve be

$$(x + a)(y - a)(x + y + a) + (bx + cy + d) = 0.$$

But it is given that the curve passes through the origin (0, 0),

therefore $\quad d = a^3$.

Thus, the equation of the curve reduces to

$$(x + a)(y - a)(x + y + a) + bx + cy + a^3 = 0$$

or $\quad xy(x + y) + a(y^2 - x^2 + xy) - 2a^2x + bx + cy = 0.$

Now equating the lowest degree terms to zero the equation of the tangent at origin is $(-2a^2 + b)x + cy = 0$.

Also given that the x-axis (*i.e.*, $y = 0$) is tangent at the origin.

∴ $\quad -2a^2 + b = 0$ or $b = 2a^2$.

$\therefore$ the equation of the curve is

$$xy(x + y) + a(y^2 - x^2 + xy) + cy = 0.$$

This equation passes through $(-2a, -2a)$. (given)

$$\therefore \quad 4a^2(-4a) + a(4a^2 - 4a^2 + 4a^2) + c(-2a) = 0 \text{ or } c = -6a^2.$$

Therefore the required equation is

$$xy(x + y) + a(y^2 - x^2 + xy) - 6a^2y = 0.$$

Example 36:

Form the equation of a curve which has $x = 0$, $y = 0$, $y = x$ and $y = -x$ four asymptotes and which passes through the point (a, b) and cuts its asymptotes again in eight points lying upon the circle $x^2 + y^2 = a^2$.

Solution:

The combined equation of the asymptotes is

$$xy(y^2 - x^2) = 0. \qquad \text{...(1)}$$

The required curve passes through the points of intersection of the asymptotes and the given circle $x^2 + y^2 - a^2 = 0$.

Let the equation of the curve be

$$xy(y^2 - x^2) + \lambda(x^2 + y^2 - a^2) = 0. \qquad \text{...(2)}$$

Now this curve also passes through the point (a, b).

$$\therefore \quad ab(b^2 - a^2) + 1(a^2 + b^2 - a^2) = 0,$$

i.e.,

$$\lambda = \left(\frac{a}{b}\right)(a^2 - b^2).$$

Substituting this value of λ in (2), we have

$$bxy(y^2 - x^2) + a(a^2 - b^2)(x^2 + y^2 - a^2) = 0$$

as the required equation of the curve.

Example 37:

Show that the four asymptotes of the curve

$$(x^2 - y^2)(y^2 - 4x^2) + 6x^3 - 5x^2y - 5x^2y - 3xy^2 + 2y^3 - x^2 + 3xy - 1 = 0$$

cut the curve in eight points which lie on the circle $x^2 + y^2 = 1$.

Solution:

The given curve is

$$P \equiv (x^2 - y^2)(y^2 - 4x^2) + 6x^3 - 5x^2y - 3xy^2 + 2y^3 - x^2 + 3xy - 1 = 0. \qquad \text{....(1)}$$

Here $\phi_4(m) = (1 - m^2)(m^2 - 4) = -4 + 5m^2 - m^4$.

The slopes of the asymptotes are given by the equation

$$\phi_4(m) = (1 - m^2)(m^2 - 4) = 0.$$

$\therefore$ $m = \pm 1, \pm 2.$

Also $\phi_3(m) = 6 - 5m - 3m^2 - 3m^2 + 2m^3,$

and $\phi'(m) = 10m - 4m^3.$

Now c is given by the equation

$$c\phi_4'(m) + \phi_3(m) = 0$$

i.e., $c(10m - 4m^3) + 6 - 5m - 3m^2 + 2m^3 = 0$

i.e., $$c = \frac{6 - 5m - 3m^2 + 2m^3}{4m^3 - 10m}.$$

When $m = 1$, $c = 0$;

when $m = -1$, $c = 1$;

when $m = 2$, $c = 0$; and

when $m = -2$, $c = 1$.

Thus, the asymptotes of the curve (1) are $y = x$, $y = -x + 1$, $y = 2x$ and $y = -2x + 1$

The combined equation of the asymptotes is

$$(y - x)(y + x - 1)(y - 2x)(y + 2x - 1) = 0$$

or $$[(y^2 - x^2) - y + x][(y^2 - 4x^2) - y + 2x] = 0$$

or $$(y^2 - x^2)(y^2 - 4x^2) - y^3 + 2xy^2 + x^2y - 2x^3 - y^3 + 4x^2y + xy^2 - 4x^3 + y^2 - 2xy - xy + 2x^2 = 0$$

or $$Q \equiv (y^2 - x^2)(y^2 - 4x^2) - 6x^3 + 5x^2y + 3xy^2 - 2y^3 + y^2 - 3xy + 2x^2 = 0 \quad ...(2)$$

Now each asymptote of (1) will cut it in 4 – 2 *i.e.*, 2 points. Therefore, the four asymptotes will cut in 4 × 2 *i.e.*, 8 points.

Now taking $\lambda = 1$, $P + Q = 0$ gives $x^2 + y^2 - 1 = 0$ *i.e.*, $x^2 + y^2 = 1$, which is the equation of a circle.

Hence the eight points of intersection of (1) and (2) lie on the circle $x^2 + y^2 = 1$.

Example 38:

Find the circular asymptote of the curve

$$r(e^\theta - 1) = a(e^\theta + 1).$$

Solution:

The given equation is

$$r = \frac{a(e^{\theta}+1)}{(e^{\theta}-1)} = f(\theta), \text{ say.}$$

$\therefore$ the circular asymptote is given by

$$r = a \lim_{\theta\to\infty} \frac{e^{\theta}+1}{e^{0}-1}, \qquad \left[\text{form } \frac{\infty}{\infty}\right]$$

$$= a \lim_{\theta\to\infty} \frac{1+e^{-\theta}}{1-e^{-\theta}} = a.$$

Hence r = a is the circular asymptote.

Example 39:

Find all asymptotes of the curve

$$3x^3 + 2x^2y - 7xy^2 + 2y^3 - 14xy + 7y^2 + 4x + 5y = 0.$$

Solution:

Putting x = 1 and y = m in the third and second degree terms separately, we have

$$\phi_3(m) = 3 + 2m - 7m^2 + 2m^3 \text{ and } \phi_2(m) = 7m^2 - 14m.$$

$$\therefore \qquad \phi_3'(m) = 2 - 14m + 6m^2.$$

Putting $\phi_3(m) = 0$, we get $3 + 2m - 7m^2 + 2m^3 = 0$

or $(m - 1)(2m^2 - 5m - 3) = 0$ (or $m - 1)(2m + 1)(m - 3) = 0$.

$$\therefore \qquad m = 1, 3, -\frac{1}{2}.$$

Also $\qquad c = -\dfrac{\phi_2(m)}{\phi_3'(m)} = \dfrac{14m - 7m^2}{2 - 14m + 6m^2}.$

Putting m = 1, we get $c = -\dfrac{7}{6}$;

when m = 3, c = – 1, and when $m = -\dfrac{1}{2}$, $c = -\dfrac{5}{4}$.

Hence asymptotes are

$$y = x = \frac{7}{6};\ y = 3x - 1 \text{ and } y = -\frac{1}{2}x - \frac{5}{6},$$

or 6x – 6y – 7 = 0, y = 3x – 1 and 6y + 3x + 5 = 0.

Example 40:

Find the asymptotes of the curve

$y^2(x^2 - a^2) = x^2(x^2 - 4a^2)$.

Solution:

The given curve is

$$y^2(x^2 - a^2) = x^2(x^2 - 4a^2)$$

or $$y^2x^2 - x^4 - a^2y^2 + 4a^2x^2 = 0. \quad ...(1)$$

Since the curve (1) is of degree 4, therefore it cannot have more than four asymptotes.

Equating to zero the coefficient of the highest power of y(*i.e.*, of y^2) the asymptotes parallel to y-axis are given by

$$x^2 - a^2 = 0 \text{ } i.e., \text{ } x = \pm a. \quad \text{(Two asymptotes)}$$

The coefficient of the highest power of x *i.e.*, of x is simply a constant and so there is no asymptote parallel to x-axis.

To find the remaining oblique asymptotes, we have put y = m and x = 1 in the highest *i.e.*, fourth degree terms in the equation (1) and we get

$$\phi_4(m) = m^2 - 1.$$

The slopes of the asymptotes are given by the equation

$$\phi_4(m) = 0 \text{ } i.e., \text{ } m^2 - 1 = 0.$$

$$\therefore \quad m = \pm 1.$$

Again putting y = m and x = 1 in the next highest *i.e.*, third degree terms, we get

$\phi_3(m) = 0$. [Note that there are no terms of degree 3 in the equation of the curve and so $\phi_3(m) = 0$.]

Now c is given by the equation $c\phi_4'(m) + \phi_3(m) = 0$

i.e., $c(2m) + 0 = 0$ *i.e.*, $2cm = 0$.

When m = 1, c = 0; and when m = – 1, c = 0.

∴ the asymptotes are $y = 1x + 0$ and $y = (-1)x + 0$

i.e., the asymptotes y = x and y = – x.

Hence all the your asymptotes of the curve are $x = \pm a$, $y = \pm x$.

Example 41:

Find the asymptotes parallel to the coordinate axes of the curve

$x^2(x - y)^2 + a^2(x^2 - y^2) - a^2xy = 0.$

Solution:

Equating to zero the coefficient of the highest power of y (*i.e.*, of y^2) the asymptotes parallel to y-axis are given by

$$x^2 - a^2 = 0 \; i.e., \; x = \pm a.$$

The coefficient of the highest power of (*i.e.*, of x^4) is merely a constant. Hence there is no asymptote parallel to x-axis.

Example 42:

Find all the asymptotes of the curve

$$y^3 - 3x^2y + xy^2 - 3x^3 + 2y^2 + 2xy + 4x + 5y + 6 = 0.$$

Solution:

Putting x = 1 and y = m in the third and second degree terms separately, we have

$$\phi_3(m) = m^3 - 3m + m^2 - 3 \text{ and } \phi_2(m) = 2(m^2 + m).$$

$$\therefore \quad \phi_3'(m) = 3m^2 + 2m - 3.$$

Equating $\phi_3(m)$ to zero, we get $m^3 + m^2 - 3m - 3 = 0$

or $m^2(m + 1) - 3(m + 1) = 0$ or $(m + 1)(m^2 - 3) = 0$

i.e., $\quad m = -1, +\sqrt{3} - \sqrt{3}.$

Also we know that $c = -\dfrac{\phi_2(m)}{\phi_3'(m)} = -\dfrac{2(m^2 + m)}{3m^2 + 2m - 3}.$

$\therefore$ when m = 1, c = – 2; when $m = \sqrt{3}$, c = – 1;

and when $m = -\sqrt{3}$, c = – 1.

Hence the asymptotes are

$$y = x - 2; \; y = \pm x\sqrt{3} - 1.$$

Example 43:

Find the asymptotes of the curve

$$y^3 - x^2y - 2xy^2 + 2x^3 - 7xy + 3y^2 + 2x^2 + 2x + 2y + 1 = 0.$$

Solution:

Putting y = mx + c, we have

$$(mx + c)^3 - x^2(mx + c) - 2x(mx + c)^2 + 2x^3 - 7x(mx + x)$$
$$+ 3(mx + c)^2 + 2x^2 + 2x + 2(mx + c) + 1 = 0$$

or $x^3(m^3 - m - 2m^2 + 2) + x^2(3m^2c - c - 4mc - 7m + 3m^2 + 2)$

$$+ \ldots = 0.$$

Equating the coefficient of the highest degree term (*i.e.*, x^3) to zero, we get

$$m^3 - m - 2m^2 + 2 = 0$$

or $(m - 1)(m + 1)(m - 2) = 0$.

Therefore, $m = 1, -1, 2$.

Now equating the coefficient of x^2 to zero, we get

$$c(3m^2 - 4m - 1) + 3m^2 - 7m + 2 = 0$$

or $c = -\dfrac{(3m^2 - 7m + 2)}{(3m^2 - 4m - 1)}$.

Therefore, when $m = 1$, $c = -1$; when $m = -1$, $c = -2$; and when $m = 2$, $c = 0$.

Substituting these pairs of values of m and c in $y = mx + c$, the required asymptotes are

$$y = x - 1,\ y = -x - 2 \text{ and } y = 2x.$$

Example 44:

Find the asymptotes of

$$x^3 - y^3 - x^2 - 5y^2 - 11x - 5y + 7 = 0.$$

Solution:

Putting $x = 1$, $y = m$ in the third degree and second degree terms separately, we have

$$\phi_3(m) = 1 - m^3 \text{ and } \phi_2(m) = -1 - 5m^2.$$

$$\therefore \quad \phi_3'(m) = -3m^2.$$

Now $\phi_3(m) = 1 - m^3 = (1 - m)(1 + m + m^2) = 0$ gives $m = 1$ as the only real root.

Also $\quad c = -\dfrac{\phi_2(m)}{\phi_3'(m)} = -\dfrac{-1-5m^2}{-3m^2}$.

Putting $m = 1$, we have $c = -2$.

Hence the only asymptote of the curve is $y = x - 2$.

Example 45:

Find the asymptotes of the curve

$$x^3 - 2y^3 + 2x^2y - xy^2 + xy - y^2 + 1 = 0.$$

Solution:

Putting y = mx + c, we have

$$x^3 - 2(mx + c)^3 + 2x^2(mx + c) - x(mx + c)^2 + x(mx + c) - (x + c)^2 + 1 = 0$$

or $x^3(-2m^3 - m^2 + 2m + 1) + x^2(-6m^2c - 2mc + 2c + m - m^2) + ... = 0.$

Equating to zero the coefficient of x^3, we have

$2m^3 + m^2 - 2m - 1 = 0$, *i.e.*, $(m^2 - 1)(2m + 1) = 0$.

$\therefore m = 1, -1, -\frac{1}{2}.$

Equating to zero the coefficient of x^2, we have

$-2(3m^2 + m - 1)c - (m^2 - m) = 0,$

i.e., $c = \dfrac{(m - m^2)}{2(3m + m - 1)}.$

Therefore, when m = 1, c = 0; when m = – 1, y = – 1; and when

$m = -\frac{1}{2}, c = \frac{1}{2}.$

$\therefore$ the asymptotes are y = x, y = – x – 1 and $y = -\frac{1}{2}x + \frac{1}{2}$;

or x – y = 0, x + y + 1 = 0 and x + 2y – 1 = 0.

Example 46:

Find the asymptotes of the curve

$$2x^3 + 3x^2y - 3xy^2 - 2y^3 + 3x^2 - 3y^2 + y - 3 = 0.$$

Solution:

Here $\phi_3(m) = 2 + 3m - 3m^2 - 2m^3$,

$\phi_3'(m) = 3 - 6m - 6m^2$ and $\phi_2(m) = 3 - 3m^2$; $\phi_3(m) = 0$, gives

$m = 1, -\frac{1}{2}, -2.$

Also $c = -\dfrac{\phi_2(m)}{\phi_3'(m)} = \dfrac{3m^2 - 3}{3 - 6m - 6m^2}.$

When m = 1, c = 0; when $m = -\frac{1}{2}$, $c = -\frac{1}{2}$;

and when m = – 2, c = – 1.

$\therefore$ $y = x$, $y = -\frac{1}{2}x - \frac{1}{2}$ and $y = -2x - 1$ are the required asymptotes.

Example 47:

Find the asymptotes of

$$4x^3 - x^2y - 4xy^2 + y^3 + 3x^2 + 2xy - y^2 - 7 = 0.$$

Solution:

Putting $x = 1$ and $y = m$ in the third degree and second degree terms separately, we have

$$\phi_3(m) = 4 - m - 4m^2 + m^3;\ \phi_2(m) = 3 + 2m - m^2.$$

Now solving the equation $\phi_3(m) = 4 - m - 4m^2 + m^3 = 0$

i.e., $(4 - m)(1 - m^2) = 0$, we get $m = 1, -1, 4$.

Also $\phi_3'(m) = -1 - 8m + 3m^2$.

$$\therefore \quad c = -\frac{\phi_2(m)}{\phi_3'(m)} = \frac{3 + 2m - m^2}{1 + 8m - 3m^2}.$$

Putting $m = 1$, we get $c = \frac{2}{3}$;

when $m = -1$, $c = 0$ and when $m = 4$, $c = \frac{1}{3}$.

Hence the asymptotes are

$$y = x + \frac{2}{3};\ y = -x \text{ and } y = 4x + \frac{1}{3}.$$

Example 48:

Find the asymptotes of

$$x^3 - 2y^3 + xy(2x - y) + y(x - 1) + 1 = 0.$$

Solution:

Putting $x = 1$, $y = m$ in the third and second degree terms separately, we have

$$\phi_3(m) = 1 - 2m^3 + 2m - m^2 \text{ and } \phi_2(m) = m.$$

$$\therefore \quad \phi_3'(m) = -6m^2 + 2 - 2m.$$

Now the slopes of the asymptotes are given by the equation $\phi_3(m) = 0$ or $(1 - 2m^3 + 2m - m^2) = 0$ or $(1 - m^2)(1 + 2m) = 0$.

$$\therefore m = 1, -1, -\frac{1}{2}.$$

Also we know that $c = -\dfrac{\phi_2(m)}{\phi_3'(m)} = \dfrac{m}{6m^2 + 2m - 2}$.

$\therefore$ when m = 1, $c = \dfrac{1}{6}$; when m = – 1, $c = -\dfrac{1}{2}$

and when $m = -\dfrac{1}{2}$, $c = \dfrac{1}{3}$.

Hence the asymptotes are

$$y = x + \frac{1}{6};\ y = -x - \frac{1}{2} \text{ and } y = -\frac{1}{2}x + \frac{1}{3}$$

i.e., $6x - 6y + 1 = 0$; $2x + 2y + 1 = 0$ and $6y + 3x - 2 = 0$.

Example 49:

Find the asymptotes of the curve $x^3 + y^3 - 3axy = 0$.

Solution:

Putting x = 1 and y = m in the third and second degree terms separately, we get

$\phi_3(m) = 1 = m^3$, and $\phi_2(m) = -3am$.

Now we solve the equation $\phi_3(m) = (1 + m)(1 - m + m^2) = 0$.

Obviously m = – 1 is the only real root of this equation.

To determine c, we have the formula

$$c = -\frac{\phi_2(m)}{\phi_3'(m)} = -\frac{3am}{3m^2} = \frac{a}{m}.$$

Putting m = 1, we get c = – a.

Hence the only asymptote of the curve is

$y = -x - a$ or $x + y + a = 0$.

Example 50(a):

Find the asymptotes of the curve $y^3 = x^3 + ax$.

Solution:

The given curve is $y^3 - x^3 - ax^2 = 0$.

Putting y = m and x = 1 in the highest *i.e.*, third degree terms in the equation of the curve, we get $\phi_3(m) = m^3 - 1$.

Solving the equation $\phi_3(m) = 0$.

i.e., $m^3 - 1 = 0$, *i.e.*, $(m - 1)(m^2 + m + 1) = 0$, we get m = 1 as the only real root.

The other two roots are imaginary.

Again putting y = m and x = 1 in the second degree terms in the equation of the curve, we get $\phi_2(m) = -a$.

Now c is given by $c\phi_3'(m) + \phi_2(m) = 0$, *i.e.*, $c(3m^2) - a = 0$.

Putting m = 1, we get $3c - a = 0$ or $c = \frac{a}{3}$.

Hence the only real asymptote of the curve is

$$y = x + \left(\frac{a}{3}\right).$$

Example 50(b):

Find the circular asymptote of the curve

$$r = \frac{3\theta^2 + 2\theta + 1}{2\theta^2 + \theta + 1}.$$

Solution:

The given equation is

$$r = \frac{3 + \left(\frac{2}{\theta}\right) + \left(\frac{1}{\theta^2}\right)}{2 + \left(\frac{1}{\theta}\right) + \left(\frac{1}{\theta^2}\right)},$$ dividing numerator and denominator by θ^2.

Taking limits when $\theta \to \infty$, we see that the circular asymptote is $r = \frac{3}{2}$.

Example 50(c):

Find the asymptotes of the curve $r \sin n\theta = a$.

Solution:

The equation of the curve can be written as

$$\frac{1}{r} = \frac{1}{a} \sin n\theta = f(\theta), \text{ say.}$$

Now $f(\theta) = 0$ if $\sin n\theta = 0$ *i.e.*, $n\theta = m\pi$ (where m is any integer).

$$\therefore \quad \theta = \left(\frac{m\pi}{n}\right) = \alpha, \text{ say.}$$

Also $$f'(\theta) = \left(\frac{1}{a}\right).n \cos n\theta.$$

$$\therefore \quad f'(\alpha) = \left(\frac{1}{a}\right).n \cos n\alpha = \left(\frac{1}{a}\right).n \cos\left(n.\frac{m\pi}{n}\right)$$

$$= \left(\frac{1}{a}\right).n.\cos m\pi.$$

$\therefore$ the asymptotes are $r \sin\left(\theta - \frac{m\pi}{n}\right) = \frac{a}{n \cos m\pi}$, where m is any integer.

Example 51:

Show that the curve $r = a \sec m\theta + b \tan m\theta$ has two sets of asymptotes; members of the first set touching one fixed circle, and those of the other, another fixed circle.

Solution:

The equation of the curve can be written as

$$\frac{1}{r} = \frac{1}{a \sec m\theta + b \tan m\theta} = \frac{\cos m\theta}{a + b \sin m\theta} = f(\theta), \text{ say.}$$

$$\therefore f'(\theta) = \frac{(a + b \sin m\theta) - (-m \sin m\theta) - \cos m\theta\,(bm \cos m\theta)}{(a + b \sin m\theta)^2}.$$

Now $f'(\theta) = 0 \Rightarrow \cos m\theta = 0$, or $m\theta = (2k + 1).\frac{1}{2}\pi$, (k is may integer).

or $\quad \theta = (2k+1).\frac{\pi}{2m} = \alpha$, say.

Obviously $\cos m\alpha = 0$.

Also $\sin ma = \sin\left\{(2k+1)\frac{1}{2}\pi\right\} = \sin\left(k\pi + \frac{1}{2}\pi\right) = \cos k\pi = (-1)^k$.

$$\therefore \quad f'(\alpha) = \frac{\{a + b(-1)^k\}\{-m(-1)^k\}}{\{a + b(-1)^k\}^2} = \frac{(-1)^{k+1}.m}{\{a + b(-1)^k\}}.$$

$\therefore$ the asymptotes are given by

$$r \sin\left[\theta - \left\{(2k+1).\frac{\pi}{2m}\right\}\right] = \frac{a + b(-1)^k}{(-1)^{k+1}.m}. \qquad ...(1)$$

Case I: When k is odd, the first set of asymptotes are

$$r \sin\left\{\theta - \frac{(2k+1)\pi}{2m}\right\} = \frac{a-b}{m}, \qquad [\because (-1)^k = -1]$$

or $$r\left[\sin\theta \cos\frac{(2k+1)\pi}{2m} - \cos\theta \sin\frac{(2k+1)\pi}{2m}\right] = \frac{a-b}{m},$$

or $$x \sin\frac{(2k+1)\pi}{2m} - y\cos\frac{(2k+1)\pi}{2m} + \frac{a-b}{m} = 0,$$

[since r cos θ = x and r sin θ = y]

and these asymptotes the circle $x^2 + y^2 = \left(\frac{a-b}{m}\right)^2$, which is fixed.

Case II: When k is even, the second set of asymptotes are

$$r\sin\left(\theta - \frac{2k+1}{2m}.\pi\right) = -\frac{a+b}{m}, \qquad [\because (-1)^k = 1]$$

or $$y\cos\left(\frac{2k+1}{2m}.\pi\right) - x\sin\left(\frac{2k+1}{2m}.\pi\right) = -\frac{a+b}{m}$$

or $$x\sin\left(\frac{2k+1}{2m}.\pi\right) - y\cos\left(\frac{2k+1}{2m}.\pi\right) - \left(\frac{a+b}{m}\right) = 0$$

and these touch another fixed circle

$$x^2 + y^2 = \left(\frac{a+b}{m}\right)^2.$$

Example 52:

Find the asymptotes of the curve

$$r\theta \cos \theta = a \cos 2\theta.$$

Solution:

The equation of the given curve can be written as

$$\frac{1}{r} = \frac{\theta \cos\theta}{a\cos 2\theta} = f(\theta), \text{ say.}$$

$$\therefore\ f'(\theta) = \frac{1}{a}\left[\frac{\cos 2\theta.\{\cos\theta - \theta.\sin\theta\} + 2\theta\cos\theta.\sin 2\theta}{\cos^2 2\theta}\right].$$

Now f(θ) = 0 ⇒ θ = 0 or cos θ = 0 *i.e.*, $\theta = (2k + 1).\frac{1}{2}\pi$.

If θ = 0 = α (say), then $f'(\alpha) = \frac{1}{a}$.

∴ r sin(θ – 0) = a

or r sin θ = a is the corresponding asymptote.

When $\theta = (2k+1).\frac{1}{2}\pi = \alpha$ (say), we have

$\cos\alpha = 0, \sin 2\alpha = 0, \sin\alpha = (-1)^k$

and cos 2α = cos (2kπ + π) = cos π = – 1.

$$\therefore \qquad f'(\alpha)=\frac{1}{a}\left[\frac{-\left\{(-1)^{k}(2k+1).\frac{1}{2}\pi\right\}}{1}\right]$$

$$=(-1)^{k+1}(2k+1).\frac{\pi}{2a}.$$

The corresponding asymptote is

$$r\sin\left\{\theta-(2k+1).\frac{\pi}{2}\right\}=\frac{2a}{(-1)^{k+1}.(2k+1)\pi}$$

or $\qquad (-1)^k\, r(\cos\,(k\pi-\theta)\}\ (2k+1)\ \pi = 2a,$

or $\qquad (-1)^{2k}\ r\cos\theta = \dfrac{2a}{\{(2k+1)\pi\}}.$

$\therefore$ the required asymptotes are $r\sin\theta = a$

and $\qquad r\cos\theta = \dfrac{2a}{\{(2k+1)\pi\}}.$

Example 53:

Find the asymptotes of the curve

$$r\theta = a.$$

Solution:

The equation of the given curve can be written as

$$\frac{1}{r}=\frac{\theta}{a}=f(\theta),\text{ say.}$$

Now $\qquad f(\theta) = 0$ if $\dfrac{\theta}{a} = 0$ *i.e.*, $\theta = 0 = a$, say.

Also $\qquad f'(\theta)=\dfrac{1}{a}$, so that

$$f'(\alpha)=f'(0)=\frac{1}{a},\ \frac{1}{f'}(\alpha)=a.$$

Now the asymptote corresponding to $\theta = \alpha$ is given by

$$r\sin(\theta-\alpha)=\frac{1}{f'}(\alpha)$$

i.e., $\qquad r\sin(\theta - 0) = a, \qquad [\because$ here $\alpha = 0]$

i.e., $\qquad r\sin\theta = a.$

Example 54:

Find the asymptotes of the curve $r = \dfrac{a}{(1-\cos\theta)}.$

Solution:

The equation to the curve can be written as

$$\frac{1}{r}=\frac{1}{a}(1-\cos\theta)=f(\theta),\text{ say.}$$

Now $f(\theta) = 0 \Rightarrow 1 - \cos\theta = 0$ or $\cos\theta = 1$

i.e., $\theta = 2k\pi$, where k is any integer.

Also $f'(\theta)=\frac{1}{a}\sin\theta$.

$\therefore$ $f'(2k\pi)=\frac{1}{a}\sin(2k\pi)=0$.

Hence $\frac{1}{f'(2k\pi)}=\infty$

and consequently there is no asymptote of the given curve.

Example 55:

Find the asymptotes of the curve

$$r=\frac{2a}{(1-2\cos\theta)}.$$

Solution:

The equation of the curve can be written as

$$\frac{1}{r}=\frac{1}{2a}(1-2\cos\theta)=f(\theta),\text{ say.}$$

Now $f(\theta) = 0$ if $1 - 2\cos\theta = 0$ *i.e.,* $2\cos\theta = 1$

or $\cos\theta = \frac{1}{2}$ *i.e.,* $\theta=2n\pi\pm\frac{1}{3}\pi$, where n is any integer.

Also $f'(\theta)=\frac{1}{2a}(2\sin\theta)=\frac{1}{a}\sin\theta$.

$$\therefore\quad f'\left(2n\pi\pm\frac{\pi}{3}\right)=\frac{1}{a}\sin\left(2n\pi\pm\frac{\pi}{3}\right)=\pm\frac{1}{a}\sin\frac{\pi}{3}=\pm\frac{\sqrt{3}}{2a}.$$

$\therefore$ the asymptotes are given by

$$r\sin\left\{\theta-\left(2n\pi\pm\frac{\pi}{3}\right)\right\}=\pm\frac{2a}{\sqrt{3}}$$

or

$$r\sin\theta\cos\left(2n\pi\pm\frac{\pi}{3}\right)-r\cos\theta.\sin\left(2n\pi\pm\frac{\pi}{3}\right)=\pm\frac{2a}{\sqrt{3}}$$

i.e.,

$$r\sin\theta\cos\frac{\pi}{3}-r\cos\theta\sin\frac{\pi}{3}=\frac{2a}{\sqrt{3}}$$

and $$r\sin\theta\cos\frac{\pi}{3}+r\cos\theta\sin\frac{\pi}{3}=-\frac{2a}{\sqrt{3}}$$

i.e., $$r\sin\left(\theta-\frac{\pi}{3}\right)=\frac{2a}{\sqrt{3}} \text{ and } r\sin\left(\theta+\frac{\pi}{3}\right)=-\frac{2a}{\sqrt{3}}.$$

Example 56:

Find the asymptotes of the curve
$r \sin \theta = 2 \cos 2\theta$.

Solution:

The equation to the curve can be written as

$$\frac{1}{r}=\frac{\sin\theta}{2\cos 2\theta}=f(\theta), \text{ say.}$$

Now $f(\theta) = 0 \Rightarrow \sin\theta = 0$ *i.e.,* $\theta = k\pi = \alpha$ (say).
Here k is any integer.

Also $$f'(\theta)=\frac{1}{2}\frac{\cos 2\theta\cos\theta-\sin\theta.(-2\sin 2\theta)}{\cos^2 2\theta}.$$

$$\therefore \quad \frac{1}{f'(\alpha)}=\frac{2\cos^2(2k\pi)}{\cos 2k\pi.\cos k\pi+2\sin k\pi.\sin 2k\pi}$$

$$=\frac{2}{\cos k\pi}, \qquad [\because \cos 2k\pi = 1]$$

$$=\frac{2}{(-1)^k}. \qquad [\because \cos k\pi = (-1)^k]$$

$\therefore$ the required asymptotes are given by

$$r\sin(\theta-k\pi)=\frac{2}{(-1)^k}$$

or $$-r\sin(k\pi-\theta)=\frac{2}{(-1)^k}$$

or $$-r\left\{(-1)^{k-1}\sin\theta\right\}=\frac{2}{(-1)^k}$$

or $r \sin\theta = 2$.

Example 57:

Find the asymptotes of the curve $r = a \operatorname{cosec}\theta + b$.

Solution:

The equation of the curve can be written as

$$\frac{1}{r} = \frac{1}{a \text{ cosec } \theta + b} = \frac{\sin \theta}{a + b \sin \theta} = f(\theta), \text{ say.}$$

Now f(θ) = 0, if sin θ = 0 *i.e.*, θ = nπ, (where n is any integer).

Also $$f'(\theta) = \frac{\cos \theta(a + b \sin \theta) - b \sin \theta \cos \theta}{(a + b \sin \theta)^2}.$$

∴ $$f'(n\pi) = \frac{\cos n\pi(a + b \sin n\pi) - b \sin n\pi \cos n\pi}{(a + b \sin n\pi)^2} \cdot \frac{\cos n\pi}{a}.$$

∴ the asymptotes are given by $r \sin (\theta - n\pi) = \dfrac{a}{\cos n\pi}$.

or $$r(\sin \theta \cos n\pi - \cos \theta \sin n\pi) = \frac{a}{\cos n\pi}$$

or $r \sin \theta \cos^2 n\pi - 0 = a$ or $r \sin \theta = a$.

Example 58:

Find the asymptotes of the curve

$$4(x^4 + y^4) - 17x^2y^2 - 4x (4y^2 - x^2) + 2(x^2 - 2) = 0$$

and show that they pass through the points of intersection of the curve with the ellipse $x^2 + 4y^2 = 4$.

Solution:

The given curve may be written as

$$(4x^4 + 4y^4 - 17x^2y^2) - 4(4xy^2 - x^3) + 2x^2 - 4 = 0. \quad ...(1)$$

Here $\phi_4(m) = 4m^4 - 17m^2 + 4 = 0$ gives $m = \pm \frac{1}{2}, \pm 2$.

Also $$c = -\frac{\phi_3(m)}{\phi'_4(m)} = \frac{4(4m^2 - 1)}{16m^3 - 34m}.$$

When $m = \frac{1}{2}$, c = 0, when $m = -\frac{1}{2}$, c = 0,

when m = 2, c = 1 and when m = – 2, c = – 1.

Thus the asymptotes of (1) are

$$y = \frac{1}{2}x, \; y = -\frac{1}{2}x, \; y = 2x + 1 \text{ and } y = -2x - 1.$$

∴ the combined equation of the asymptotes is

$$(x - 2y)(x + 2y)(2x - y + 1)(2x + y + 1) = 0$$

or $$(4x^4 + 4y^4 - 17x^2y^2) - 4(4xy^2 - x^3) + (x^2 - 4y^2) = 0. \quad ...(2)$$

Subtracting (2) from (1), we get

$$x^2 + 4y^2 - 4 = 0. \quad ...(3)$$

Also the four asymptotes cut the curve into n(n – 2) *i.e.*, 4(4 – 2) = 8 points. These 8 points of intersection lie on the ellipse (3).

Example 59:

Find the equation of the cubic which has the same asymptotes as the curve $x^3 - 6x^2y + 11xy^2 - 6y^3 + x + y + 1 = 0$ *and which touches the axis of y at the origin and passes through the point (3, 2).*

Solution:

The given curve is

$$(x^3 - 6x^2y + 11xy^2 - 6y^3) + (x + y) + 1 = 0.$$

Here $\phi_3(m) - 1 - 6m + 11m^2 - 6m^3$

$$= (1 - m)(1 - 2m)(1 - 3m).$$

Therefore, $\phi_3(m) = 0$ gives $m = 1, \frac{1}{2}, \frac{1}{3}$. Also $\phi_2(m) = 0$.

$\therefore$ $$c = -\frac{\phi_2(m)}{\phi_3'(m)} = 0, \text{ for all the three values of m.}$$

$\therefore$ the asymptotes of the given curve are

$$y = x;\ y = \frac{1}{2}x \text{ and } y = \frac{1}{3}x.$$

Hence the combined equation of the asymptotes is

$$(x - y)(x - 2y)(x - 3y) = 0. \quad ...(1)$$

Now, the most general equation of any curve, having these asymptotes, is of the form

$$(x - y)(x - 2y)(x - 3y) + F_1 = 0,$$

where F_1 is of the first degree in x and y (say $F_1 = ax + by + c$).

If the curve $(x - y)(x - 2y)(x - 3y) + ax + by + c = 0$, passes through the origin (0, 0), then c = 0 and the equation of the curve becomes

$$(x - y)(x - 2y)(x - 3y) + ax + by = 0. \quad ...(2)$$

Equating to zero, the lowest degree terms in (2), we get ax + by = 0, as the equation of the tangent at origin.

But x = 0 *i.e.*, y-axis is given to be the tangent at origin.

Therefore, b = 0 and hence the equation of the curve is

$$(x - y)(x - 2y)(x - 3y) + ax = 0 \quad ...(3)$$

It passes through the point (3, 2), (given).

$$\therefore \quad (3 - 2)(3 - 4)(3 - 6) + 3a = 0; \text{ or } a = -1.$$

Therefore, the required equation of the curve is

$$(x - y)(x - 2y)(x - 3y) - x = 0, \quad [\text{putting } a = -1 \text{ in } (3)]$$

or $$x^3 - 6x^2y + 11xy^2 - 6y^3 - x = 0.$$

Example 60:

Show that the eight points of the curve

$$x^4 - 5x^2y^2 + 4y^4 + x^2 - y^2 + x + y + 1 = 0$$

and its asymptotes lie on a rectangular hyperbola.

Solution:

The equation of the given curve is

$$x^4 - 5x^2y^2 + 4y^4 + x^2 - y^2 + x + y + 1 = 0 \quad ...(1)$$

or $(x^2 - y^2)(x^2 - 4y^2) + x^2 - y^2 + x + y + 1 = 0$

or $(x - y)(x + y)(x - 2y)(x + 2y) + x^2 - y^2 + x + y + 1 = 0.$

$\therefore$ by inspection, the combined equation of the asymptotes of (1) is

$$(x - y)(x + y)(x - 2y)(x + 2y) = 0$$

or $$x^4 - 6x^2y^2 + 4y^4 = 0. \quad ...(2)$$

Now each asymptote of (1) will cut it in 4 – 2 *i.e.*, 2 points. Therefore, the four asymptotes will cut it in 4 × 2 *i.e.*, 8 points.

Subtracting (2) from (1), we get

$$x^2 - y^2 + x + y + 1 = 0. \quad ...(3)$$

The curve (3) passes through the eight points of intersection of (1) and (2). Also the conic (3) is a rectangular hyperbola because in its equation the sum of the coefficients of x^2 and y^2 is zero.

Hence the eight points of intersection of (1) and (2) lie on a rectangular hyperbola.

Example 61:

Show that asymptotes of the cubic

$$x^3 - 2y^3 + xy(2x - y) + y(x - y) + 1 = 0$$

cut the curve again in three points which lie on the straight line

$$x - y + 1 = 0.$$

Solution:

The given curve is

$$x^3 - 2y^3 + xy(2x - y) + y(x - y) + 1 = 0. \qquad ...(1)$$

$\therefore \quad \phi_3(m) = 1 - 2m^3 + 2m - m^2 = 0$ gives $m = 1, -1, -\frac{1}{2}$.

Also $\quad \phi_3'(m) - 6m^2 + 2 - 2m$ and $\phi_2(m) = m - m^2$.

$$\therefore \quad c = -\frac{\phi_2(m)}{\phi_3'} = -\frac{(m - m^2)}{(2 - 2m - 6m^2)}.$$

When $m = 1$, $c = 0$; when $m = -1$, $c = -1$;

and when $m = -\frac{1}{2}, c = \frac{1}{2}$.

$\therefore$ the asymptotes of (1) are $y = x$, $y = -x - 1$

and $\quad y = -\frac{1}{2}x + \frac{1}{2}$.

Hence the combined equation of the asymptotes of (1) is

$$(x - y)(x + y + 1)(x + 2y - 1) = 0$$

or $\quad x^3 - 2y^3 + 2x^2y - xy^2 + xy - y^2 - x + y = 0.$

Subtracting (2) form (1), we get

$$x - y + 1 = 0,$$

which shows that the points of intersection of the curve and its asymptotes lie on the straight line $x - y + 1 = 0$.

Also the three asymptotes cut the cubic in $n(n - 2)$, *i.e.*, $3(3 - 2) = 3$ points. As shown above, these, three points lie on the straight line

$$x - y + 1 = 0.$$

Example 62:

A thin closed rectangular box is to have one rectangular edge n times the length of another edge and the volume of the box is given to be S. Prove that the least surface S is given by

$$nS^3 = 54(n + 1)^2 V^2.$$

Solution:

Let the lengths of the edges be x, nx and y. Then V = the volume of the rectangular box = $x.nx.y = nx^2y$ = constant. ...(1)

Differentiating (1) w.r.t. x, we get

$$0 = nx^2\left(\frac{dy}{dx}\right) + 2nxy \text{ of } \left(\frac{dy}{dx}\right) = -\frac{2y}{x}. \qquad \text{...(2)}$$

Now S = the surface of the box = $2nx^2 + 2(1 + n)\,xy$. ...(3)

Here S is a function of x and y. But y is connected with x by (1). So we can regard S as a function of x only. For a maximum or a minimum of S, we have $\frac{dS}{dx} = 0$.

Now $\frac{dS}{dx} = 4nx + 2(1+n)\left(x\frac{dy}{dx} + y\right) = 4nx - 2(1 + n)y$, from (2).

$\therefore$ $\frac{dS}{dx} = 0$ gives $2nx = (1 + n)y$.

Also $\frac{d^2S}{dx^2} = 4n - 2(1+n)\frac{dy}{dx} = 4n + \{4(1 + n)y\}/x$, from (2) = + ive.

Therefore, S is least when $2nx = (1 + n)y$. For this least S, on putting $y = 2nx/(1 + n)$ in (1) and (3), we get

$$V = \frac{2n^2x^3}{1+n} \text{ and } S = 2nx^2 + 2(1 + n)x.\frac{2nx}{1+n} = 6nx^2.$$

$\therefore$ $S^3 = 6^3n^3x^6 = 6^3n^3[(1 + n)^2V^2/(4n^4)]$.

Hence $nS^3 = 54(n + 1)^2V^2$, when S is least.

2

Expansions of Functions

NORMAL EXPANSIONS OF FUNCTIONS

Maclaurin's Theorem

Let f(x) be a function of x which possesses continues derivatives of all orders in the interval [0, x]. Assuming that f(x) can be expanded as an infinite power series in x, we have

$$f(x) = f(0) + +\frac{x}{1!}f'(0)+\frac{x^2}{2!}f''(0)+...+\frac{x^n}{n!}f^{(n)}(0)+ ...$$

Proof:

Suppose $f(x) = A_0 + A_1x + A_2x^2 + A_3x^3 + ...$ ...(1)

Let the expansion (1) be differentiable term by term any number of times. Then by successive differentiation, we have

$$f'(x) = A_1 + 2A_2 + 3A_3x^2 + 4A_4x^3 + ...,$$

$$f''(x) = 2.1\ A_2 + 3.2\ A_3x + 4.3\ A_4x^2 + ...,$$

$$f'''(x) = 3.2.1\ A_3 + 4.3.2\ A_4x + ..., \text{ and so on.}$$

Putting x = 0 in each of these relations, we get

$$f(0) = A_0,\ f'(0) = A_1,\ f''(0) = 2\ !\ A_2,\ f'''(0) = 3\ !\ A_3, ...$$

Substituting theses values of A_0, A_1, A_2, ... in (1), we get

$$f(x) = f(0) + xf' + \frac{x^2}{2!}f^{(n)}(0) + ...$$

This is **Maclaurin's Theorem:** If we denote f(x) by y, then Maclaurin's theorem can be written in either of the following ways:

$$y = (y)_0 + \frac{x}{1!}(y_1)_0 + \frac{x^2}{2!}(y_2)_0 + \frac{x^3}{3!}(y_3)_0 + ... + \frac{x^n}{n!}(y_n)_0 + ...$$

or $y = y(0) + \frac{x}{1!} y_1(0) + \frac{x^2}{2!} y_2(0) + \ldots + \frac{x^n}{n!} y_n(0) + \ldots$

Taylor's Theorem

Let f(x) be a function of x which possesses continuous derivatives of all orders in the interval [a, a + h]. Assuming that f(a + h) can be expanded as an infinite power series in h, we have

$$f(a + h) = f(a) + hf'(a) + \frac{h^2}{2!} f''(a) + \ldots + \frac{h^n}{n!} f^{(n)}(a) + \ldots$$

Proof:

Suppose $f(a + h) = A_0 + A_1h + A_2h^2 + A_3h^3 + \ldots$ ...(1)

Let the expansion (1) be differentiable term by term any number of times w.r.t. h. Then by successive differentiation w.r.t. h, we have

$f'(a + h) = A_1 + 2A_2h + 3A_3h^2 + \ldots,$

$f''(a + h) = 2.1A_2 + 3.2\, A_3h + \ldots,$

$f'''(a + h) = 3.2.1\, A_3 + \ldots,$ and so on.

Putting h = 0 in each of the above relations, we get

$f(a) = A_0,\ f'(a) = A_1, f''(a)$

$= 2!\, A_2,\ f'''(a) = 3!\, A_3,$ and so on.

$\therefore\ A_0 = f(a),\ A_1,\ f'(a),\ A_2$

$= \frac{1}{2!} f''(a),\ A_3 = \frac{1}{3!} f'''(a)$, and so on.

Substituting these values of $A_0, A_1, A_2, A_3, \ldots$ in (1), we get

$$f(a + h) = f(a) + hf'(a) + \frac{h^2}{2!} f'''(a) + \ldots + \frac{h^n}{n!} f^{(n)}(a) + \ldots$$

This is **Taylor's Theorem.** Another useful form is obtained on replacing h by (x – a). Thus

$$f(x) = f(a) + (x - a) f'(a) + \frac{(x-a)^2}{2!} f''(a) + \ldots + \frac{(x-a)^n}{n!} f^{(n)}(a) + \ldots,$$

which is an expansion of f(x) as a power series in (x – a).

Note: If we expand f(x + h), by Taylor's theorem, as a power series in h, then the result is as follows:

$$f(x + h) = f(x) + hf'(x) + \frac{h^2}{2!} f''(x) + \ldots + \frac{h^n}{n!} f^{(n)}(x) + \ldots$$

SOME IMPORTANT EXPANSIONS

***Expansion of e^x:* (Exponential Series)**

Let $f(x) = e^x$. Then $f(0) = e^0 = 1$;

$f^{(n)}(x) = e^x$ so that $f^{(n)}(0) = e^0 = 1$, where $n = 1, 2, 3, ...$

Substituting these values in Maclaurin's series.

TAYLOR'S SERIES

Suppose f(x) possesses continuous derivatives of all orders in the interval [a, a + h]. Then for every positive integral value of n, we have

$$f(a + h) = f(a) + hf'(a) + \frac{h^2}{2!}f''(a)+...+\frac{h^{n-1}}{(n-1)!}f(n-1)(a)+R_n, \quad ...(1)$$

where $R_n = \frac{h^n}{n!}$ f(n) (a + θh), (0 < θ < 1).

Suppose $R_n \to 0$,a s $n \to \infty$. Then taking limits of both sides of equation (1) when $n \to \infty$, we get

$$f(a + h) = f(a) = hf'(a) = \frac{h^2}{2!}f''(a) + ... + \frac{h^n}{n!}f^{(n)}(a)+... \quad ...(2)$$

The series given in (2) is known as **Taylor's Infinite Series** for the expansion of f(a + h) as a **Power Series** in h.

A more useful form is obtain from the above on replacing ah by (x – a) Thus

$$f(x) = f(a) + (x - a) f'(a) + \frac{(x-a)^2}{2!} f''(a)+\frac{(x-a)^3}{3!}f'''(a)$$

$+ ... + \frac{(x-a)^n}{n!}f^n(a) + ...$ which is an expansion in powers of (x – a).

MACLAURIN'S SERIES

Let f(x) be a function of x which can be expanded in power of x and let the expansion be differentiable term by term any number of times.

$$f(x) = f(0) + xf' + \frac{x^2}{2!}f''(0) ... + \frac{x^{n-1}}{(n-1)!}f^{(n-1)}(0)+R_n, \quad ...(1)$$

where $R_n = \frac{x^n}{n!}f^{(n)}(\theta x), (0<\theta<1)$.

Suppose $R_n \to 0$, as $n \to \infty$. Then taking limits of both sides of (1) when $n \to \infty$, we get

$$f(x) = f(0) + xf'(0) + \frac{x^2}{2!}f''(0) + \dots + \frac{x^n}{n!}f^{(n)}(0) + \dots \qquad \dots(2)$$

The series given in (2) is known as **Maclaurin's Infinite Series** for the expansion of f(x) as a **Power Series** in x. Maclaurin's series is a particular case of Taylor's series. If in Taylor's series we put a = 0 and h = x we get Maclaurin's series.

Note: Maclaurin's Expansion of f(x) fails if any of the functions f(x), f'(x), f''(x), ... becomes infinite or discontinuous at any point of the interval [0, x] or if R_n does not tend to zero as $n \to \infty$.

SOLVED EXAMPLES

Example 1:

Explained by Maclaurin's Theorem $e^x/(1 + e^x)$ as far as the term x^3.

Solution:

Let $$y = \frac{e^x}{1+e^x} = \frac{1+e^x-1}{1+e^x} = 1 - \frac{1}{1+e^x}.$$

Then $$(y)_0 = \frac{e^0}{1+e^0} = \frac{1}{2},$$

$$y_1 = 0 + \frac{e^x}{(1+e^x)^2} = \frac{e^x}{1+e^x}\,\frac{1}{1+e^x} = y(1-y) = y - y^2 \text{ so that}$$

$$(y_1)_0 = \frac{1}{2} - \frac{1}{4} = \frac{1}{4},$$

$$y_2 = y_1 - 2yy_1 \text{ so that } (y_2)_0 = \frac{1}{4} - 2.\frac{1}{2}.\frac{1}{4} = 0,$$

$$y_3 = y_2 - 2y_1^2 - 2yy_2 \text{ so that } (y_3)_0 = 0 - 2.(1/4)^2 - 0 = -1/8 \text{ and so on.}$$

Substituting these values in Maclaurin's theorem, we get

$$\frac{e^x}{1+e^x} = \frac{1}{2} + x.\frac{1}{4} + \frac{x^2}{2!}.0 + \frac{x^3}{3!}.(-1/8) + \dots$$

$$= \frac{1}{2} + \frac{x}{4} - \frac{1}{48}x^3 + \dots$$

Example 2:

Expand the following functions by Maclaurin's theorem:

(i) $e^{\sin x}$

(ii) $e^{x \cos x}$

(iii) $e^x \log (1 + x)$

(iv) $\log (1 + \sin x)$.

Solution:

(i) Let $y = e^{\sin x}$. Then $(y)_0 = e^{\sin 0} = e^0 = 1$,

$y_1 = e^{\sin x} \cos x = y \cos x$ so that $(y_1)_0 = (y)_0 \cos 0 = 1 \times 1 = 1$,

$y_2 = y_1 \cos x - y \sin x$ so that $(y_2)_0 = 1 \times 1 - 1 \times 0 = 1$,

$y_3 = y_2 \cos x - y_1 \sin x - y_1 \sin x - y \cos x$

$= y_2 \cos x - 2y_1 \sin x - y \cos x$ so that $(y_3)_0 = 1 - 0 - 1 = 0$,

$y_4 = y_3 \cos x - 3y_2 \sin x - 3y_1 \cos x + y \sin x$ so that $(y_4)_0 = -3$, and so on.

Now by Maclaurin's Theorem, we have

$$y = (y)_0 + x(y_1)_0 + \frac{x^2}{2!}(y_2)_0 + \frac{x^3}{3!} + \frac{x^4}{4!}(y_4)_0 + \ldots$$

$$\therefore\ e^{\sin x} = 1 + x \,.\, 1 + \frac{x^2}{2!}.1 + \frac{x^3}{3!}.0 + \frac{x^4}{4!}.(-3) + \ldots$$

$$= 1 + x + \frac{x^2}{2} - \frac{x^4}{8} + \ldots$$

(ii) Let $y_1 = e^{x \cos x}$. Then $(y)_0 = e^0 = 1$,

so that $(y_1)_0 = 1.(1 - 0) = 1$,

$y_2 = y_1(\cos x - x \sin x) + (-2 \sin x - x \cos x)$ giving

$(y_2)_0 = 1.(1 - 0) + 1.0 = 1$,

$y_3 = y_2 (\cos x - x \sin x) + y_1(-2 \sin x - x \cos x)$

$+ y_1(-2 \sin x - x \cos x) + y (-3 \cos x + x \sin x)$

$= y_2(\cos x - \sin x) + 2y_1 (-2 \sin x - x \cos x)$

$+ y (-3 \cos x + x \sin x)$

so that $(v_3)_0 = 1.(1 - 0) + 0 + 1.(-3) = -2$,

$y_4 = y_3 (\cos x - x \sin x) + 3y_2 (-2 \sin x - x \cos x)$

$+ 3y_1 (-3 \cos x + x \sin x) + y(4 \sin x + x \cos x)$

giving $(y_4)_0 = -2.(1 - 0) + 0 + 3.1.(-3 + 0) + 0 - 11$,

$y_5 = y_4 (\cos x - x \sin x) + 4y_3 (-2 \sin x - x \cos x)$

$+ 6y_2 (-3 \cos x + x \sin x) + 4y_1 (4 \sin x + x \cos x)$

$+ y (5 \cos x - x \sin x)$

so that $(y_5)_0 = -11 + 6.1.(-3) + 5 = -24$, and so on.

Substituting these values in Maclaurin's theorem, we get

$$e^{x\cos x} = 1 + x.1 + \frac{x^2}{2!}.1 + \frac{x^3}{3!}.(-2) + \frac{x^4}{4!}.(-11) + \frac{x^5}{5!}.(-24) + ...$$

$$= 1 + x + \frac{x^2}{2} - \frac{x^3}{3} - \frac{11x^4}{24} - \frac{x^5}{5} + ...$$

(iii) Let $y = e^x \log(1 + x)$. Then $(y)_0 = e^0 \log 1 = 1 \times 0 = 0$,

$y_1 = e^x \log(1 + x)^{-1} = y + e^x(1 + x)^{-1}$ so that $(y_1)_0 = 0 + 1 = 1$,

$y_2 = y_1 + e^x(1 + x)^{-1} - e^x(1 + x)^{-2}$ so that $(y_2)_0 = 1 + 1 - 1 = 1$,

$y_3 = y_2 + e^x(1 + x)^{-1} - e^x(1 + x)^{-2} - e^x(1 + x)^{-2} + 2e^x(1 + x)^{-3}$

$= y_2 + e^x(1 + x)^{-1} - 2e^x(1 + x)^{-2} + 2e^x(1 + x)^{-3}$

so that $(y_3)_0 = 1 + 1 - 2 + 2 = 2$,

$y_4 = y_3 + e^x(1 + x)^{-1} - 3e^x(1 + x)^{-2} + 6e^x(1 + x)^{-3}$

$- 6e^x(1 + x)^{-4}$ so that $(y_4)_0 = 2 + 1 - 3 + 6 - 6 = 0$,

$y_5 = y_4 + e^x(1 + x)^{-1} - 4e^x(1 + x)^{-2} + 12e^x(1 + x)^{-3}$

$- 24e^x(1 + x)^{-4} + 24e^x(1 + x)^{-5}$ so that

$(y_5)_0 = 0 + 1 - 4 + 12 - 24 + 24 = 9$, and so on.

Substituting these values in Maclaurin's Theorem, we get

$$e^x \log(1 + x) = 0 + x.1 + \frac{x^2}{2!}.1 + \frac{x^3}{3!}.2 + \frac{x^4}{4!}.0 + \frac{x^5}{5!}.9 + ...$$

$$= x + \frac{x^2}{2!} + \frac{2x^3}{3!} + \frac{9x^5}{5!} + ...$$

(iv) Let $y = \log(1 + \sin x)$. Then $(y)_0 = 0$,

$y_1 = \dfrac{\cos x}{1 + \sin x}$ so that $(y_1)_0 = 1$,

$$y_2 = \frac{-\sin x(1 + \sin x) - \cos^2 x}{(1 + \sin x)^2} = \frac{-(1 + \sin x)}{(1 + \sin x)^2}$$

$= -\dfrac{1}{1 + \sin x}$ so that $(y_2)_0 = -1$,

$$y_3 = \frac{\cos x}{(1 + \sin x)^2} = \frac{\cos x}{1 + \sin x} \cdot \frac{1}{1 + \sin x} = -y_1 y_2$$

so that $(y_3)_0 = -1.(-1) = 1$,

$y_4 = -y_1y_3 - y_2^2$ so that $(y_4)_0 = -1.1 - (-1)^2 = -1 - 1 = -2$,

$y_5 = -y_1y_4 - y_2y_3 - 2y_2y_3 = -y_1y_4 - 3y_2y_3$ so that

$(y_5)_0 = -1.(-2) - 3.(-1).1 = 2 + 3 = 5$, and so on.

Substituting these values in Maclaurin's theorem, we get

$\log(1 + \sin x) = 0 + x.1 + (x^2/2\,!).(-1) + (x^3/3\,!).1$

$+ (x^4/4\,!).(-2) + (x^5/5\,!).5 + \ldots$

$$= x - \frac{x^2}{2} + \frac{x^3}{6} - \frac{x^4}{12} + \frac{x^5}{24} + \ldots$$

Example 3:

Apply Maclaurin's Theorem to find the expansion in ascending powers of x of $\log_e(1 + e^x)$ to the term containing x^4.

Solution:

Let $y = \log_e(1 + e^x)$. Then $(y)_0 = \log_e(1 + e^0) = \log_e 2$,

$$y_1 = \frac{e^x}{1+e^x} = \frac{(1+e^x)-1}{1+e^x} = 1 - \frac{1}{1+e^x} \text{ so that } (y_1)_0 = 1 - \frac{1}{2} = \frac{1}{2},$$

$$y_2 = 0 + \frac{e^x}{(1+e^x)^2} = \frac{e^x}{(1+e^x)}.\frac{1}{1+e^x} = y_1(1-y_1) = y_1 - y_1^2$$

so that $(y_2)_0 = \dfrac{1}{2} - \left(\dfrac{1}{2}\right)^2 = \dfrac{1}{4}$,

$y_3 = y_2 - 2y_1y_2$ so that $(y_3)_0 = \dfrac{1}{4} - 2.\dfrac{1}{2}.\dfrac{1}{4} = 0$,

$y_4 = y_3 - 2y_2^2 - 2y_1y_3$ so that $(y_4)_0 = 0 - 2.\left(\dfrac{1}{4}\right)^2 - 0 = -\dfrac{1}{8}$, and so on.

Substituting these values in Maclaurin's Expansion.

$$y = (y)_0 + x(y_1)_0 + \frac{x^2}{2!}(y_2)_0 + \frac{x^3}{3!}(y_3)_0 + \frac{x^4}{4!}(y_4)_0 + \ldots$$

$\therefore \log(2 + e^x)$

$$= \log 2 + x.\frac{1}{2} + \frac{x^2}{2!}.\frac{1}{4} + \frac{x^3}{3!}.0 + \frac{x^4}{4!}.\left(-\frac{1}{8}\right) + \ldots$$

$$= \log 2 + \frac{x}{2} + \frac{x^2}{8} - \frac{x^4}{192} + \ldots$$

Example 4:

Expand $e^{a \sin^{-1} x}$ *by Maclaurin's theorem and find the general term. Hence show that*

$$e^{\theta} = 1 + \sin\theta + \frac{1}{2!}\sin^2\theta + \frac{2}{3!}\sin^3\theta + \ldots$$

Solution:

Let $y = e^{a \sin^{-1} x}$. We have

$y(0) = 1,\ y_1(0) = a,\ y_2(0) = a^2,$

and $y_{n+2}(0) = (n^2 + a^2)y_n(0).$ ...(1)

Putting n = 1, 2, 3, 4, ... in (1), we get

$$y_3(0) = (1^2 + a^2)\, y_1(0) = (1^2 + a^2)\, a,\ y_4(0) = (2^2 + a^2)\, y_2(0)$$
$$= (2^2 + a^2)\, a^2,\ y_5(0) = (3^2 + a^2)\, y_3(0) = (3^2 + a^2)(1^2 + a^2)a$$
$$y_6(0) = (4^2 + a^2)\, y_4(0) = (4^2 + a^2)(2^2 + a^2)\, a^2, \text{ etc.}$$

In general,

$$y_n(0) = \begin{cases} a\left(1^2+a^2\right)\left(3^2+a^2\right)\ldots\left[(n-2)^2+a^2\right] \text{ if n is odd} \\ a^2\left(2^2+a^2\right)\left(4^2+a^2\right)\ldots\left[(n-2)^2+a^2\right] \text{ if n is even.} \end{cases}$$

Substituting these values in Maclaurin's Expansion

$$y = y(0) + xy_1(0) + \frac{x^2}{2!}y_2(0) + \ldots + \frac{x^n}{n!}y_n(0) + \ldots,$$

we get

$$e^{a \sin^{-1} x} + 1 + ax + \frac{a^2}{2!}x^2 + \frac{a\left(1^2+a^2\right)}{3!}x^3 + \frac{a^2\left(2^2+a^2\right)}{4!}x^4 + \ldots \quad \ldots(2)$$

The general term is $(x^n/n!)\, y_n(0)$, where $y_n(0)$ is as given above.

Now putting $x = \sin\theta$ and $a = 1$ in (2), we get

$$e^{\theta} = 1 + \sin\theta + \frac{1}{2!}\sin^2\theta + \frac{2}{3!}\sin^3\theta + \ldots$$

Example 5:

Applying Maclaurin's theorem to obtain terms into x^4 *in the expansion of* $\log(1 + \sin^2 x)$.

Solution:

Let $y = \log(1 + \sin^2 x)$. Then $(y)_0 = 0$.

Now $e^y = 1 + \sin^2 x$.

Differentiating, we get

$$e^y y_1 = 2 \sin x \cos x = \sin 2x \quad ...(1)$$

Putting x = 0 in (1), we get

$$e^0 (y_1)_0 = 0 \text{ or } (y_1)_0 = 0.$$

Differentiating (1), we get

$$e^y\left(y_1^2 + y_2\right) = 2 \cos 2x \quad ...(2)$$

Putting x = 0 in (2), we get

$$(y_2)_0 = 2.$$

Differentiating (2), we get

$$e^y\left[\left(y_1^2 + y_2\right)y_1 + 2y_1y_2 + y_3\right] = -4 \sin 2x$$

or $$e^y\left(y_1^3 + 2y_1y_2 + y_3\right) = -4\sin 2x. \quad ...(3)$$

Putting x = 0 in (3), we get

$$(y_3)_0 = 0.$$

Differentiating (3), we get

$$e^y y_1\left(y_1^3 + 3y_1y_2 + y_3\right) + e^y\left(3y_1^2y_2 + 3y_2^2 + 3y_1y_3 + y_4\right) = -8 \cos 2x$$

or $$e^y\left(y_1^4 + 6y_1^2y_2 + 4y_1y_3 + 3y_2^2 + y_4\right) = -8 \cos 2x. \quad ...(4)$$

Putting x = 0 in (4), we get

$$3.2^2 + (y_4)_0 = -8 \text{ or } (y_4)_0 = -20.$$

Now substituting these values in Maclaurin's theorem, we get

$$\log\left(1+\sin^2 x\right) = 0 + x.0 + \frac{x^2}{2!}.2 + \frac{x^3}{3!}.0 + \frac{x^4}{4!}(-20) + ...$$

$$= x^2 - \frac{5}{6}x^4 + ...$$

Example 6:

Expand log $\left\{x + \sqrt{\left(1+x^2\right)}\right\}$ *in ascending powers of x and find the general term.*

Solution:

Let y log = $\left\{x + \sqrt{\left(1-x^2\right)}\right\}$, then

$$y_1 + \frac{1}{x+\sqrt{(1+x^2)}}.\left\{1+\frac{2x}{2\sqrt{(x^2+1)}}\right\} = \frac{1}{\sqrt{(x^2+1)}}$$

Therefore $y_1^2(x^2+1)-1=0$.

Differentiating again, we get

$$(x^2 + 1)\, 2y_1\, y_2 + 2xy_1^2 = 0$$

or $(x^2 + 1)y_2 + xy_1 = 0$, since $2y_1 \neq 0$.

Now differentiating n times by Leibnitz's theorem, we get

$$(x^2 + 1)y_{n+2} + (2n + 1)xy_{n+1} + n^2y_n = 0.$$

Putting x = 0 in the above relations, we get

$(y)_0 = 0$, $(y_1)_0 = 1$, $(y_2)_0 = 0$, and

$(y_{n+2})_0 = -n^2(y_n)_0$. ...(1)

Now putting n = 1, 3, 5, ... in (1), we get

$(y_3)_0 = -1^2(y_1)_0 = -1^2$, $(y_5)_0 = (-3^2)(y_3)_0 = (-3^2)(-1^2) = 3^2.1^2$,

$(y_7)_0 = (-5^2)(y_5)_0 = (-5^2)(-3^2)(-1^2) = -5^2.3^2.1^2$, etc.

In general, if n is odd, we have

$(y_n)_0 = \{-(n-2)^2\}\{-(n-4)^2 \ldots (-5^2)(-3^2)(-1^2)$

$= (-1)^{(n-1)/2}(n-2)^2(n-4)^2 \ldots 5^2.3^2.1^2$. ...(2)

Again, putting n = 2, 4, 6 ... in (1), we get

$(y_4)_0 = -2^2\,(y_2)_0 = 0$, $(y_6)_0 = -4^2\,(y_4)_0 = 0$, etc.

Thus, if n is even, we have $(y_n)_0 = 0$.

Now by Maclaurin's theorem, we have

$$\log\left[x+\sqrt{(1+x^2)}\right] = (y)_0 + x(y_1)_0 + \frac{x^2}{2!}(y_2)_0 + \frac{x^3}{3!}(y_3)_0 + \ldots$$

$$= 0 + x.1 + \frac{x^2}{2!}.0 + \frac{x^3}{3!}(-1^2) + \frac{x^4}{4!}.0 + \frac{x^5}{5!}(3^2.1^2)$$

$$= x - \frac{x^3}{3!}1^2 + \frac{x^5}{5!}(3^2.1^2) - \frac{x^7}{7!}(5^2.3^2.1^2) + \ldots$$

The general term = $(x^n/n!)\,(y_n)_0$, where $(y_n)_0$ is given by (2) when n is odd and $(y_n)_0 = 0$, when n is even.

Putting 2n – 1 in place of n in (2), we find that

$(x_{2n-1})_0 = (-1)^{n-1}(2n-3)^2 \ldots 5^2.3^2.1^2.$

$\log\left[x+\sqrt{(1+x^2)}\right]$

$$= x.1^2.\frac{x^3}{3!}+1^2.3^2\frac{x^5}{5!}-1^2.3^2.5^2.\frac{x^7}{7!}+\ldots$$

$$+(-1)^{(n-1)}1^2.3^2.5^2.(2n-3)^2.\frac{x^{2n-1}}{(2n-1)^2}$$

Example 7:

Find the first four terms in the expansion of log (1 + tan x) in powers of x.

Solution:

Let $y = \log(1 + \tan x)$. Then $(y)_0 = \log(1 + \tan 0) = 0$.

Now $e^y = 1 + \tan x$. Differentiating both sides w.r.t. x, we get

$$e^y y_1 = \sec^2 x. \qquad \ldots(1)$$

Putting $x = 0$ on both sides of (1), we get

$$e^0 (y_1)_0 = 1 \text{ or } (y_1)_0 = 1.$$

Differentiating (1), we get

$$e^y y_1^2 + e^y y_2 = 2\sec^2 x \tan x$$

or $$e^y\left(y_1^2+y_2\right)=2\sec^2 x \tan x. \qquad \ldots(2)$$

Putting $x = 0$ in (2), we get

$$1 + (y_2)_0 = 0 \text{ or } (y_2)_0 = 1.$$

Differentiating (2), we get

$$e^y y_1\left(y_1^2+y_2\right)+e^y\left(2y_1y_2+y_3\right) = 2\sec^4 x + 4\sec^2 x \tan^2 x$$

or $$e^y\left(y_1^3+3y_1y_2+y_3\right)=2\sec^4 x+4\sec^2 x \tan^2 x. \qquad \ldots(3)$$

Putting $x = 0$ in (3), we get

$$1 + 3.1.(-1) + (y_3)_0 = 2 \text{ or } (y_3)_0 = 4.$$

Differentiating (3), we get

$$e^y y_1\left(y_1^3+3y_1y_2+y_3\right)+e^y\left(3y_1^2y_2+3y_2^2+3y_1y_3+y_4\right)$$

$$= 8\sec^4 x \tan x + 8\sec^2 x \tan^3 x + 8\sec^4 x \tan x$$

or $\quad e^y\left(y_1^4+6y_1^2y_2+4y_1y_3+3y_2^2+y_4\right)$

$$= 16 \sec^4 x \tan x + 8 \sec^2 x \tan^3 x. \qquad ...(4)$$

Putting x = 0 in (4), we get

$$1 + 6.1.(-1) + 4.1.4 + 3.(-1)^2 + (y_4)_0 = 0$$

or $\quad (y_4)_0 + 14 = 0$ or $(y_4)_0 = -14.$

Now substituting these values in Maclaurin's theorem, we get

$$\log(1+\tan x) = 0 + x.1 + \frac{x^2}{2!}.(-1)+\frac{x^3}{3!}.4+\frac{x^4}{4!}.(-14)+...$$

$$= x-\frac{1}{2}x^2+\frac{2}{3}x^3-\frac{7}{12}x^4+...$$

Example 8:

Show that

(i) $e^x \cos x = 1 + x - \frac{2x^3}{3!}-\frac{2^2x^4}{4!}-\frac{2^2x^5}{5!}+\frac{2^3x^7}{7!}+...$

$$+2^{n/2}\cos\frac{n\pi}{4}.\frac{x^n}{n!}+...$$

(ii) $e^x \sin x = x + x^2 - \frac{2}{3!}x^3-\frac{2^2}{5!}x^5-...+\sin\left(\frac{1}{4}n\pi\right)\frac{2^{n/2}}{n!}x^n+...$

Solution:

(i) Let $y = e^x \cos x$. Then $(y)_0 = e^0 \cos 0 = 1$,

$y_1 = e^x \cos x - e^x \sin x = e^x (\cos x - \sin x)$

so that $(y_1)_0 = 1(1-0) = 1$,

$y_2 = e^x (\cos x - \sin x) + e^x (-\sin x - \cos x) = -2e^x \sin x$

so that $(y_2)_0 = 0$,

$y_3 = -2e^x \sin x - 2e^x \cos x = -2e^x (\sin x + \cos x)$

so that $(y_3)_0 = -2$,

$y_4 = -2e^x(\sin x + \cos x) - 2e^x (\cos x - \sin x)$

$= -4e^x \cos x = -2^2y$ so that $(y_4)_0 = -2^2$,

$y_5 = -2^2y_1$ so that $(y_5)_0 = -2^2$,

$y_6 = -2^2y_2$ so that $(y_6)_0 = 0$,

$y_7 = -2^2y_3$ so that $(y_7)_0 = 2^3$, and so on.

In general

$y_n = (1 + 1)^{n/2} \cos (x + n \tan^{-1} 1) = 2^{n/2} \cos (x + n\pi/4)$

so that $(y_n)_0 = 2^{n/2} \cos \left(\frac{1}{4} n\pi\right)$.

Now by Maclaurin's theorem, we have

$$y = (y)_0 + x(y_1)_0 + \frac{x^2}{2!}(y_2)_0 + \ldots + \frac{x^n}{n!}(y_n)_0 + \ldots$$

$$= 1 + x.1 + \frac{x^2}{2!}.0 + \frac{x^3}{3!}.(-2) + \frac{x^4}{4!}.(-2^2) + \frac{x^5}{5!}.(-2^2)$$

$$+ \frac{x^6}{6!}.0 + \frac{x^7}{7!}.2^3 + \ldots + \frac{x^n}{n!} 2^{n/2} \cos\left(\frac{1}{4} n\pi\right) + \ldots$$

$$= 1 + x - \frac{2x^3}{3!} - \frac{2^2 x^4}{4!} - \frac{2^2 x^5}{5!} + \frac{2^3 x^7}{7!} + \ldots + 2^{n/2} \cos\left(\frac{1}{4} n\pi\right)\frac{x^n}{n!} + \ldots$$

(ii) Proceed as in part (i).

Example 9(a):

Apply Maclaurin's theorem to prove that

(i) $$e^{ax} \sin bx = bx + abx^2 + \frac{3a^2 b - b^3}{3!} x^3 + \ldots$$

$$\left. + \frac{(a^2 + b^2)^{n/2}}{n!} x^n \sin\left(n \tan^{-1}(b/a)\right)\right\} + \ldots$$

(ii) $$e^{ax} \cos bx = 1 + ax + \frac{a^2 - b^2}{2} x^2 + \frac{a(a^2 - 3b^2)}{3!} x^3 + \ldots$$

$$+ \frac{(a^2 + b^2)^{n/2}}{n!} x^n \cos\left\{n \tan^{-1}(b/a)\right\} + \ldots$$

Hence deduce that

$$e^{x \cos \alpha} \cos (x \sin \alpha) = 1 + x \cos a + \frac{x^2}{2!} \cos 2\alpha + \frac{x^3}{3!} \cos 3\alpha + \ldots$$

Solution:

(i) Let $y = e^{ax} \sin bx$. Then $(y)_0 = e^0 \sin 0 = 0$,

$y_1 = ae^{ax} \sin bx + be^{ax} \cos bx = ay + be^{ax} \cos bx$

so that $(y_1)_0 = b$,

$y_2 = ay_1 + abe^{ax} \cos bx - b^2e^{ax} \sin bx$

$= ay_1 - b^2y + abe^{ax} \cos bx$ so that $(y_2)_0 = ab - 0 + ab = 2ab,$

$y_3 = ay_2 - b^2y_1 + a^2be^{ax} \cos bx - ab^2e^{ax} \sin bx$

$= ay_2 - b^2y_1 - ab^2y + a^2be^{ax} \cos bx$

so that $(y_3)_0 = 2a^2b - b^3 + a^2b = 3a^2 b - b^3$, and so on.

In general

$$y_n = (a^2 + b^2)^{n/2} \sin \{bx + n \tan^{-1} (b/a)\}$$

so that $(y_n)_0 = (a^2 + b^2)^{n/2} \sin \{n \tan^{-1} (b/a)\}$.

Now by Maclaurin's theorem, we have

$$y = (y)_0 + \frac{x}{1!}.(y_1)_0 + \frac{x^2}{2!}(y_2)_0 + \frac{x^3}{3!}(y_3)_0 + \ldots + \frac{x^n}{n!}(y_n)_0 + \ldots$$

$$= 0 + \frac{x}{1!}.b + \frac{x^2}{2!}.(2ab) + \frac{x^3}{3!}(3a^2b - b^3) + \ldots$$

$$+ \frac{x^n}{n!}(a^2 + b^2)^{n/2} \sin \{n \tan^{-1}(b/a)\} + \ldots$$

$$= bx + abx^2 + \frac{3a^2b - b^3}{3!}x^3 + \ldots$$

$$+ \frac{(a^2 + b^2)^{n/2}}{n!} x^n \sin \{n \tan^{-1}(b/a)\} + \ldots$$

(ii) Let $y = e^{ax} \cos bx$. Then $(y)_0 = e^0 \cos 0 = 1$,

$y_1 = ae^{ax} \cos bx - be^{ax} \sin bx = ay - be^{ax} \sin bx$ so that $(y_1)_0 = a$,

$y_2 = ay_1 - abe^{ax} \sin bx - b^2e^{ax} \cos bx = ay_1 - b^2y - abe^{ax} \sin bx$

so that $(y_2)_0 = a^2 - b^2$,

$y_3 = ay_2 - b^2y_1 - a^2be^{ax} \sin bx - ab^2e^{ax} \cos bx$

$= ay_2 - b^2y_1 - ab^2y - a^2be^{ax} \sin bx$

so that $(y_3)_0 = a(a^2 - b^2) - b^2a - ab^2 = a(a^2 - 3b^2)$, and so on.

In general, $y_n = (a^2 + b^2)^{n/2} \cos \{bx + n \tan^{-1} (b/a)\}$ so that

$(y_n)_0 = (a^2 + b^2)^{n/2} \cos \{n \tan^{-1} (b/a)\}$.

Substituting these values in Maclaurin's theorem, we get

$$e^{ax} \cos bx = 1 + ax + \frac{a^2 - b^2}{2!}x^2 + \frac{a(a^2 - 3b^2)}{3!}x^3 + \ldots$$

$$+ \frac{(a^2 + b^2)^{n/2}}{n!} x^n \cos \{n \tan^{-1}(b/a)\} + \ldots$$

Deduction: Putting a = cos α and b = sin α, we get

$(y_n)_0 = (\cos^2 \alpha + \sin^2 \alpha)^{n/2} \cos (n \tan^{-1} \tan \alpha) = \cos n\alpha$ so that

$(y_1)_0 = \cos \alpha$, $(y_2)_0 = \cos 2\alpha$, $(y_3)_0 = \cos 3\alpha$, etc.

$\therefore\ e^{x \cos \alpha} \cos (x \sin \alpha)$

$$= 1 + x \cos \alpha + (x^2/2\,!) \cos 2\alpha + (x^3/3\,!) \cos 3\alpha + \ldots$$

Example 9(b):

Expand log sin x in powers of (x – a).

Solution:

Let f(x) = log sin x. We can write f(x) = f [a + (x – a)]. Expanding f [a + (x – a)] by Taylor's theorem in powers of (x – a), we get

$$f(x) = f(a) + (x - a) f'(a) + (1/2\,!)(x - a)^2 f''(a) + (1/3\,!)(x - a)^3 f'''(a) + \ldots \qquad \ldots(1)$$

Now f(x) = log sin x. Therefore f(a) = log sin a,

$f'(a) = (1/\sin x).\cos x = \cot x$, giving $f'(c) = \cot a$

$f''(x) = -\operatorname{cosec}^2 x$ so that $f''(a) = -\operatorname{cosec}^2 a$,

$f'''(x) = 2 \operatorname{cosec}^2 x \cot x$ so that $f'''(a) = 2 \operatorname{cosec}^2 a \cot a$, and so on.

Substituting these values in (1), we get

$$\log \sin x = \log \sin a + (x - a) \cot a - \frac{(x-a)^2}{2!} \operatorname{cosec}^2 a + \frac{(x-a)^3}{3!} 2 \operatorname{cosec}^2 a \cot a + \ldots$$

Example 10:

Expand log {1 – log (1 – x)} in powers of x by Maclaurin's theorem as far as the term x^3.

By substituting x/(1 + x) for x deduce the expansion of g{1 + log (1 + x)} as far as the term in x^3.

Solution:

Let y = log{1 – log (1 – x)}. Then $(y)_0 = 0$.

Now $e^y = 1 - \log (1 - x)$. Differentiating, we get

$$e^x y_1 = (1 - x)^{-1}. \qquad \ldots(1)$$

Putting x = 0 in (1), we get $(y_1)_0 = 1$.

Differentiating (1), we get

$$e^y y_1^2 + e^y y_2 = (1-x)^{-2}$$

or $$e^y\left(y_1^2 + y_2\right) = (1-x)^{-2} \qquad ...(2)$$

Putting x = 0 in (2), we get

$$1 + (y_2)_0 = 1 \text{ or } (y_2)_0 = 0. \qquad ...(3)$$

Differentiating (2), we get

$$e^y\left(y_1^3 + 3y_1y_2 + y_3\right) = 2(1-x)^{-3}$$

Putting x = 0 in (3), we get

$$1 + (y_3)_0 = 2 \text{ or } (y_3)_0 = 1.$$

Substituting these values in Maclaurin's theorem, we get

$$\log\{1 - \log(1-x)\} = 0 + x.1 + \frac{x^2}{2!}.0 + \frac{x^3}{3!}.1 + ...$$

$$= x + \left(\frac{x^3}{6}\right) + ... \qquad ...(A)$$

Now substituting x/(1 + x) for x on both sides of (A), we get

$$\log\left\{1 - \log\left(1 - \frac{x}{1+x}\right)\right\} = \frac{x}{1+x} + \frac{1}{6}\left(\frac{x}{1+x}\right) + ...$$

or $$\log\{1 + \log(1+x)\} = x(1+x)^{-1} + \left(\frac{1}{6}\right)x^3(1+x)^{-3} + ...$$

$$= x\left\{1 + (-1)x + \frac{(-1)(-2)}{1.2}x^2 + ...\right\} + \frac{1}{6}x^3\{1 + (-3)x + ...\}$$, on expanding by binomial theorem

$$= (x - x^2 + x^3 + ..) + \left(\frac{1}{6}x^3 + ...\right)$$

$$= x - x^2 + \frac{7}{6}x^3 + ...$$

Example 11:

If $y = \sin\log(x^2 + 2x + 1)$, *prove that*

$$(x+1)^2 y_{n+2} + (2n+1)(x+1)y_{n+1} + (n^2+4)y_n = 0.$$

Hence or otherwise expand y in ascending powers of x as for as x^6.

Solution:

Here $y = \sin\log(x^2 + 2x + 1) = \sin\log(x+1)^2$ ...(1)

$$\therefore\ y_1 = [\cos\log(x+1)^2].\frac{1}{(x+1)^2}.2(x+1)$$

$$= [\cos\log(x+1)^2].\frac{2}{x+1}. \qquad ...(2)$$

Squaring both sides of (2), we get

$$(x+1)^2 y_1^2 = 4\cos^2\log(x+1)^2 = 4[1 - \sin^2\log(x+1)^2]$$

$$= 4(1 - y^2) \qquad ...(3)$$

or $(x+1)^2 y_1^2 + 4y^2 - 4 = 0$.

Differentiating equation (3), we get

$$(x+1)^2\, 2y_1y_2 + 2(x+1)y_1^2 + 8yy_1 = 0$$

or $\quad 2y_1[(x+1)^2y_2 + (x+1)y_1 + 4y] = 0$ ...(4)

or $\quad (x+1)^2y_2 + (x+1)y_1 + 4y = 0,$

since $\quad 2y \neq 0.$

Differentiating (4) n times by Leibnitz's theorem, we get

$$(x+1)^2y_{n+2} + {}^nC_1.y_{n+1}.2(x+1) + {}^nC_2.y_n.2 + (x+1)y_{n+1} + {}^nC_1.y_n.1 + 4y_n = 0$$

or $(x+1)^2y_{n+2} + (2n+1)(x+1)y_{n+1} + (n^2+4)y_n = 0$ ...(5)

Putting x = 0 in (5), we get

$$(y)_0 = 0,\ (y_1)_0 = 2,\ (y_2)_0 + (y_1)_0 + 4(y)_0 = 0$$

or $\quad (y_2)_0 = -2$

Also putting x = 0 in (5), we get

$$(y_{n+2})_0 + (2n+1)(y_{n+1})_0 + (n^2+4)(y_n)_0 = 0$$

or $\quad (y_{n+2})_0 = -[(2n+1)(y_{n+1})_0 + (n^2+4)(y_n)_0].$ ...(6)

Now putting n = 1, 2, 3, 4 in (6), we get

$(y_3)_0 = -[3(y_2)_0 + 5(y_1)_0] = -[3.(-2) + 5.2] = -4,$

$(y_4)_0 = -[5(y_3)_0 + 8(y_2)_0] = -[5.(-4) + 8.(-2)] = 36,$

$(y_5)_0 = -[7(y_4)_0 + 13(y_3)_0] = -[7.(36) + 13.(-4)] = -200,$

and $(y_6)_0 = -[9(y_5)_0 + 20(y_4)_0] = -[9.(-200) + 20.(36)] = 1080.$

Now by Maclaurin's theorem, we have

$$y=(y)_0+\frac{x}{1!}(y_1)_0+\frac{x^2}{2!}(y_0)+\frac{x^3}{3!}(y_3)_0+\frac{x^4}{4!}(y_4)_0+\frac{x^5}{5!}(y_5)_0+\ldots$$

$$=0+\frac{x}{1}.2+\frac{x^2}{2}.(-2)+\frac{x^3}{1.2.3}.(-4)+\frac{x^4}{1.2.3.4}.36$$

$$+\frac{x^5}{1.2.3.4.5}.(-200)+\frac{x^6}{1.2.3.4.5.6}.(1080)+\ldots$$

$$=2x-x^2-\frac{2}{3}x^3+\frac{3}{2}x^4-\frac{5}{3}x^5+\frac{3}{2}x^6+\ldots$$

Example 12:

Use Maclaurin's theorem to show that

$$e^{m\cos 3^{-1}x}=e^{mp/2}\left[1-mx+\frac{m^2}{2!}x^2-\frac{m(1^2+m^2)}{3!}x^3+\frac{m^2(2^2+m^2)}{4!}x^4-\ldots\right]$$

Solution:

Do yourself.

Example 13:

Expand sin (m sin^{-1} x) by Maclaurin's theorem as far as h^5. Hence expand sin mθ in powers of sin θ.

Solution:

Let $y = \sin(m \sin^{-1} x)$, we get

$y(0) = 0,\ y_1(0) = m,\ y_2(0) = 0,$

and $y_{n+2}(0) = (n^2 - m^2)y_n(0).$ ...(1)

Putting $n = 1, 2, 3, \ldots$ in (1), we get

$y_3(0) = (1^2 - m^2)y_1(0) = (1^2 - m^2)\,m,$

$y_4(0) = (2^2 - m^2)y_2(0) = 0,$

$y_5(0) = (3^2 - m^2)y_3(0) = (3^2 - m^2)(1^2 - m^2)\,m$, etc.

Substituting these values in Maclaurin's theorem, we get

$$\sin(m\sin^{-1}x) = y(0) + xy_1(0) + \frac{x^2}{2!}y_2(0)+\ldots$$

$$=mx+\frac{m(1^2-m^2)}{3!}x^3+\frac{m(1^2-m^2)(3^2-m^2)}{5!}x^5+\ldots$$

Putting x = sin θ on both sides, we get

$$\sin m\theta = m\sin\theta + \frac{m(1^2 - m^2)}{3!}\sin^3\theta + \frac{m(1^2 - m^2)(3^2 - m^2)}{5!}\sin^5\theta + \ldots$$

Example 14(a):

(a) *If* $y = (\sin^{-1} x)/\sqrt{(1-x)^2}$, *where* $-1 < x < 1$

and $-\pi/2 < \sin^{-1} x < \pi/2$

prove that $(1 - x^2)y_{n+1} - (2n + 1)xy_n - n^2y_{n-1} = 0.$

Also if $y = a_0 + a_1x + a_2x^2 + \ldots + a_nx^n + \ldots,$

prove that $(n + 1)a_{n+1} = n\, a_{n-1}$ *and hence obtain the general term of the expansion.*

(b) *Expand* $(\sin^{-1} x)/\sqrt{(1-x^2)}$ *in powers of x upto three terms.*

Solution:

(a) Here $y = (\sin^{-1} x)\sqrt{(1-x^2)}$.

$\therefore \quad y^2(1 - x^2) = (\sin^{-1} x)^2.$

Differentiating w.r.t. x, we ger

$$2yy_1(1 - x^2) - 2xy^2 = 2(\sin^{-1} x)/\sqrt{(1-x^2)} = 2y.$$

Since $2y \neq 0$, therefore $y_1(1 - x^2) - xy = 1$

i.e., $\quad y_1(1 - x^2) - xy - 1 = 0.$ ...(2)

Differentiating (2) n times by Leibnitz's theorem, we get

$$y_{n+1}(1 - x^2) + ny_n(-2x) + \frac{n(n-1)}{2!}y_{n-1}\cdot(-2) - xy_n - ny_{n-1} = 0$$

or $\quad (1 - x^2)y_{n+1} - (2n + 1)xy_n - n^2y_{n-1} = 0.$ ...(3)

Now putting x = 0 in (1), (2) and (3), we get

$$(y)_0 = 0,\ (y_1)_0 = 1 \text{ and } (y_{n+1})_0 = n^2(y_{n-1})_0$$

By Maclaurin's theorem, we have

$$y = (y)_0 + x(y_1)_0 + (x^2/2!)\,(y_2)_0 + \ldots + (x^n/n!\,(y_n)_0 + \ldots$$

Also we are given that

$$y = a_0 + a_1x + a_2x^2 + \ldots + a_nx^n + \ldots$$

Comparing the coefficients of x^n in the two expansions for y, we get $a_n = (y_n)_0/n!$.

$$\therefore \quad \frac{a_{n+1}}{a_{n-1}} = \frac{(y_{n+1})_0}{(n+1)!} \div \frac{(y_{n-1})_0}{(n-1)!} = \frac{(y_{n+1})_0}{(y_{n-1})_0} \cdot \frac{(n-1)!}{(n+1)!}$$

$$= n^2 \cdot \frac{1}{n(n+1)}, \qquad \text{from (4)}$$

$$= \frac{n}{n+1}.$$

$$\therefore \quad (n+1)a_{n+1} = n\, a_{n-1}$$

or $$a_{n+1} = \frac{n}{n+1} a_{n-1} \qquad ...(5)$$

Example 14(b):

(a) *By Maclaurin's theorem or otherwise find the expansion of* $y = \sin(e^x - 1)$ *upto and including the term in* x^4.

(b) *Also show that* $x = y - \frac{1}{2}y^2 + ...$

Solution:

(a) Let $y = \sin(e^x - 1)$. Then $(y)_0 = \sin 0 = 0$,

$y_1 = [\cos(e^x - 1)].e^x$ so that $(y_1)_0 = (\cos 0).e^0 = 1$,

$y_2 = [\cos(e^x - 1)].e^x - [\sin(e^x - 1)].e^{2x} = y_1 - ye^{2x}$

so that $(y_2)_0 = (y_1)_0 - (y)_0\, e^0 = 1 - 0 = 1$,

$y_3 = y_2 - y_1\, e^{2x} - 2y\, e^{2x}$ so that $(y_3)_0 = 1 - 1 - 0 = 0$,

$y_4 = y_3 - y_2\, e^{2x} - 4y_1\, e^{2x} - 4y\, e^{2x}$ so that $(y_4)_0 = -5$, etc.

Hence by Maclaurin's theorem, we get

$$\sin(e^x - 1) = (y)_0 + x(y_1)_0 + \frac{x^2}{2!}(y_2)_0 + \frac{x^3}{3!}(y_3)_0 + ...$$

$$= 0 + x.1 + \frac{x^2}{2!}.1 + \frac{x^3}{3!}.0 + \frac{x^4}{4!}.(-5) + ...$$

$$= x + \frac{1}{2}x^2 - \frac{5}{24}x^4 + ...$$

(b) We have

$y = \sin(e^x - 1) \Rightarrow e^x - 1 = \sin^{-1}y \Rightarrow e^x = 1 + \sin^{-1} y.$ ...(1)

Differentiating (1) w.r.t. 'y', we get

$$e^x.x_1 = 1/\sqrt{(1-y^2)}, \text{ where } x_1 = \frac{dx}{dy}. \qquad ...(2)$$

From equation (2), we get $(1 - y^2)x_1^2 = e^{-2x}$.

Differentiating it w.r.t 'y', we get

$$\left(1-y^2\right)2x_1x_2-2yx_1^2=e^{-2x}\left(-2x_1\right)$$

or $(1 - y^2)x_2 - yx_1 = - e^{-2x}$, since $2x_1 \neq 0$. ...(3)

From (1), we have $x = \log(1 + \sin^{-1} y)$. ...(4)

Now putting $y = 0$ in (4), (2) and (3), we get

$(x)_0 = \log(1 + 0) = 0, e^0.(x_1)_0 = 1/\sqrt{(1-0)}$ giving $(x_1)_0 = 1$,

$(x_2)_0 = - e^0 = - 1$. [Note that $(x)_{y=0} = 0$]

Hence by Maclaurin's theorem, we get

$$x = (x)_0 + y(x_1)_0 + (y^2/2\,!)\,(x_2)_0 + \ldots$$

$$= 0 + y.1 + (y_2/2\,!)\,(-1) + \ldots = y - \frac{1}{2}y^2 + \ldots$$

Now $a_0 = (y)_0 = 0$, $a_1 = (y_1)_0 = 1$. Putting $n = 1, 3, 5, \ldots$ in (5), we get $a_2 = \frac{1}{2}a_0 = 0$, $a_4 = \frac{3}{4}a_2 = 0$, $a_6 = \frac{5}{6}a_4 = 0$, etc.

Thus, $a_n = 0$ if n is even i.e., $a_{2m} = 0$.

Again putting $n = 2, 4, 6, \ldots$ in (5), we get

$$a_3 = \frac{2}{3}a_1 = \frac{2}{3},\ a_5 = \frac{4}{5}.a_3 = \frac{4}{5}.\frac{2}{3},\ a_7 = \frac{6}{7}a_5 = \frac{6}{7}.\frac{4}{5}.\frac{2}{3}, \text{ etc.}$$

In general, if n is odd, we have

$$a_n = \frac{n-1}{n}.\frac{n-3}{n-2}\ldots\frac{4}{5}.\frac{2}{3}.$$

Thus $a_{2m+1} = \frac{2m}{2m+1}.\frac{2m-2}{2m-1}\ldots\frac{4}{5}.\frac{2}{3}$.

(b) As found in part (a), we have $a_0 = 0$, $a_1 = 1$, $a_2 = 0$, $a_3 = \frac{2}{3}$, $a_4 = 0$, $a_5 = \frac{2}{3}.\frac{4}{5}$, etc.

$$\therefore \frac{\sin^{-1} x}{\sqrt{\left(1-x^2\right)}} = x + \frac{2}{3}x^3 + \frac{2.4}{3.5}x^5 + \ldots$$

Example 14(c):

Show that

$$(\sin^{-1} x)^2 = \frac{2}{2!}x^2 + \frac{2.2^2}{4!}x^4 + \frac{2.2^2.4^2}{6!}x^6 + \ldots$$

Deduce that

$$\theta^2 = 2.\frac{\sin^2\theta}{2!} + 2^2.\frac{2\sin^4\theta}{4!} + 2^2.4^2\frac{2\sin^6\theta}{6!} + \ldots$$

Solution:

Let $y = (\sin^{-1} x)^2$, we get

$y(0) = 0, y_1(0) = 0,\ y_2(0) = 2,$

and $y_{n+2}(0) = n^2 y_n(0).$ ...(1)

Putting n = 1, 2, 3, 4, ... in (1), we get

$y_3(0) = 1^2 y_1(0) = 0, y_4(0) = 2^2 y_2(0) = 2^2.2,$

$y_5(0) = 3^2 y_5(0) = 0,\ y_6(0) = 4^2 y_4(0) = 4^2.2^2.2$, etc.

Hence by Maclaurin's theorem, we get

$$(\sin^{-1} x)^2 = y(0) + xy_1(0) + \frac{x^2}{2!}y_2(0) + \ldots$$

$$= \frac{2}{2!}x^2 + \frac{2.2^2}{4!}x^4 + \frac{2.2^2.4^2}{6!}x^6 + \ldots$$

Now putting $\sin^{-1} x = \theta$ or $x = \sin\theta$ on both sides, we get the required expansion for θ^2.

Example 15:

If $y = \sin^{-1} x = a_0 + a_1 x + a_2 x^2 + \ldots$, *prove that* $(n + 1)(n + 2)a_{n+2} = n^2 a_n$.

Solution:

Let $y = \sin^{-1} x.$ ... (1)

Then $y_1 = \frac{1}{\sqrt{(1-x^2)}}.$ ...(2)

$\therefore$ $(1-x^2)y_1^2 - 1 = 0.$

Differentiating again, we get

$(1-x^2)2y_1y_2 - 2xy_1^2 = 0$ or $2y_1[(1 - x^2)y_2 - xy_1] = 0$

or $(1 - x^2)y_2 - xy_1 = 0,$ ...(3)

since $2y_1 \neq 0.$

Now differentiating (3) n times by Leibnitz's theorem, we get

$$(1 - x^2)y_{n+2} + n.y_{n+1}.(-2x) + \frac{n(n-1)}{1.2}y_n.(-2) - y_{n+1}.x - ny_n.\ 1 = 0$$

or $$(1 - x^2)y_{n+2} - (2n + 1) - n^2y_n = 0. \qquad ...(4)$$

Putting x = 0 in (4), we get

$$(y_{n+2})_0 = n^2(y_n)_0. \qquad ...(5)$$

By Maclaurin's theorem, we have

$$y = (y)_0 + \frac{x}{1!}(y_1)_0 + \frac{x^2}{2!}(y_2)_0 + \frac{x^3}{3!}(y_3)_0 + ... + \frac{x^n}{n!}(y_n)_0 + ...$$

Also we are given that

$$y = \sin^{-1} x = a_0 + a_1x + a_2x^2 + ... + a_nx^n + ...$$

Equating the coefficients of x^n in the two expansions for y, we get

$$a_n = \frac{(y_n)_0}{n!}.$$

$$\therefore \frac{a_{n+2}}{a_n} = \frac{(y_{n+2})_0}{(n+2)!}.\frac{n!}{(y_n)_0} = \frac{(y_{n+2})_0}{(y_n)_0}.\frac{1}{(n+2)(n+1)}$$

$$= \frac{n^2}{(n+2)(n+1)},$$

substituting for $\frac{(y_{n+2})_0}{(y_n)_0}$ from (5).

Hence $(n + 1)(n + 2)a_{n+2} = n^2a_n$.

Example 16:

Expand $\sin^{-1}(x + h)$ in powers of x as far as the term in x^3.

Solution:

First we observe that we are to expand $\sin^{-1}(x + h)$ in ascending powers of x. So let $f(h) = \sin^{-1} h$. Then

$$f(h + x) = \sin^{-1}(h + x).$$

Thus, we are to expand f(h + x) in powers of x. So by Taylor's theorem, we have

$$f(h + x) = f(h) + xf'(h) + \frac{x^2}{2!}f''(h) + \frac{x^3}{3!}f'''(h) + ... \qquad ...(1)$$

Now f(h) = $\sin^{-1}h$. Therefore

$$f'(h) = \frac{1}{\sqrt{(1-h^2)}} = (1-h^2)^{-1/2}$$

$f''(h) = h(1 - h^2)^{-3/2}$,

$f''(h) = h(1 - h^2)^{-3/2}$,

$f'''(h) = (1 - h^2)^{-3/2} + h(-3/2)(1 - h^2)^{-5/2}(-2h)$

$= (1 - h^2)^{-3/2} + 3h^2(1 - h^2)^{-5/2} = (1 - h^2)^{-5/2}[(1 - h^2) + 3h^2]$

$= (1 - h^2)^{-5/2}(1 + 2h^2)$, etc.

Substituting these values in (1), we have

$\sin^{-1}(h + x) = \sin^{-1}h + (1 - h^2)^{-1/2}x + (x^2/2\,!)\,h\,(1 - h^2)^{-3/2}$

$+ (x^3/3\,!)(1 - h^2)^{-5/2}(1 + 2h^2) + \ldots$

Example 17:

Expand $2x^3 + 7x^2 + x - 1$ 9n powers of $(x - 2)$.

Solution:

Let $f(x) = 2x^3 + 7x^2 + x - 1$. We can write

$f(x) = f[2 + (x - 2)]$.

Now expanding $f[2 + (x - 2)]$ by Taylor's Theorem in powers of $x - 2$, we get

$f(x) = f[2 + (x - 2)] = f(2) + (x - 2)f'(2)$

$+ (1/2\,!)(x - 2)^2 f''(2) + \ldots$ ...(1)

Now $f(x) = 2x^3 + 7x^2 + x - 1$

so that $f(2) = 2.2^3 + 7.2^2 + 2 - 1$

$= 45$, $f'(x) = 6x^2 + 14x + 1$

so that $f'(2) = 53$, $f''(x)$

$= 12x + 14$ so that $f''(2) = 38$,

$f'''(x) = 12$, so that $f'''(2) = 12$,

$f^{iv}(x) = 0$ so that $f^{iv}(2) = 0$. Obviously $f^n(2) = 0$ when $n \geq 4$. Now substituting these values in (1), we get

$$f(x) = 45 + (x - 2).53 + \frac{(x-2)^2}{2!}.38 + \frac{(x-2)^3}{3!}.12$$

$= 45 + 53(x - 2) + 19(x - 2)^2 + 2(x - 2)^3.$

Example 18:

Show that $\log(x + h) = \log h + \frac{x}{h} - \frac{x^2}{2h^2} + \frac{x^3}{3h^3} - \dots$

Solution:

First we observe that we are to expand log (x + h) in ascending powers of x. So let f(h) = log h. Then

$$f(h + x) = \log(h + x).$$

Here $f(h) = \log h,\ f'(h)$

$$= 1/h,\ f''(h) = -\frac{1}{h^2},$$

$$f'''(h) = \frac{2}{h^3},\ \dots \text{ etc.}$$

Substituting these values in Taylor's expansion, we get

$$\log(h + x) = \log h + \frac{x}{h} - \frac{x^2}{2h^2} + \frac{x^3}{3h^3} - \dots$$

Example 19:

Expand $\tan^{-1} x$ *in powers of* $\left(x - \frac{1}{4}\pi\right)$.

Solution:

Let $f(x) = \tan^{-1} x$. Then, we have

$$\tan^{-1} x = f(x) = f\left[\frac{1}{4}\pi + \left(x - \frac{1}{4}\pi\right)\right],$$

$$\left[\because \text{ we have to expand } f(x) \text{ in powers of } \left(x - \frac{1}{4}\pi\right)\right]$$

$$= f\left(\frac{\pi}{4}\right) + \left(x - \frac{1}{4}\pi\right)f'\left(\frac{\pi}{4}\right) + \frac{1}{2!}\left(x - \frac{1}{4}\pi\right)^2 f''\left(\frac{\pi}{4}\right) + \dots,$$

on expanding $f\left[\frac{1}{4}\pi + \left(x - \frac{1}{4}\pi\right)\right]$ by Taylor's theorem in powers of $\left(x - \frac{1}{4}\pi\right)$.

Now $f(x) = \tan^{-1}x$. Therefore

$$f\left(\frac{\pi}{4}\right) = \tan^{-1}\left(\frac{\pi}{4}\right),\ f'(x) = \frac{1}{1 + x^2}$$

so that $f'\left(\frac{\pi}{4}\right) = 1/\left(1+\frac{\pi^2}{16}\right)$, $f''(x) = -2x/(1+x^2)^2$ so that

$$f''\left(\frac{\pi}{4}\right) = -\pi/\{2(1+\pi^2/16)^2\} \text{ and so on.}$$

Substituting these values in the above expansion, we get

$$\tan^{-1} x = \tan^{-1}\left(\frac{\pi}{4}\right) + \left(x-\frac{1}{4}\pi\right)/\left(1+\frac{\pi^2}{16}\right)$$

$$-\pi\left(x-\frac{1}{4}\pi\right)^2 / \left\{4\left(1+\frac{\pi^2}{16}\right)^2\right\} + \ldots$$

Example 20:

Expand log sin (x + h) in powers of h by Taylor's theorem.

Solution:

First we observe that we are to expand log sin (x + h) in powers of h. So let f(x) = log sin x. Then

$$f(x+h) = \log \sin (x+h).$$

Expanding f(x + h) by Taylor's theorem in powers of h, we have

$$f(x+h) = f(x) = h$$

$$f'(x) + \frac{h^2}{2!}f''(x) + \frac{h^3}{3!}f'''(x) + \ldots \qquad \ldots(1)$$

Now f(x) = log sin x. Therefore f '(x)

$$= \frac{1}{\sin x}.\cos x = \cot x,$$

$$f''(x) = -\operatorname{cosec}^2 x,$$

$$f'''(x) = 2\operatorname{cosec}^2 x \cot x, \text{ etc.}$$

Substituting these values in (1), we get

$$\log \sin (x+h) = \log \sin x + h \cot x - (h^2/2!)\operatorname{cosec}^2 x$$
$$+ (2h^3/3!)\operatorname{cosec}^2 x \cot x + \ldots$$

Example 21:

Expand sin x in powers of $\left(x-\frac{1}{2}\pi\right)$.

Solution:

Let f(x) = sin x. We want to expand f(x) in powers of $x-\frac{1}{2}\pi$. We can write f(x) = $f\left[\frac{1}{2}\pi+\left(x-\frac{1}{2}\pi\right)\right]$.

Now expanding $f\left[\frac{1}{2}\pi+\left(x-\frac{1}{2}\pi\right)\right]$ by Taylor's theorem in powers of $\left(x-\frac{1}{2}\pi\right)$, we get

$$f(x)=f\left[\frac{1}{2}\pi+\left(x-\frac{1}{2}\pi\right)\right]=f\left(\frac{\pi}{2}\right)+\left(x-\frac{1}{2}\pi\right)f'\left(\frac{\pi}{2}\right)$$

$$+\frac{1}{2!}\left(x-\frac{1}{2}\pi\right)^2 f''\left(\frac{\pi}{2}\right)+\frac{1}{3!}\left(x-\frac{1}{2}\pi\right)^3 f'''\left(\frac{\pi}{2}\right)+\ldots \qquad \ldots(1)$$

Now f(x) = sin x. Therefore $f\left(\frac{\pi}{2}\right)=\sin\frac{\pi}{2}=1$,

f '(x) = cos x giving $f'\left(\frac{\pi}{2}\right)$

$= \cos\frac{\pi}{2} = 0$,

f "(x) = – sin x so that $f''\left(\frac{\pi}{2}\right)$

$= -\sin\frac{\pi}{2} = -1$,

f "'(x) = – cos x so that $f'''\frac{\pi}{2}$

$= -\cos\frac{\pi}{2} = 0$, etc.

Substituting these values in (1), we get

$$\sin x=1+\left(x-\frac{1}{2}\pi\right).0+\frac{1}{2!}\left(x-\frac{1}{2}\pi\right)^2.(-1)+\frac{1}{3!}\left(x-\frac{1}{2}\pi\right)^3.0+\ \ldots$$

$$=1-\left(\frac{1}{2!}\right)\left(x-\frac{1}{2}\pi\right)^2+\ldots$$

Example 22:

Prove that

$$sin(x + h) = sin\ x + h\ cos\ x - \frac{h^2}{2!} sin - ...$$

Solution:

First we observe that we are to expand sin(x + h) in powers of h. So let f(x) = sin x. Then

$$f(x + h) = \sin(x + h).$$

Expanding f(x + h) by Taylor's theorem in powers of h, we have

$$\sin(x + h) = f(x + h) = f(x) + hf'(x) + \frac{h^2}{2!} f''(x) + \frac{h^3}{3!} f'''(x) + ...$$

$$= \sin x + h \cos x + \frac{h^2}{2!}(-\sin x) + \frac{h^3}{3!}(-\cos x) + ...$$

$$= \sin x + h \cos - \frac{h^2}{2!} \sin x - \frac{h^3}{3!} \cos x + ...$$

3

Successive Differentiation

LEIBNITZ'S THEOREM

This theorem helps us to find the nth differential coefficient of the product of two functions. The statement of the theorem is as follows:

If u and v are any two functions of x such that all their desired differential coefficients exist, then the nth differential coefficient of their product is given by

$$D^n(uv) = (D^n u).v + {}^nC_1\, D^{n-1}\, u\, Dv + {}^nC_2\, D^{n-2}\, u.D^2v + \ldots + {}^nC_r\, D^{n-r}\, u.\, D^rv + \ldots + u\, D^nv.$$

Proof:

We shall prove the theorem by mathematical induction. By actual differentiation, we have

$$D(uv) = (Du).v + u.Dv. \qquad \ldots(1)$$

From (1) we see that the theorem is true for n = 1.

Now assume that the theorem is true for a particular value of n. Then we have

$$D^n\,(uv) = (D^n u).v + {}^nC_1 D^{n-1}\, u\, Dv = {}^nC_2\, D^{n-2}\, u.D^{2v} + \ldots + {}^nC_r D^{n-r}\, u\, D^rv + {}^nC_{r+1}\, D^{n-r-1}\, u\, D^{r+1}\, v + \ldots + u\, D^nv. \quad \ldots(2)$$

Differentiating both sides of (2) with respect to x, we get

$$D^{n+1}\,(uv) = \{(D^{n+1}u).v + D^nu\, Dv\} + \{{}^nC_1\, D^nu.Dv + {}^nC_1\, D^{n-1}\, u.D^2v\} + \{{}^nC_2 D^{n-1}\, u.D^2v + {}^nC_2 D^{n-2} u.D^3v\} + \ldots + \{{}^nC_r D^{n-r+1} u.D^rv + {}^nC_r D^{n-r} u.D^{r+1}v\} + \{{}^nC_{r+1}\, D^{n-r} u.D^{r+1}\, v + {}^nC_{r+1}\, D^{n-r-1} u.D^{r+2}v\} + \ldots + \{Du\, D^nv + u\, D^{n+1}\, v\}.$$

Rearranging the terms, we get

$$D^{n+1}(uv) = (D^{n+1}u).v + (1 + {}^nC_1)(D^n u\, Dv)$$
$$+ ({}^nC_1 + {}^nC_2) D^{n-1} u.D^2v$$
$$+ \ldots + ({}^nC_r + {}^nC_{r+1}(D^{n-r} u\, D^{r+1}v) + \ldots + u\, D^{n+1} v. \quad \ldots(3)$$

But from algebra, ${}^nC_r + {}^nC_{r+1} = {}^{n+1}C_{r+1}$. Therefore

$1 + {}^nC_1 = {}^nC_0 + {}^nC_1 = {}^{n+1}C_1$, ${}^nC_1 + {}^nC_2 = {}^{n+1}C_2$, and so on.

Hence equation (3) gives

$$D^{n+1}(uv) = (D^{n+1} u).v + {}^{n+1}C_1 (D^n u).Dv$$
$$+ {}^{n+1}C_2(D^{n-1} u).(D^2v) + \ldots$$
$$+ {}^{n+1}C_{r+1}D^{n-r} u.D^{r+1}v + \ldots + u.D^{n+1} v. \quad \ldots(4)$$

From (4) we see that if the theorem is true for any value of n it is also true for the next value of n. But we have already seen that the theorem is true for n = 1. Hence it must be true for n = 2 and so for n = 3, and so on. Thus the theorem is true for all positive integral values of n.

Important Note: While applying Leibnitz's theorem if we observe that one of the two functions is such that all its differential coefficients after a certain stage become zero, then we should take that function as the second function.

Example 1:

If $y = x^2 \tan^{-1} x$, *find* y_n.

Solution:

Let $u = \tan^{-1} x$ and $v = x^2$. Then $y = uv$. Differentiating n times by Leibnitz's theorem, we get

$$y_n = D^n(uv) = (D^n u).v + {}^nC_1(D^{n-1} u).Dv + {}^nC_2(D^{n-2}u).(D^2v) + \ldots$$

Now $\quad D^n u = D^n (\tan^{-1} x) = (-1)^{n-1} (n-1)! \sin^n \theta \sin n\theta,$

where $\quad \theta = \tan^{-1} (1/x)$.

Also $Dv = Dx^2 = 2x$, $D^2v = 2$ = constant and so the third and the higher derivatives of v all vanish.

Hence
$$y_n = [(-1)^{n-1} (n-1)! \sin^n \theta \sin n\theta].x^2$$
$$+ {}^nC_1[(-1)^{n-2} (n-2)! \sin^{n-1} \theta \sin (n-1)\theta].2x$$
$$+ {}^nC_2[(-1)^{n-3} (n-3)! \sin^{n-2} \theta \sin (n-2)\theta].2$$
$$= (-1)^{n-1} (n-3)! [(n-1)(n-2) x^2 \sin^n \theta \sin n\theta$$
$$- 2nx (n-2) \sin^{n-1} \theta \sin (n-1)\theta$$
$$+ n(n-1) \sin^{n-2} \theta \sin (n-2)\theta], \text{ where } \theta = \tan^{-1} (1/x).$$

Example 2:

If $y = (\sin^{-1} x)\sqrt{(1-x^2)}$, *prove that*

$(1 - x^2)y_{n+1} - (2n + 1)xy_n - n^2y_{n-1} = 0.$

Solution:

We have $y\sqrt{(1-x^2)} = \sin^{-1} x.$

Differentiating both sides, we get

$$y_1\sqrt{(1-x^2)} + y.\frac{1}{2}\frac{-2x}{\sqrt{(1-x^2)}} = \frac{1}{\sqrt{(1-x^2)}}$$

or $\quad y_1(1 - x^2) - yx - 1 = 0.$ ...(1)

Differentiating (1) n times by Leibnitz's theorem, we get

$$y_{n+1}(1 - x^2) + {}^nC_1y_n(-2x) + {}^nC_2y_{n-1}(-2)$$
$$- y_n.x - {}^nC_1\, y_{n-1}.1 = 0$$

or $(1 - x^2)y_{n+1} - (2n + 1)xy_n - n^2y_{n-1} = 0.$

Definition: It y be a function of x, etc., differential coefficient $\frac{dy}{dx}$, will be in general a function of x which can be differentiated. The differential co-efficient of y. Similarly, the differential co-efficient of the second differential co-efficient is called the third differential co-efficient, and so on the successive differential co-efficients of y are denoted by

$$\frac{dy}{dx}, \frac{d^2y}{dx^2}, \frac{d^3y}{dx^3} \ldots$$

The nth differential co-efficient of y being $\frac{d^2y}{dx^n}$.

Alternative methods of writing the nth differential co-efficient are

$$\left(\frac{d}{dx}\right)^n y, D^ny, y_n, \frac{d^2y}{dx^n}, y(x).$$

In the last case the first second etc. differential co-efficients would be written as y', y'', etc.

The value of a differential co-efficient at x = a is usually indicated by adding a suffix. Thus, $(y_n)x \to a$ or $(y_n)a$. It $y = f(x)$. The same thing can also be indicated by $f^{(n)}(a)$.

STANDARD RESULTS

(i) If $y = (ax + b)^m$, then $y_1 = ma\,(ax + b)^{m-1}$;

$y_2 = m(m-1)\,a^2(ax+b)^{m-2}$, $y_3 = m(m-1)(m-2)\,a^3(ax+b)^{m-3}$, and so on.

In general,

$$y_n = m(m-1)(m-2)\ldots\{m-(n-1)\}\,a^n(ax+b)^{m-n}.$$

Thus,

$D^n(ax+b)^m = m(m-1)(m-2)\ldots(m-n+1)\,a^n(ax+b)^{m-n}$.

If m is a positive integer, the above result can be written in a compact form by using the factorial notation. Thus in this case $D^n(ax+b)^m$

$$= \frac{m(m-1)\ldots(m-n+1)(m-n)(m-n-1)\ldots 1}{(m-n)(m-n-1)\ldots 2.1}\,a^n(ax+b)^{m-n}$$

$$= \frac{m!}{(m-n)!}a^n(ax+b)^{m-n}.$$

Note 1: If m is a + ive integer, then $D^m(ax+b)^m$

$= (m!/0!)\cdot a^m(ax+b)^0 = m!\,a^m$. In particular, $D^m x^m$ - m!.

Note 2: If m is a + ive integer, the mth diff. co-eff. of $(ax+b)^m$ is constant. Therefore the $(m+1)^{th}$ and all the higher differential coefficients of $(ax+b)^m$ are zero.

Note 3: If m is a negative integer, say $m = -p$, where p is a positive integer, then

$$D^n(ax+b)^{-p} = (-p)(-p-1)(-p-2)\ldots\{-p-(n-1)\}\,a^n(ax+b)^{-p-n}$$

$$= (-1)^n p(p+1)\ldots(p+n-1)\,a^n(ax+b)^{-p-n}$$

$$= (-1)^n\frac{(p+n-1)!}{(p-1)!}a^n(ax+b)^{-p-n}.$$

(ii) If $y = e^{ax}\sin(bx+c)$, then

$y_1 = ae^{ax}\sin(bx+c) + be^{ax}\cos(bx+c)$

$= e^{ax}[a\sin(bx+c) + b\cos(bx+c)]$.

Putting $a = r\cos\phi$ and $b = r\sin\phi$, we get

$$y_1 = re^{ax}\sin(bx+c+\phi),$$

where $r^2 = a^2 + b^2$ and $\phi = \tan^{-1}(b/a)$.

Similarly, $y_2 = r^2e^{ax}\sin(bx+c+2f)$, and so on.

Thus, $D^n\{e^{ax}\sin(bx+c)\} = r^n e^{ax}\sin(bx+c+n\phi)$, where

$r = (a^2 + b^2)^{1/2}$ and $\phi = \tan^{-1}(b/a)$.

Similarly, $D^n\{e^{ax} \cos bx + c)\} = r^n e^{ax} \cos(bx + c + n\phi)$, where

$r = (a^2 + b^2)^{1/2}$ and $\phi = \tan^{-1}(b/a)$.

(iii) If $y = (ax + b)^{-1}$, then $y_1 = (-1)\, a\, (ax + b)^{-2}$,
$y^2 = (-1)(-2)\, a^2 (ax + b)^{-3}$, $y_3 = (-1)(-2)(-3)\, a^3 (ax + b)^{-4}$, and so on.

In general, $y_n = (-1)(-2)(-3) \ldots (-n)\, a^n (ax + b)^{-(n+1)}$.

Thus, $D^n (ax + b)^{-1} = (-1)^n n!\, a^n (ax + b)^{-n-1}$.

(iv) If $y = e^{ax+b}$, then $y_1 = a\, e^{ax+b}$, $y_2 = a^2 e^{ax+b}$, $y_3 = a^3 e^{ax+b}$, and so on.

Thus, $D^n e^{ax+b} = a^n e^{ax+b}$. In particular, $D^n e^{ax} = a^n e^{ax}$.

(v) If $y = a^x$, then $y_1 = a^x \log a$, $y_2 = a^x (\log a)^2$, $y_3 = a^x (\log a)^3$, and so on. Thus, $D^m a^x = a^x (\log)^n$

(vi) If $y = \log(ax + b)$, then $y_1 = a/(ax + b) = a\,(ax + b)^{-1}$,.
$y_2 = (-1)\, a^2 (ax + b)^{-2}$, $y_3 = (-1)(-2)\, a^3 (ax + b)^{-3}$, and so on.
In general, $y_n = (-1)(-2) \ldots \{-(n-1)\}\, a^n (ax + b)^{-n}$.

Thus, $D^n \log(ax + b) = \dfrac{(-1)^{n-1}(n-1)!\, a^n}{(ax+b)^n}$.

In particular, $D^n \log x = \dfrac{(-1)^{n-1}(n-1)!}{x^n}$.

(vii) If $y = \sin(ax + b)$, then

$y_1 = a \cos(ax + b) = a \sin\left(ax + b + \frac{1}{2}\pi\right)$. Comparing y with y_1, we find that

$y_2 = a^2 \sin\left(ax + b + \frac{1}{2}\pi + \frac{1}{2}\pi\right) = a^2 \sin\left(ax + b + \frac{2}{2}\pi\right)$,

$y_3 = a^3 \sin\left(ax + b + \frac{3}{2}\pi\right)$, and so on.

Thus $D^n \sin(ax + b) = a^n \sin\left(ax + b + \frac{1}{2}n\pi\right)$.

(viii) Similarly, $D^n \cos(ax + b) = a^n \cos\left(ax + b + \frac{1}{2}n\pi\right)$.

SOLVED EXAMPLES

Example 1:

(i) $\dfrac{1}{(x-1)^3(x-2)}$

(ii) $\dfrac{x^4}{(x-1)(x-2)}$

(iii) $\dfrac{1}{x^2-a^2}$

(iv) $\dfrac{1}{a^2-x^2}$

(v) $\dfrac{x^2}{(x+2)(2x+3)}$

Solution:

(i) Let $y=\dfrac{1}{(x-1)^3(x-2)}$. To resolve y into partial fractions we put $x - 1 = z$. Then $y=\dfrac{1}{z^3}.\dfrac{1}{z-1}=\dfrac{1}{z^3}.\dfrac{1}{-1+z}$, arranging the numerator and the denominator both in ascending powers of z.

Now dividing the numerator 1 by the denominator $-1+z$ till z^3 is a common factor in the remainder, we get

$$y=\frac{1}{z^3}\left[-1-z-z^2+\frac{z^3}{-1+z}\right]=-\frac{1}{z^3}-\frac{1}{z^2}-\frac{1}{z}+\frac{1}{-1+z}$$

$$=-\frac{1}{(x-1)^3}-\frac{1}{(x-1)^2}-\frac{1}{x-1}+\frac{1}{x-2}$$

$$= -(x-1)^{-3}-(x-1)^{-2}-(x-1)^{-1}+(x-2)^{-1}.$$

Now $D^n(ax+b)^{-p} = (-p)(-p-1)(-p-2)\ldots$

$\ldots\{(-p-(n-1)\}\,a^n(ax+b)^{-p-n}$

$$=\frac{(-1)^n(p+n-1)!}{(p-1)!}a^n(ax+b)^{-p-n}.$$

$$\therefore\ y_n=-\frac{(-1)^n(3+n-1)!}{(3-1)!}(x-1)^{-3-n}-\frac{(-1)^n(2+n-1)!}{(2-1)!}.(x-1)^{-2-n}$$

$- (-1)^n\, n!\, (x-1)^{-n-1} + (-1)^n\, n!\, (x-2)^{-n-1}$

$$= (-1)^n \left[-\frac{(n+2)!}{2!}(x-1)^{-3-n} - \frac{(n+1)!}{1!}(x-1)^{-2-n} \right.$$

$$\left. - n!(x-1)^{-n-1} + n! + n(x-2)^{-n-1} \right]$$

$$= (-1)^{n+1} n! \left[\frac{(n+2)(n+1)}{2(x-1)^{n+3}} + \frac{n+1}{(x-1)^{n+2}} + \frac{1}{(x-1)^{n+1}} - \frac{1}{(x-2)^{n+1}} \right].$$

(ii) Let $y = \dfrac{x^4}{(x-1)(x-2)} = \left[x^2 + 3x + 7 + \dfrac{15x-14}{(x-1)(x-2)} \right]$,

dividing N^r by the D^r

$= \left[x^2 + 3x + 7 + \dfrac{16}{x-2} - \dfrac{1}{x-1} \right]$, on resolving into partial fractions.

Then $y_n = 16\,(-1)^n . n!\, (x-2)^{-1-n} - (-1)^n\, n!\, (x-1)^{-1-n}$, if $n > 2$

$= (-1)^n . n!\, [16\,(x-2)^{-n-1} - (x-1)^{-n-1}]$.

(iii) Let $y = \dfrac{1}{x^2 - a^2} = \dfrac{1}{(x-a)(x+a)} = \dfrac{1}{2a}\left[\dfrac{1}{x-a} - \dfrac{1}{x+a} \right]$

$= \dfrac{1}{2a}\left[(x-a)^{-1} - (x+a)^{-1} \right]$.

Then $y_n = \dfrac{1}{2a}\,(-1)^n\, n!\, \{(x-a)^{-n-1} - (x+a)^{-n-1}\}$.

(iv) Let $y = \dfrac{1}{a^2 - x^2} = \dfrac{1}{(a-x)(a+x)} = \dfrac{1}{2a}\left[\dfrac{1}{a-x} + \dfrac{1}{a+x} \right]$

$= \dfrac{1}{2a}\left[\dfrac{1}{x+a} - \dfrac{1}{x-a} \right] = \dfrac{1}{2a}\left[(x+a)^{-1} - (x-a)^{-1} \right]$

Then $y_n = \dfrac{1}{2a}(-1)^n\, n!\ \{(x+a)^{-n-1} - (x-a)^{-n-1}\}$.

(v) Let $y = x^2/[(x+2)(2x+3)]$.

The given fraction is not a proper one. If we divide the numerator by the denominator, we observe orally that the quotient will be 1/2. So let

$$\frac{x^2}{(x+2)(2x+3)} \equiv \frac{1}{2} + \frac{A}{x+2} + \frac{B}{2x+3}.$$

Then A = – 4 and B = 9/2.

Hence $y = \frac{1}{2} - \frac{4}{x+2} + \frac{9}{2(2x+3)}$

$$= \frac{1}{2} - 4(x+2)^{-1} + \frac{9}{2}(2x+3)^{-1}.$$

$\therefore y_n = -4(-1)^n \, n! \, (x+2)^{-n-1}$

$$+ \frac{9}{2} \cdot (-1)^n \, n! \cdot 2^n \, (2x+3)^{-n-1}$$

$$= (-1)^n n! \left[\frac{9.2^{n-1}}{(2x+3)^{n+1}} - \frac{4}{(x+2)^{n+1}} \right].$$

Example 2:

Find the nth derivatives of

(i) $\frac{x}{1+3x+2x^2}$

(ii) $\frac{1}{1-5x-6x^2}$

(iii) $\frac{17x^2+26x-42}{6x^3-25x^2-29x+20}$

(iv) $\frac{x^2}{(x-a)(a-b)}$.

Solution:

(i) Let $y = \frac{x}{1+3x+2x^2} = \frac{x}{(1+x)(1+2x)}$

Resolving into partial fractions, we get

$$y = \frac{-1}{(1-2)(1+x)} + \frac{-\frac{1}{2}}{\left(1-\frac{1}{2}\right)(1+2x)} = \frac{1}{x+1} - \frac{1}{2x+1}.$$

Now using the standard result for $D^n (ax+b)^{-1}$, we get

$$y = \frac{(-1)^n n!}{(x+1)^{n+1}} - \frac{(-1)^n n! \, 2^n}{(2x+1)^{n+1}}$$

$$=(-1)^n n!\left[\frac{1}{(x+1)^{n+1}}-\frac{2^n}{(2x+1)^{n+1}}\right].$$

(ii) Let $y=\dfrac{1}{6x^2-5x+1}=\dfrac{1}{(3x-1)(2x-1)}$

$=\dfrac{2}{2x-1}-\dfrac{3}{3x-1}$, (on resolving into partial fractions)

$= 2(2x - 1)^{-1} - 3(3x - 1)^{-1}$.

Now $D^n(ax + b)^{-1} = (-1)^{-n}\, n!\, a^n (ax + b)^{-n-1}$.

$\therefore\ y_n = 2\,(-1)^n\, n!\, 2^n (2x - 1)^{-n-1}$

$\qquad - 3\,(-1)^n\, n!\, 3^n (3x - 1)^{-n-1}$

$= (-1)^n\, n!\, [2^{n+1} (2x - 1)^{-n-1} - 3^{n+1} (3x - 1)^{-n-1}]$.

(iii) Let y = the given traction. Resolving into partial fractions, we get

$$y=\frac{1}{2x-1}-\frac{2}{3x+4}+\frac{3}{x-5}.$$

$$\therefore\ y_n=\frac{(-1)^n n!\,2^n}{(2x-1)^{n+1}}-\frac{2(-1)^n n!\,3^n}{(3x+4)^{n+1}}+\frac{3(-1)^n.n!}{(x-5)^{n+1}}$$

$$=(-1)^n.n!\left\{\frac{2n}{(2x-1)^{n+1}}-\frac{2.3^n}{(3x+4)^{n+1}}+\frac{3}{(x-5)^{n+1}}\right\}.$$

(iv) Let $y=\dfrac{x^2}{(x-a)(x-b)}$. Since the given fraction is not a proper one, therefore we should first divide the numerator by the denominator before resolving it into partial fractions. Here we observe orally that the quotient will be 1. So let

$$\frac{x^2}{(x-a)(x-b)}\equiv 1+\frac{A}{x-a}+\frac{B}{x-b}.$$

Then $A=\dfrac{a^2}{a-b}$ and $B=\dfrac{b^2}{b-a}$

Hence $y=1+\dfrac{a^2}{(a-b)(x-a)}+\dfrac{b^2}{(b-a)(x-b)}$.

$$= 1 + \frac{a^2}{(a-b)}(x-a)^{-1} - \frac{b^2}{(a-b)}(x-b)^{-1}.$$

Now differentiating both sides n times, we get

$$y_n = \frac{a^2}{(a-b)}(-1)^n n!\,(x-a)^{-n-1} - \frac{b^2}{(a-b)}(-1)^n n!\,(x-b)^{-n-1}$$

$$= \frac{(-1)^n n!}{(a-b)}\left[\frac{a^2}{(x-a)^{n+1}} - \frac{b^2}{(x-b)^{n+1}}\right].$$

Example 3:

(a) If $y = \sin^{-1}\{2x/(1+x^2)\}$, prove that
$y_n = 2\,(-1)^{n-1}\,(n-1)!\,\sin^n \sin n\theta$, *where* $\theta = \cot^{-1} x$.

(b) If $y = \tan^{-1}\left\{\frac{\sqrt{(1+x^2)}-1}{x}\right\}$, *show that*

$$y_n = \frac{1}{2}(-1)^{n-1}\,(n-1)!\,\sin^n\theta\,\sin n\theta,$$

where $\theta = \cot^{-1} x$.

Solution:

(a) Put $x = \tan\phi$. Then

$$y \sin^{-1}\left(\frac{2\tan\phi}{1+\tan^2\phi}\right) = \sin^{-1}\sin 2\phi = 2\phi$$

$= 2\tan^{-1} x$.

(b) Let $y = \tan^{-1}\left\{\frac{\sqrt{(1+x^2)}-1}{x}\right\}$.

Put $x = \tan\phi$. Then

$$y = \tan^{-1}\left[\frac{\sqrt{(1+\tan^2\phi)}-1}{\tan\phi}\right] = \tan^{-1}\frac{\sec\phi - 1}{\tan\phi}$$

$$= \tan^{-1}\frac{1-\cos\phi}{\sin\phi} = \tan^{-1}\frac{2\sin^2(\phi/2)}{2\sin(\phi/2)\cos(\phi/2)}$$

$= \tan^{-1} \tan (\phi/2) = \phi/2 = 1/2\tan^{-1} x.$

$\therefore\ y_1 = 1/\{2(1 + x^2)\}.$

We get

$$y_n = \frac{1}{2} (-1)^{n-1} (n-1)! \sin^n \theta \sin n\theta, \text{ where } \theta = \cot^{-1} x.$$

Example 4:

Prove that the value of the nth differential coefficient of $x^3/(x^2 - 1)$ for $x = 0$, is zero if n is even, and is $-$ n! if n is odd and greater than 1.

Solution:

$$\text{Let } y = \frac{x^3}{x^2 - 1} = x + \frac{x}{x^2 - 1} = x + \frac{x}{(x-1)(x+1)}$$

$$= x + \frac{1}{(1+1)(x-1)} + \frac{-1}{(-1-1)(x+1)}$$

$$= x + \frac{1}{2(x-1)} + \frac{1}{2(x+1)}.$$

$\therefore$ When $n > 1$, we have

$$y_n = \frac{(-1)^n n!}{2}\left[\frac{1}{(x-1)^{n+1}} + \frac{1}{(x+1)^{n+1}}\right].$$

Putting $x = 0$ in the expression for y_n, we get

$$(y_n)_0 = \frac{(-1)^n n!}{2}\left[\frac{1}{(-1)^{n+1}} + \frac{1}{1}\right] = \frac{(-1)^n n!}{2}\left[\frac{1}{(-1)^{n+1}} + 1\right].$$

When n is even, $(y_n)_0 = \dfrac{n!}{2}\left[\dfrac{1}{(-1)} + 1\right] = \dfrac{n!}{2}.0 = 0.$

When n is odd, $(y_n)_0 = -\dfrac{n!}{2}[1+1] = -n!.$

Example 5:

Find the nth derivative of $\tan^{-1}\left(\dfrac{1+x}{1-x}\right)$.

Solution:

Let

$$y = \tan^{-1}\left(\frac{1+x}{1-x}\right) = \tan^{-1}\left\{\frac{1+x}{1-1.x}\right\} = \tan^{-1} 1 + \tan^{-1} x.$$

Then $y_1 = 1/(1 + x^2)$. We have

$y_n = (-1)^{n-1} (n-1)! \sin^n \phi \sin n\phi$, where $\phi = \tan^{-1}(1/x)$.

Example 6:

If $y = x(a^2 + x^2)^{-1}$, prove that

$y_n = (-1)^n n! a^{-n-1} \sin^{n+1} \phi \cos(n+1) f$, where $\phi = \tan^{-1}(a/x)$.

Solution:

We have $y = \dfrac{x}{a^2 + x^2} = \dfrac{x}{(x - ia)(x + ia)}$

$= \dfrac{1}{2}\left[\dfrac{1}{(x - ia)} + \dfrac{1}{(x + ia)}\right]$, on resolving into partial fractions.

Differentiating both sides n times w.r.t 'x', we get

$$y_n = \frac{1}{2}[(-1)^n n! (x - ia)^{-n-1} + (-1)^n n! (x + ia)^{-n-1}]$$

$$= \frac{1}{2} (-1)^n n! [(x - ia)^{-n-1} + (x + ia)^{-n-1}].$$

Putting $x = r \cos \phi$ and $a = r \sin \phi$, we get

$$y_n = \frac{1}{2}(-1)^n n! [r^{-n-1} (\cos \phi - i \sin \phi)^{-n-1} + r^{-n-1} (\cos \phi + i \sin \phi)^{-n-1}]$$

$$= \frac{1}{2} (-1)^n n! r^{-n-1} [\{\cos (n+1) \phi + i \sin (n+1) \phi\} + \{\cos (n+1) \phi - i \sin (n+1) \phi\}]$$

$$= \frac{1}{2} (-1)^n n! r^{-n-1} 2 \cos (n+1) \phi$$

$= (-1)^n n! (a/\sin \phi)^{-n-1} \cos (n+1) \phi$, since $r = a/\sin \phi$

$= (-1)^n.n! a^{-n-1} \sin^{n+1} \phi \cos (n+1) \phi$,

where $\phi = \tan^{-1}(a/x)$.

Example 7:

Find the nth differential coefficient of $1/(x^2 + a^2)$.

Solution:

We have $y = \dfrac{1}{x^2 + a^2} = \dfrac{1}{(x + ia)(x - ia)}$

$$= \frac{1}{2ia}\left\{\frac{1}{x - ia} - \frac{1}{x + ia}\right\}.$$

$$\therefore\ y_n = \frac{1}{2ia}.(-1)^n n!\left\{\frac{1}{(x-ia)^{n+1}} - \frac{1}{(x+ia)^{n+1}}\right\}.$$

Let $x = r\cos\phi$ and $a = r\sin\phi$, so that $\phi = \tan^{-1}(a/x)$. Then

$$y_n = \frac{(-1)^n n!}{2ia}\left\{\frac{1}{(r\cos\phi - ir\sin\phi)^{n+1}} - \frac{1}{(r\cos\phi + ir\sin\phi)^{n+1}}\right\}$$

$$= \frac{(-1)^n n!}{2ia\, r^{n+1}}\{(\cos\phi - i\sin\phi)^{-(n+1)} - (\cos\phi + i\sin\phi)^{-(n+1)}\}$$

$$= \frac{(-1)^n n!}{2ia\, r^{n+1}}[\{\cos(n+1)\phi + i\sin(n+1)\phi\}$$

$$- \{\cos(n+1)\phi - i\sin(n+1)\phi\}],$$

(by De-Moivre's Theorem)

$$= \frac{(-1)^n n!}{2ia\, r^{n+1}} 2i\sin(n+1)\phi = \frac{(-1)^n n!}{a.a^{n+1}}\sin(n+1)\phi\sin^{n+1}\phi,$$

since $r = a/\sin\phi$

$$= (-1)^n n!\, a^{-(n+2)}\sin(n+1)\phi\sin^{n+1}\phi,$$

where $\phi = \tan^{-1}(a/x)$.

Example 8:

(a) *If $y = \sin^{-1} x$, prove that*

$$(1 - x^2)(d^2y/dx^2) = x.(dy/dx).$$

(b) *Find d^2y/dx^2 for the cycloid whose equation is*

$$x = a(\theta - \sin\theta),\ y = a(1 - \cos\theta).$$

(c) *If $x = a(t - \sin t)$ and $y = a(1 + \cos t)$, prove that*

$$\frac{d^2y}{dx^2} = \frac{1}{4a}\operatorname{cosec}^4\left(\frac{t}{2}\right).$$

(d) *If $x = a(\cos\theta + \theta\sin\theta)$, $y = a(\sin\theta - \theta\cos\theta)$, find d^2y/dx^2.*

(e) *If $x = 2\cos t - \cos 2t$, $y = 2\sin t - \sin 2t$, find the value of (d^2y/dx^2) at $t = \pi/2$.*

(f) *If $y = \sin(\sin x)$, prove that*

$$(d^2y/dx^2) + (dy/dx)(\tan x + y\cos^2 x = 0.$$

Solution:

(a) We have y = $\sin^{-1} x$; $\therefore$ dy/dx = $1/\sqrt{(1-x^2)}$

or $(1 - x^2)(dy/dx)^2 = 1$.

Differentiating again with respect to x, we get

$$\left(1-x^2\right).2\frac{dy}{dx}.\frac{d^2y}{dx^2}-2x\left(\frac{dy}{dx}\right)^2=0$$

or $\left(1-x^2\right)\frac{d^2y}{dx^2}=x\frac{dy}{dx}$. $\left[\because \frac{dy}{dx}\neq 0\right]$

(b) We have x = a (θ – sin θ);

$\therefore$ (dy/dθ) = a (1 – cos θ).

Also y = a(1 – cos θ);

$\therefore$ (dy/dθ) = a sin θ.

$$\text{Now } \frac{dy}{dx}=\frac{dy/d\theta}{dx/d\theta}=\frac{a\sin\theta}{a(1-\cos\theta)}=\frac{2\sin\theta/2\cos\theta/2}{2\sin^2\theta/2}=\cot\frac{\theta}{2}.$$

$$\therefore \frac{d^2y}{dx^2}=\frac{d}{dx}\left(\frac{dy}{dx}\right)=\frac{d}{dx}(\cot\theta/2)=\left[\frac{d}{d\theta}(\cot\theta/2)\right]\frac{d\theta}{d}$$

$$=-\frac{1}{2}\operatorname{cosec}^2\frac{\theta}{2}.\frac{1}{a(1-\cos\theta)}$$

$$=-\frac{1}{2}\operatorname{cosec}^2\frac{\theta}{2}.\frac{1}{2a\sin^2\theta/2}=-\frac{1}{4a}\operatorname{cosec}^4\frac{\theta}{2}.$$

(c) Proceed as in (b).

(d) We have x = a (cos θ + θ sin θ).

$\therefore$ dx/dθ = a (– sin θ + sin θ + θ cos θ) = aθ cos θ.

Also y = a (sin θ – θ cos θ).

$\therefore$ dy/dθ = a (cos θ – cos θ + θ sin θ) = aθ cos θ.

$$\text{Now } \frac{dy}{dx}=\frac{dy/d\theta}{dx/d\theta}=\frac{a\theta\sin\theta}{a\theta\cos\theta}=\tan\theta.$$

$$\therefore \frac{d^2y}{dx^2}=\frac{d}{dx}\left(\frac{dy}{dx}\right)=\frac{d}{dx}(\tan\theta)=\left(\frac{d}{d\theta}(\tan\theta)\right)\frac{d\theta}{dx}$$

$$=\sec^2\theta.\frac{1}{a\theta\cos\theta}=\frac{1}{a}\frac{\sec^3\theta}{\theta},$$

(e) We have x = 2 cos t – cos 2t;

$\therefore \dfrac{dx}{dt} = -2 \sin t + 2 \sin 2t.$

Also y = 2 sin t – sin 2t;

$\therefore \dfrac{dy}{dt} = 2 \cos t - 2 \cos 2t.$

Now $\dfrac{dy}{dx} = \dfrac{dy/dt}{dx/dt} = \dfrac{2\cos t - 2\cos 2t}{-2 \sin t + 2 \sin 2t} = \dfrac{\cos t - \cos 2t}{\sin 2t - \sin t}$

$$= \frac{2 \sin \frac{3t}{2} . \sin \frac{t}{2}}{2 \cos \frac{3t}{2} . \sin \frac{t}{2}} = \tan \frac{3t}{2}.$$

$$\therefore \frac{d^2y}{dx^2} = \frac{d}{dx}\left(\frac{dy}{dx}\right) = \frac{d}{dx}\left(\tan \frac{3t}{2}\right) = \left[\frac{d}{dt}\left(\tan \frac{3t}{2}\right)\right]\frac{dt}{dx}$$

$$= \frac{3}{2} \sec^2 \frac{3t}{2} . \frac{1}{2 \sin 2t - 2 \sin t}.$$

When $t = \dfrac{\pi}{2}, \dfrac{d^2y}{dx^2} = \dfrac{3}{2} \sec^2 \dfrac{3\pi}{4} . \dfrac{1}{2 \sin \pi - 2 \sin \frac{\pi}{2}}$

$$\therefore \left(\frac{d^2y}{dx^2}\right)_{at\, t=\pi/2} = \frac{3}{2} . 2 . \frac{1}{(0-2)} = -\frac{3}{2}.$$

(f) We have y = sin (sin x). ...(1)

$\therefore \dfrac{dy}{dx} = [\cos (\sin x)] \cos x.$

Differentiating again, we have

$(d^2y/dx^2) = [-\sin (\sin x)] \cos^2 x - [\cos (\sin x)] \sin x$

$= - y \cos^2 x - [\cos (\sin x)] \cos x . (\sin x / \cos x),$

[∵ from (1), y = sin (sin x)]

$= - y \cos^2 x - \left(\dfrac{dy}{dx}\right) . \tan x.$

$\left[\because \text{from (2), } \dfrac{dy}{dx} = \{\cos (\sin x)\} \cos x\right]$

$$\therefore \frac{d^2y}{dx^2} + \frac{dy}{dx} \tan x + y \cos^2 x = 0.$$

Example 9:

Find the nth derivative of $\tan^{-1}\{2x/(1 - x^2)\}$.

Solution:

Let $y = \tan^{-1} \{2x/(1 - x^2)\} = 2 \tan^{-1} x$.

Then $y_1 = \dfrac{2}{1+x^2} = \dfrac{2}{(x-i)(x+i)} = \dfrac{1}{i}\left[\dfrac{1}{x-i} - \dfrac{1}{x+i}\right]$, on resolving into partial fractions.

Now differentiating both sides (n – 1) times w.r.t. ‘x’, we have

$$y_n = (-1)^{n-1} (n - 1)! [(x - i)^{-n} - (x + i)^{-n}],$$

Putting $x = r \cos \phi$ and $1 = r \sin \phi$ we have

$$y_n = \frac{(-1)^{n-1}(n-1)!}{i} [r^{-n} (\cos \phi - i \sin \phi)^{-n} - r^{-n}(\cos\phi + i \sin \phi)^{-n}]$$

$$= \frac{(-1)^{m-1}(n-1)!}{i} (r^{n} [(\cos n\phi + i \sin n\phi) - (\cos n\phi - i \sin n\phi)]$$

$$= 2(- 1)^{n-1} (n - 1)! r^{-n} \sin n\phi$$

$$= 2(- 1)^{n-1} (n - 1)! (1/\sin \phi)^{-n} \sin n\phi, \text{ since } r = 1/\sin \phi$$

$$= 2(- 1)^{n-1} (n - 1)! \sin^n \phi \sin n\phi, \text{ where } \phi = \tan^{-1} (1/x).$$

Example 10:

(i) *If $y = A \sin mx + B \cos mx$, prove that*

$$y_2 + m^2y = 0.$$

(ii) *If $y = e^{ax} \sin bx$, prove that*

$$y_2 - 2ay_1 + (a^2 + b^2) y = 0.$$

(iii) *If $y = \sin nx + \cos nx$, show that*

$$y_r = n^r\{1 + (- 1)^r \sin 2nx\}^{1/2}$$

Solution:

(i) We have, $y = A \sin mx + B \cos mx$. ...(1)

Differentiating both sides w.r.t. ‘x’ we get

$$y_1 = Am \cos mx - Bm \sin mx.$$

Again $y_2 = - Am^2 \sin mx - Bm^2 \cos mx$

$$= - m^2 (A \sin mx + B \cos mx) = - m^2y, \text{ from (1).}$$

$\therefore \quad y_2 = m^2y = 0$

(ii) Here $y = e^{ax} \sin bx$. ...(1)

$\therefore \quad y_1 = e^{ax}\, b \cos bx + a \,.\, e^{ax} \sin bx.$

Replacing $e^{ax} \sin bx$ by y, we get

$$y_1 = be^{ax} \cos bx + ay. \qquad ...(2)$$

Differentiating both sides (2) w.r.t. 'x', we get

$y_2 = abe^{ax} \text{ cox } bx - b^2 e^{ax} \sin bx + ay_1$

$= a(y_1 - ay) - b^2y + ay_1$, [$\because$ from (2), $be^{ax} \cos bx = y_1 - ay$ and from (1), $e^{ax} \sin bx = y$]

$= 2ay_1 - (a^2 + b^2)y.$

$\therefore y_2 - 2ay_1 + (a^2 + b^2)y = 0.$

(iii) Here y = sin nx + cos nx.

$$\therefore y_r = n^r \sin\left(nx + \frac{1}{2}r\pi\right) + n^r \cos\left(nx + \frac{1}{2}r\pi\right)$$

$$= n^r\left[\left\{\sin\left(nx + \frac{1}{2}r\pi\right) + \cos\left(nx + \frac{1}{2}r\pi\right)\right\}^2\right]^{1/2}$$

$$= n^r\left[\sin^2\left(nx + \frac{1}{2}r\pi\right) + \cos^2\left(nx + \frac{1}{2}r\pi\right)\right.$$

$$\left. + 2\sin\left(nx + \frac{1}{2}r\pi\right)\cos\left(nx + \frac{1}{2}r\pi\right)\right]^{1/2}$$

$$= n^r\left[1 + \sin 2\left(nx + \frac{1}{2}r\pi\right)\right]^{1/2} = n^r\left[1 + \sin\left(2nx + r\pi\right)\right]^{1/2}$$

$= n^r[1 + (-1)^r \sin 2nx]$, since $(r\pi + \theta) = (-1)^r \sin\theta$.

Example 11(a):

Find the nth differential coefficient of $\tan^{-1} (x/a)$.

Solution:

Let $y = \tan^{-1}\left(\frac{x}{a}\right)$. Then $y_1 = \frac{a}{x^2 + a^2} = \frac{a}{(x + ia)(x - ia)}$

$$= \frac{a}{(-ia - ia)(x + ia)} + \frac{a}{(ia + ia)(x - ia)}$$

$$= -\frac{1}{2i(x + ia)} + \frac{1}{2i(x - ia)} = \frac{1}{2i}\left[(x - ia)^{-1} - (x + ia)^{-1}\right].$$

Now differentiating both sides (n – 1) times w.r.t. 'x', we get

$$y_n = \frac{1}{2i}[(-1)^{n-1}(n-1)!\,(x - ia)^{-n} - (-1)^n (n-1)!\,(x + ia)^{-n}]$$

$$= \frac{(-1)^{n-1}(n-1)!}{2i} [(x - ia)^{-n} - (x + ia)^{-n}].$$

Put $x = r \cos \phi$ and $a = r \sin \phi$. Then

$$y_n = \frac{(-1)^{n-1}(n-1)!}{2i} [(r \cos \phi - ir \sin \phi)^{-n} - (r \cos \phi + ir \sin \phi)^{-n}]$$

$$= \frac{(-1)^{n-1}(n-1)!}{2i} r^{-n} [(\cos \phi - i \sin \phi)^{-n} - (\cos \phi + i \sin \phi)^{-n}]$$

$$= \frac{(-1)^{n-1}(n-1)!}{2i} r^{-n} [(\cos n\phi + i \sin n\phi) - (\cos n\phi - i \sin n\phi)]$$

$$= \frac{(-1)^{n-1}(n-1)!}{2i} r^{-n} 2i \sin n\phi = (-1)^{n-1} (n-1)!\, r^{-n} \sin n\phi$$

$= (-1)^{n-1} (n-1)!\, (a/\sin \phi)^{-n} \sin n\phi$, since $r = a/\sin \phi$

$= (-1)^{n-1} (n-1)!\, a^{-n}\sin^n \phi \sin n\phi$, where $\phi = \tan^{-1} (a/x)$.

Example 11(b):

Prove that the value when $x = 0$ of $D^n (\tan^{-1} x)$ is 0; $(n - 1)!$ or $-(n - 1)!$ according as n is of the form 2p, 4p + 1 or 4p + 3 respectively.

Solution:

Let $y = \tan^{-1} x$. Then

$$y_1 = \frac{1}{1+x^2} = \frac{1}{(x-i)(x+i)} = \frac{1}{2i}\left[\frac{1}{x-i} - \frac{1}{x+i}\right]$$

Differentiating both sides $(n - 1)$ times w.r.t. 'x', we get

$$y_n = \frac{(-1)^{n-1}(n-1)!}{2i} [(x - i)^{-n} - (x + i)^{-n}].$$

Putting $x = 0$ in the expression for y_n, we get

$$(y_n)_{x=0} = \frac{(-1)^{n-1}(n-1)!}{2i}\left[(-i)^{-n} - i^{-n}\right]$$

$$= \frac{(-1)^{n-1}(n-1)!}{2i} [(-1)^{-n} i^{-n} - i^{-n}]$$

$$= \frac{(-1)^{n-1}(n-1)!}{2i} i^{-n} [(-1)^{-n} - 1].$$

Now we shall consider the three cases.

(i) If n is of the form 2p, we have

$$(y_n)_0 = \frac{(-1)^{2p-1}(n-1)!}{2i} i^{-2p}\left[(-1)^{-2p} - 1\right]$$

$$= -\frac{(n-1)!}{2i} i^{-2p}[1-1] = -\frac{(n-1)!}{2i} i^{-2p}.0 = 0.$$

(ii) If n is of the form 4p + 1, we have

$$(y_n)_0 = \frac{(-1)^{4p}(n-1)!}{2i} i^{-(4p+1)}\left[(-1)^{-(4p+1)} - 1\right]$$

$$= \frac{(n-1)!}{2.i^{4p+2}}[-1-1],$$

$[\because (-1)^{-(4p+1)} = 1/(-1)^{4p+1} = 1/(-1) = -1]$

$$= \frac{(n-1)!}{2.i^{4p+2}}(-2) = -\frac{(n-1)!}{i^{4p}.i^2} = -\frac{(n-1)!}{(i^4)^p i^2} = -\frac{(n-1)!}{4p(-1)}$$

$[\because i^4 = 1 \text{ and } i^2 = -1]$

$= (n-1)!$.

(iii) If n is of the form 4p + 3, we have

$$(y_n)_0 = \frac{(-1)^{4p+2}(n-1)!\, i^{-(4p+3)}}{2i} [(-1)^{-(4p+3)} - 1]$$

$$= \frac{(n-1)!}{2.i^{4p+4}}[-1-1] = \frac{(n-1)!}{2.i^{4p+4}}(-2)$$

$$= -\frac{(n-1)!}{(i^4)^{p+1}} = -\frac{(n-1)!}{1^{p+1}} = -(n-1)!.$$

Example 12:

If $p^2 = a^2 \cos^2 \theta + b^2 \sin^2 q$, prove that

$$p + \frac{d^2 p}{d\theta^2} = \frac{a^2 b^2}{p^3}.$$

Solution:

We have $p^2 = a^2 \cos^2 \theta + b^2 \sin^2 \theta$. ...(1)

Differentiating both sides of (1) w.r.t. θ, we have

$$2p\left(\frac{dp}{d\theta}\right) = -2a^2 \cos\theta \sin\theta + 2b^2 \sin\theta \cos\theta$$

or $$p\left(\frac{dp}{d\theta}\right) = (b^2 - a^2) \sin\theta \cos\theta. \qquad ...(2)$$

Now differentiating both sides of (2) w.r.t. θ, we have

$$p\frac{d^2p}{d\theta^2} + \left(\frac{dp}{d\theta}\right)^2 = (b^2 - a^2)(\cos^2\theta - \sin^2\theta)$$

$$= (b^2\cos^2\theta + a^2\sin^2\theta) - (a^2\cos^2\theta + b^2\sin^2\theta)$$

$$= (b^2\cos^2\theta + a^2\sin^2\theta) - p^2. \qquad ...[\text{from } (1)]$$

$$\therefore\ p\frac{d^2p}{d\theta^2} + p^2 = \left(b^2\cos^2\theta + a^2\sin^2\theta\right) - \left(\frac{dp}{d\theta}\right)^2$$

$$= \left(b^2\cos^2\theta + a^2\sin^2\theta\right) - \frac{\left(b^2 - a^2\right)^2 \sin^2\theta\cos^2\theta}{p^2}$$

substituting for $\frac{dp}{d\theta}$ from (2)

$$= \frac{1}{p^2}\left[p^2(b^2\cos^2\theta + a^2\sin^2\theta) - (b^2 - a^2)\sin^2\theta\cos^2\theta\right]$$

$$= \frac{1}{p^2}\left[(a^2\cos^2\theta + b^2\sin^2\theta)(b^2\cos^2\theta + a^2\sin^2\theta) - (b^2 - a^2)^2 \sin^2\theta\cos^2\theta\right]$$

$$= \frac{1}{p^2}\left[a^2b^2\cos^4\theta + a^2b^2\sin^4\theta + 2a^2b^2\sin^2\theta\cos^2\theta\right]$$

$$= \frac{a^2b^2}{p^2}\left(\cos^2\theta + \sin^2\theta\right)^2 = \frac{a^2b^2}{p^2}.$$

Thus, $$p\frac{d^2p}{d\theta^2} + p^2 = \frac{a^2b^2}{p^2}.$$

Dividing both sides by p, we have

$$\frac{d^2p}{d\theta^2} + p = \frac{a^2b^2}{p^3}.$$

Example 13:

If $y = (\sin^{-1} x)^2$, prove that

$$\left(1 - x^2\right)\frac{d^2y}{dx^2} - x\frac{dy}{dx} - 2 = 0.$$

Differentiate the above equation n times w.r.t. 'x'.

Solution:

We have $y = (\sin^{-1} x)^2$.

Therefore $y_1 = 2(\sin^{-1} x)/\sqrt{(1-x^2)}$,

or $$y_1^2 = \frac{4(\sin^{-1} x)^2}{1-x^2} = \frac{4y}{1-x^2}, \qquad [\because (\sin^{-1} x)^2 = y]$$

or $y_1^2(1 - x^2) - 4y = 0$. Differentiating again, we get

$$2y_1y_2(1 - x^2) - 2xy_1^2 - 4y_1 = 0$$

or $$2y_1[y_2(1 - x^2) - xy_1 - 2] = 0.$$

Cancelling $2y_1$, since $2y_1 \neq 0$, we get

$$y_2(1 - x^2) - y_1x - 2 = 0$$

proving the first result

Defferentiating (1) n times by Leibnitz's theorem, we get

$$y_{n+2}(1 - x^2) + {}^nC_1y_{n+1}.(-2x) + {}^nC_2y_n.(-2) - y_{n+1}.x - {}^nC_1y_n.1 = 0$$

or $(1 - x^2)y_{n+2} - (2n + 1)xy_{n+1} - n^2y_n = 0$.

Example 14:

Find the nth differential coefficient of x^3 cos x.

Solution:

Since the fourth and higher derivatives of x^3 are all zero, therefore for the sake of convenience we shall take x^3 as the second function. Applying Leibnitz's Theorem, we have

$$D^n[(\cos x)x^3] = (D^n \cos x).x^3 + {}^nC_1(D^{n-1} \cos x).(Dx^3) + {}^nC_2(D^{n-2} \cos x).(D^2x^3) + {}^nC_3(D^{n-3} \cos x).(D^3x^3),$$

since all other terms become zero

$$= \left[\cos\left(x + \frac{1}{2}n\pi\right).x^3 + n\left[\cos\left\{x + \frac{1}{2}(n-1)\pi\right\}\right].3x^2\right.$$

$$+ \frac{n(n-1)}{1.2}\left[\cos\left\{x + \frac{1}{2}(n-2)\right\}\pi\right].6x$$

$$+ \frac{n(n-1)(n-2)}{1.2.3}\left[\cos\left\{x + \frac{1}{2}(n-3)\pi\right\}\right].6$$

$$= x^3 \cos\left(x+\frac{1}{2}n\pi\right) + 3nx^2 \sin\left(x+\frac{1}{2}n\pi\right)$$

$$- 3n(n-1)x \cos\left(x+\frac{1}{2}n\pi\right) - n(n-1)(n-2) \sin\left(x+\frac{1}{2}n\pi\right)$$

$$= [x^3 - 3n(n-1)x] \cos\left(x+\frac{1}{2}n\pi\right)$$

$$+ \left[3x^2 n - n(n-1)(n-2)\right] \sin\left(x+\frac{1}{2}n\pi\right).$$

Example 15:

If $y = x^2 e^x \cos x$, find y_n.

Solution:

Let $u = e^x \cos x$ and $v = x^2$. Then $y = uv$. Differentiating n times by Leibnitz's Theorem, we get

$$y_n = D^n(uv) = (D^n u)v + {}^nC_1 (D^{n-1}u).(Dv) + {}^nC_2 (D^{n-2}u).(D^2 v) + \ldots$$

Now $D^n u = D^n(e^x \cos x) = r^n e^x \cos(x + n\phi)$, where

$$r = \sqrt{(1+1)} = \sqrt{2}$$

and $\phi = \tan^{-1}(1/1) = \tan^{-1} 1 = \frac{\pi}{4}.$

Thus, $D^n u = 2^{n/2} e^x \cos\left(x+\frac{1}{4}n\pi\right).$

Also $Dv = Dx^2 = 2x$,

$D^2 v = 2$ = constant and so the third and the higher derivatives of v all vanish. Hence

$$y_n = \left[2^{n/2} e^x \cos\left(x+\frac{1}{4}n\pi\right)\right].x^2 + n\left[2^{(n-1)/2} e^x \cos\left\{x+\frac{1}{4}(n-1)\pi\right\}\right].2x$$

$$+ \frac{1}{2}n(n-1)\left[2^{(n-2)/2} e^x \cos\left\{x+\frac{1}{4}(n-2)\pi\right\}\right].2$$

$$= e^x\left[2^{n/2} x^2 \cos\left(x+\frac{1}{4}n\pi\right) + 2nx\, 2^{(n-1)/2} \cos\left(x+\frac{1}{4}(n-1)\pi\right]\right.$$

$$\left. + n(n-1)\, 2^{(n-2)/2} \cos\left\{x+\frac{1}{4}(n-2)\pi\right\}\right].$$

Example 16:

Find the nth differential coefficient of $x^{n-1} \log x$.

Solution:

Let $y = x^{n-1} \log x$ ...(1)

Then $y_1 = x^{n-1}(1/x) + (n-1).x^{n-2}.\log x.$

Multiplying both sides by x, we have

$xy_1 = x^{n-1} + (n-1)\, x^{n-1} \log x$

or $xy_1 = x^{n-1} + (n-1)\, y.$ ...(2)

[$\because$ from (1), $y = x^{n-1} \log x$]

Differentiating both sides of (2) (n – 1) times, we have

$D^{n-1}(y_1 x) = D^{n-1} x^{n-1} + (n-1)\, D^{n-1} y$

or $(D^{n-1} y_1).x + {}^{n-1}C_1\, (D^{n-2} y_1).1 = (n-1)! + (n-1)\, y_{n-1}$

or $xy_n + (n-1)y_{n-1} = (n-1)!\ (n-1)y_{n-1}$

or $xy_n = (n-1)!$ or $y_n = (n-1)!/x.$

Example 17:

(a) If $y = \sin^{-1} \{2x/(1+x^2)\}$, prove that

$y_n = 2\,(-1)^{n-1}\,(n-1)!\,\sin^n \sin n\theta$, *where* $\theta = \cot^{-1} x.$

(b) If $y = \tan^{-1}\left\{\dfrac{\sqrt{(1+x^2)}-1}{x}\right\}$, *show that*

$$y_n = \frac{1}{2}(-1)^{n-1}\,(n-1)!\,\sin^n \theta \sin n\theta,$$

where $\theta = \cot^{-1} x.$

Solution:

(a) Put $x = \tan\phi$. Then

$$y \sin^{-1}\left(\frac{2\tan\phi}{1+\tan^2\phi}\right) = \sin^{-1} \sin 2\phi = 2\phi$$

$= 2\tan^{-1} x.$

(b) Let $y = \tan^{-1}\left\{\dfrac{\sqrt{(1+x^2)}-1}{x}\right\}.$

Put $x = \tan\phi$. Then

$$y = \tan^{-1}\left[\frac{\sqrt{(1+\tan^2\phi)}-1}{\tan\phi}\right] = \tan^{-1}\frac{\sec\phi - 1}{\tan\phi}$$

$$= \tan^{-1}\frac{1-\cos\phi}{\sin\phi} = \tan^{-1}\frac{2\sin^2(\phi/2)}{2\sin(\phi/2)\cos(\phi/2)}$$

$= \tan^{-1}\tan(\phi/2) = \phi/2 = 1/2\tan^{-1} x.$

$\therefore y_1 = 1/\{2(1 + x^2)\}.$

We get

$$y_n = \frac{1}{2}(-1)^{n-1}(n-1)!\sin^n\theta\sin n\theta, \text{ where } \theta = \cot^{-1} x.$$

Example 18:

Find the nth derivative of $1/(1 + x + x^2 + x^3)$ and show that for $x = 0$ it is zero when n is of the form $4p + 2$ or $4p + 3$ and is n! or – n! according as n is of the form $4p + 1$.

Solution:

Let $y = \dfrac{1}{1+x+x^2+x^3} = \dfrac{1}{(1+x)(x^2+1)}$

$$= \frac{1}{2(x+1)} - \frac{x-1}{2(x^2+1)}, \text{ (on resolving into partial fractions)}$$

$$= \frac{1}{2(x+1)} - \frac{x-1}{2(x+i)(x-i)}$$

$$= \frac{1}{2(x+1)} - \frac{1}{2}\left[\frac{-i-1}{(-i-i)(x+i)} + \frac{i-1}{(i+i)(x-i)}\right]$$

$$= \frac{1}{2(x+1)} - \frac{1+i}{4i(x+i)} + \frac{1-i}{4i(x-i)}.$$

Differentiating n times w.r.t. ‘x’, we get

$$y_n = (-1)^n n!\left[\frac{1}{2(x+1)^{n+1}} - \frac{1+i}{4i(x+i)^{n+1}} + \frac{1-i}{4i(x-i)^{n+1}}\right]$$

Putting $x = 0$ in the expression for y_n, we get

$$(y_n)_0 = (-1)^n n!\left[\frac{1}{2} - \frac{1+i}{4.i^{n+2}} + \frac{1-i}{4(-1)^{n+1}\, i^{n+2}}\right].$$

Now we shall discuss the **different cases.**

(i) When $n = 4p + 2$, we have $(-1)^n = (-1)^{4p+2} = 1$,
$(-1)^{n+1} = (-1)^{4p+3} = -1$,
$i^{n+2} = i^{4p+4} = i^{4(p+1)} = 1^{p+1} = 1$.

$$\therefore (y_n)_0 = n!\left[\frac{1}{2} - \frac{1+i}{4} - \frac{1-i}{4}\right] = (n!).0 = 0.$$

(ii) When $n = 4p + 3$, we have $(-1)^n = -1$, $(-1)^{n+1} = 1$,
$i^{n+2} = i^{4p+5} = i^{4p+4}\, i = i$.

$$\therefore (y_n)_0 = -n!\left[\frac{1}{2} - \frac{1+i}{4i} + \frac{1-i}{4i}\right] = -(n!).0 = 0.$$

(iii) When $n = 4p$, we have $(-1)^n = 1$, $(-1)^{n+1} = -1$,
$i^{n+2} = i^{4p+2} = i^{4p}\, i^2 = -1$.

$$\therefore \qquad (y_n)_0 = n!\left[\frac{1}{2} + \frac{1+i}{4} + \frac{1-i}{4}\right] = n!.$$

(iv) When $n = 4p + 1$, we have $(-1)^n = -1$, $(-1)^{n+1} = 1$,
$i^{n+2} = i^{4p+3} = i^{4p}\, i^3 = i^3 = -i$.

$$\therefore (y_n)_0 = -n!\left[\frac{1}{2} + \frac{1+i}{4i} - \frac{1-i}{4i}\right] = -n!.$$

Example 19:

Find the nth derivative of $x^2e^x \cos^3 x$.

Solution:

Let $\quad y = x^2e^x \cos^3x$. Since $\cos 3x = 4\cos^3x - 3\cos x$,

therefore $\cos^3 x = \dfrac{1}{4}(3\cos x + \cos 3x)$.

Hence $\quad y = \dfrac{1}{4}x^2 e^x (3\cos x + \cos 3x)$

$$= \frac{3}{4}\left(e^x \cos x\right)x^2 + \frac{1}{4}\left(e^x \cos 3x\right).x^2.$$

Now differentiating n times by Leibnitz's theorem, we get

$$y_n = \frac{3}{4}[\{r^n e^x \cos(x + n\phi)\}.x^2 + {}^nC_1\,\{r^{n-1} e^x \cos(x + n-1\phi)\}.2x$$

$$+ {}^nC_2\,\{r^{n-2} e^x \cos(x + n - 2\phi)\}.2] + \frac{1}{4}\left[r_1^n e^x \cos\left(3x + n\phi_1\right).x^2\right.$$

$$+{}^{n}C_1\left\{r_1^{n-1}e^x\cos\left(3x+n-1\phi_1\right)\right\}.2x+{}^{n}C_2\left\{r_1^{n-2}e^x\cos\left(3x+n-2\phi_1\right)\right\}.2\Big],$$

where $r = \sqrt{2}$, $\phi = \tan^{-1} = \dfrac{\pi}{4}$; $r_1 = \sqrt{10}$, $\phi_1 = \tan^{-1} 3$.

EXERCISES

1. If $y^{1/m} + y^{-1/m} = 2x$ prove that
   $(x^2 - 1)\, y_{n+2} + (2n + 1)\, xy_{n+1} + (x^2 - m^2)y_n = 0$ where y_n denotes the nth derivative of y.
2. If $u = \sin nx + \cos nx$ show that
   $$u_r = n^r[1 + (-1)^r \sin 2m]^{1/3}$$
   where u_r denotes the rth differential co-efficient of u with respect to x.
3. If $y = (x^2 - 1)^n$ prove that
   $$(x^2 - 1)y_{n+2} + 2xy_{n+1} - n(x + 1)\, y_n = 0$$
   Hence, if $P_n = \dfrac{d^n}{dx^n}(x^2 - 1)^n$, show that
   $$\frac{d}{dx}\left[\left(1-x^2\right)\frac{dP_n}{dx}\right]+n(n+1)\,P_n=0.$$
4. If $y = \dfrac{b+cx}{a+2bx+cx_2}$, where $ac > b^2$ prove that $y_n = (-1)^n\, (n)!$
   $$\left(\frac{C}{a+2bx+cx^2}\right)^{(n+1)/2}\cos\left[(n+1)\tan^{-1}\frac{\sqrt{\left(ac-b^2\right)}}{b+cx}\right].$$
5. Show that the 2nd differential co-efficient of $\cos^{2m}x.-(1)^n2 - 2^{m+1}$ $[(2m)^{2n}$ co cos $2m + (2m - 2)^{2n}\, C_1.\cos(2m - 2)^x + \ldots\, 2^{2n}\,(m - 2)\cos 2x]$ where C_r is the co-efficient of x^r in the expansion of $(1 + x)^{2m}$ in powers of x.
6. If $p^2 = a^2 \cos 2\theta + b^2 \sin^2 \theta$, prove that
   $$p+\frac{d^2p}{d\theta^2}=\frac{a^2b^2}{p^2}.$$

4

Jacobians

CASE OF FUNCTIONS

We shall establish the formula for two variables and the result can be easily extended to any number of variables.

Theorem:

If u_1, u_2 are functions of y_1, y_2 and y_1, y_2 are functions of x_1, x_2, then

$$\frac{\partial(u_1, u_2)}{\partial(x_1, x_2)} = \frac{\partial(u_1, u_2)}{\partial(y_1, y_2)} \cdot \frac{\partial(y_1, y_2)}{\partial(x_1, x_2)}.$$

Proof:

We have

$$\left.\begin{aligned} \frac{\partial u_1}{\partial x_1} &= \frac{\partial u_1}{\partial y_1}\frac{\partial y_1}{\partial x_1} + \frac{\partial u_1}{\partial y_2}\frac{\partial y_2}{\partial x_1}, \quad \frac{\partial u_1}{\partial x_2} = \frac{\partial u_1}{\partial y_1}\frac{\partial y_1}{\partial x_2} + \frac{\partial u_1}{\partial y_2}\frac{\partial y_2}{\partial x_2}, \\ \frac{\partial u_1}{\partial x_1} &= \frac{\partial u_2}{\partial y_1}\frac{\partial y_1}{\partial x_1} + \frac{\partial u_2}{\partial y_2}\frac{\partial y_2}{\partial x_1}, \quad \frac{\partial u_2}{\partial x_2} = \frac{\partial u_2}{\partial y_1}\frac{\partial y_1}{\partial x_2} + \frac{\partial u_2}{\partial y_2}\frac{\partial y_2}{\partial x_2}. \end{aligned}\right\} \quad ...(1)$$

Now $\dfrac{\partial(u_1, u_2)}{\partial(y_1, y_2)} \cdot \dfrac{\partial(y_1, y_2)}{\partial(x_1, x_2)}$

$$= \begin{vmatrix} \frac{\partial u_1}{\partial y_1} & \frac{\partial u_1}{\partial y_2} \\ \frac{\partial u_2}{\partial y_1} & \frac{\partial u_2}{\partial y_2} \end{vmatrix} \times \begin{vmatrix} \frac{\partial y_1}{\partial x_1} & \frac{\partial y_1}{\partial x_2} \\ \frac{\partial y_2}{\partial x_1} & \frac{\partial y_2}{\partial x_2} \end{vmatrix}$$

$$= \begin{vmatrix} \frac{\partial u_1}{\partial y_1}\frac{\partial u_1}{\partial y_2}\frac{\partial y_2}{\partial x_1} & \frac{\partial u_1}{\partial y_1}\frac{\partial y_1}{\partial x_2} + \frac{\partial u_1}{\partial y_2}\frac{\partial y_2}{\partial x_2} \\ \frac{\partial u_2}{\partial y_1}\frac{\partial y_1}{\partial x_1} + \frac{\partial u_2}{\partial y_2}\frac{\partial y_2}{\partial x_1} & \frac{\partial u_2}{\partial y_1}\frac{\partial y_1}{\partial x_2} + \frac{\partial u_2}{\partial y_2}\frac{\partial y_2}{\partial x_2} \end{vmatrix},$$

applying row-by-column multiplication rule

$$= \begin{vmatrix} \frac{\partial u_1}{\partial x_1} & \frac{\partial u_1}{\partial x_2} \\ \frac{\partial u_2}{\partial x_1} & \frac{\partial u_2}{\partial x_2} \end{vmatrix}, \text{ using the relations (1)}$$

$$= \frac{\partial(u_1, u_2)}{\partial(x_1, x_2)}.$$

Note: The above formula resembles very much with the formula

$$\frac{df}{dx} = \frac{df}{dt} \cdot \frac{dt}{dx},$$

for the derivative of the function of a function.

Generalization of the Above Formula: *If $u_1, u_2, ..., u_n$ are functions of $y_1, y_2, ..., y_n$ and $y_1, y_2, ..., y_n$ are functions of $x_1, x_2, ..., x_n$, then*

$$\frac{\partial(u_1, u_2, ..., u_n)}{\partial(x_1, x_2, ..., x_n)} = \frac{\partial(u_1, u_2, ..., u_n)}{\partial(y_1, y_2, ..., y_n)} \cdot \frac{\partial(y_1, y_2, ..., y_n)}{\partial(x_1, x_2, ..., x_n)}.$$

The proof may be easily extended as in the case of two variables and has been left as an exercise for the students.

NECESSARY AND SUFFICIENT CONDITION FOR A JACOBIAN TO VANISH

Theorem: *Let $u_1, u_2, ..., u_n$ be functions of n independent variables $x_1, x_2, ..., x_n$. In order that these n functions may not be independent, i.e., there may exist between these n functions a relation*

$$F(u_1, u_2, ..., u_n) = 0, \qquad ...(1)$$

it is necessary and sufficient that the Jacobian $\frac{\partial(u_1, u_2, ..., u_n)}{\partial(x_1, x_2, ..., x_n)}$ should vanish identically.

Proof: *The condition is necessary i.e.*, if there exists between $u_1, u_2, ..., u_n$ a relation

$$F(u_1, u_2, ..., u_n) = 0. \qquad ...(1)$$

their Jacobian is necessarily zero.

Differentiating eqn. (1) partially with respect to $x_1, x_2, ..., x_n$, we get

$$\frac{\partial F}{\partial u_1}\frac{\partial u_1}{\partial x_1} + \frac{\partial F}{\partial u_2}\frac{\partial u_2}{\partial x_1} + ... + \frac{\partial F}{\partial u_n}\frac{\partial u_n}{\partial x_1} = 0,$$

$$\frac{\partial F}{\partial u_1}\frac{\partial u_1}{\partial x_2}+\frac{\partial F}{\partial u_2}\frac{\partial u_2}{\partial x_2}+\ldots+\frac{\partial F}{\partial u_n}\frac{\partial u_n}{\partial x_2}=0,$$

....

$$\frac{\partial F}{\partial u_1}\frac{\partial u_1}{\partial x_n}+\frac{\partial F}{\partial u_2}\frac{\partial u_2}{\partial x_n}+\ldots+\frac{\partial F}{\partial u_n}\frac{\partial u_n}{\partial x_n}=0.$$

Eliminating $\frac{\partial F}{\partial u_1}, \frac{\partial F}{\partial u_2}, \ldots, \frac{\partial F}{\partial u_n}$ from these equations, we get

$$\begin{vmatrix} \frac{\partial u_1}{\partial x_1} & \frac{\partial u_2}{\partial x_1} & \cdots & \frac{\partial u_n}{\partial x_1} \\ \frac{\partial u_1}{\partial x_2} & \frac{\partial u_2}{\partial x_2} & \cdots & \frac{\partial u_n}{\partial x_2} \\ \cdots & \cdots & \cdots & \cdots \\ \frac{\partial u_1}{\partial x_n} & \frac{\partial u_2}{\partial x_n} & \cdots & \frac{\partial u_n}{\partial x_n} \end{vmatrix}=0$$

or
$$\frac{\partial(u_1, u_2, \ldots, u_n)}{\partial(x_1, x_2, \ldots, x_n)}=0.$$

The Condition is Sufficient, *i.e.,* if the Jacobian $J(u_1, u_2, \ldots, u_n)$ is zero, then there must exist a relation between $u_1, u_2, \ldots, u_n$.

The equations connecting the functions $u_1, u_2, \ldots, u_n$ and the variables $x_1, x_2, \ldots, x_n$ are always capable of being put into the following form :

$\phi_1(x_1, x_2, \ldots, x_n, u_1) = 0$

$\phi_2(x_2, x_3, \ldots, x_n, u_1, u_2) = 0$

....

$\phi_r(x_r, x_{r+1}, \ldots, x_n, u_1, u_2, \ldots, u_r) = 0$

$\phi_n(x_n, u_1, u_2, \ldots, u_n) = 0.$

Then, we have

$$J=\frac{\partial(u_1, u_2, \ldots, u_n)}{\partial(x_1, x_2, \ldots, x_n)}=(-1)^n\frac{\dfrac{\partial(\phi_1, \phi_2, \ldots, \phi_n)}{\partial(x_1, x_2, \ldots, x_n)}}{\dfrac{\partial(\phi_1, \phi_2, \ldots, \phi_n)}{\partial(u_1, u_2, \ldots, u_n)}}$$

$$=(-1)^n\frac{\dfrac{\partial\phi_1}{\partial x_1}\dfrac{\partial\phi_2}{\partial x_2}\cdots\dfrac{\partial\phi_n}{\partial x_n}}{\dfrac{\partial\phi_1}{\partial u_1}\dfrac{\partial\phi_2}{\partial u_2}\cdots\dfrac{\partial\phi_n}{\partial u_n}}.$$

Now, if J = 0, we have

$$\frac{\partial \phi_1}{\partial x_1} \cdot \frac{\partial \phi_2}{\partial x_2} \cdots \frac{\partial \phi_r}{\partial x_r} \cdots \frac{\partial \phi_n}{\partial x_n} = 0$$

i.e., $\frac{\partial \phi_r}{\partial x_r} = 0$ for *some value* of r between 1 and n.

Hence, for that particular value of r the function ϕ_r must not contain x_r; and accordingly the corresponding equation is if the form

$$\phi_r(x_{r+1}, ..., x_n, u_1, u_2, ..., u_r) = 0.$$

Consequently between this and the remaining equations $\phi_{r+1} = 0$, $\phi_{r+2} = 0, ..., \phi_n = 0$, the variables $x_{r+1}, x_{r+2}, ..., x_n$ can be eliminated so as to give a final equation between $u_1, u_2, ..., u_n$ alone.

Hence the theorem is establish.

JACOBIAN OF IMPLICIT FUNCTIONS

Theorem:

Suppose $u_1, u_2, ..., u_n$ instead of being given explicitly in terms of $x_1, x_2, ..., x_n$ are connected with them by equations as

$$F_1(u_1, u_2, ..., u_n, x_1, x_2, ..., x_n) = 0,$$
$$F_2(u_1, u_2, ..., u_n, x_1, x_2, ..., x_n) = 0,$$
$$\cdots \cdots \cdots \cdots \cdots \cdots$$
$$F_n(u_1, u_2, ..., u_n, x_1, x_2, ..., x_n) = 0.$$

Then, we have

$$\frac{\partial(u_1, u_2, ..., u_n)}{\partial(x_1, x_2, ..., x_n)} = (-1)^n \frac{\dfrac{\partial(F_1, F_2, ..., F_n)}{\partial(x_1, x_2, ..., x_n)}}{\dfrac{\partial(F_1, F_2, ..., F_n)}{\partial(u_1, u_2, ..., F_n)}}.$$

Proof:

Here also we shall establish the result for two variables and the proof can be extended easily for n variables. The students should themselves write the proof for n variables on the basis of the proof given below for two variables.

In the case of two variables, the connecting relations are

$$\left.\begin{aligned} F_1(u_1, u_2, x_1, x_2) = 0, \\ F_2(u_1, u_2, x_1, x_2) = 0. \end{aligned}\right\} \quad ...(1)$$

From relations (1), we have by differentiation

$$\left.\begin{aligned}
&\frac{\partial F_1}{\partial x_1}+\frac{\partial F_1}{\partial u_1}\frac{\partial u_1}{\partial x_1}+\frac{\partial F_1}{\partial u_2}\frac{\partial u_2}{\partial x_1}=0,\\
&\frac{\partial F_1}{\partial x_2}+\frac{\partial F_1}{\partial u_1}\frac{\partial u_1}{\partial x_2}+\frac{\partial F_1}{\partial u_2}\frac{\partial u_2}{\partial x_2}=0,\\
&\frac{\partial F_2}{\partial x_1}+\frac{\partial F_2}{\partial u_1}\frac{\partial u_1}{\partial x_1}+\frac{\partial F_2}{\partial u_2}\frac{\partial u_2}{\partial x_1}=0,\\
&\frac{\partial F_2}{\partial x_2}+\frac{\partial F_2}{\partial u_1}\frac{\partial u_1}{\partial x_1}+\frac{\partial F_2}{\partial u_2}\frac{\partial u_2}{\partial x_2}=0.
\end{aligned}\right\} \quad \text{...(2)}$$

Now $\dfrac{\partial(F_1, F_2)}{\partial(u_1, u_2)}\cdot\dfrac{\partial(u_1, u_2)}{\partial(x_1, x_2)}$

$$=\begin{vmatrix}\dfrac{\partial F_1}{\partial u_1} & \dfrac{\partial F_1}{\partial u_2}\\[2mm] \dfrac{\partial F_2}{\partial u_1} & \dfrac{\partial F_2}{\partial u_2}\end{vmatrix}\times\begin{vmatrix}\dfrac{\partial u_1}{\partial x_1} & \dfrac{\partial u_1}{\partial x_2}\\[2mm] \dfrac{\partial u_2}{\partial x_1} & \dfrac{\partial u_2}{\partial x_2}\end{vmatrix}$$

$$=\begin{vmatrix}\dfrac{\partial F_1}{\partial u_1}\dfrac{\partial u_1}{\partial x_1}+\dfrac{\partial F_1}{\partial u_2}\dfrac{\partial u_2}{\partial x_1} & \dfrac{\partial F_1}{\partial u_1}\dfrac{\partial u_1}{\partial x_2}+\dfrac{\partial F_1}{\partial u_2}\dfrac{\partial u_2}{\partial x_2}\\[2mm] \dfrac{\partial F_2}{\partial u_1}\dfrac{\partial u_1}{\partial x_1}+\dfrac{\partial F_2}{\partial u_2}\dfrac{\partial u_2}{\partial x_1} & \dfrac{\partial F_2}{\partial u_1}\dfrac{\partial u_1}{\partial x_2}+\dfrac{\partial F_2}{\partial u_2}\dfrac{\partial u_2}{\partial x_2}\end{vmatrix},$$

applying row-by-column multiplication rule

$$=\begin{vmatrix}\dfrac{-\partial F_1}{\partial x_1} & \dfrac{-\partial f_1}{\partial x_2}\\[2mm] \dfrac{-\partial F_2}{\partial x_1} & \dfrac{-\partial F_2}{\partial x_2}\end{vmatrix}, \text{ using the relations (2)}$$

$$=(-1)^2\frac{\partial(F_1, F_2)}{\partial(x_1, x_2)}.$$

Accordingly, we have $\dfrac{\partial(u_1, u_2)}{\partial(x_1, x_2)}=(-1)^2\dfrac{\dfrac{\partial(F_1, F_2)}{\partial(x_1, x_2)}}{\dfrac{\partial(F_1, F_2)}{\partial(u_1, u_2)}}$.

Definition

If $u_1, u_2, \ldots, u_n$ are functions of n independent variables $x_1, x_2, \ldots, x_n$ then the determinant

$$\begin{vmatrix} \frac{\partial u_1}{\partial x_1} & \frac{\partial u_1}{\partial x_2} & \dots\dots\dots & \frac{\partial u_1}{\partial x_n} \\ \frac{\partial u_1}{\partial x_1} & \frac{\partial u_2}{\partial x_2} & \dots\dots\dots & \frac{\partial u_2}{\partial x_n} \\ \dots\dots & \dots\dots & \dots\dots\dots & \dots\dots \\ \frac{\partial x_n}{\partial x_1} & \frac{\partial u_n}{\partial x_2} & \dots\dots\dots & \frac{\partial u_n}{\partial x_n} \end{vmatrix}$$

is called the Jacobian of $u_1, u_2, ..., u_n$ with respect to $x_1, x_2, ..., x_n$ and is denoted either by $\frac{\partial(u_1, u_2, ..., u_n)}{\partial(x_1, x_2, ..., x_n)}$ or by $J(u_1, u_2, ..., u_n)$. The second notation is used when there is no doubt as regards the independent variables.

Thus, if u and v are functions of two independent variables x and y, we have

$$\frac{\partial(u, v)}{\partial(x, y)} = \begin{vmatrix} \frac{\partial u}{\partial x} & \frac{\partial u}{\partial y} \\ \frac{\partial v}{\partial x} & \frac{\partial v}{\partial y} \end{vmatrix} = J(u, v)$$

Similarly, if u, v and w are functions of three independent variables x, y and z, we have

$$\frac{\partial(u,v,w)}{\partial(x,y,z)} = \begin{vmatrix} \frac{\partial u}{\partial x} & \frac{\partial u}{\partial y} & \frac{\partial u}{\partial z} \\ \frac{\partial v}{\partial x} & \frac{\partial v}{\partial y} & \frac{\partial v}{\partial z} \\ \frac{\partial w}{\partial x} & \frac{\partial w}{\partial y} & \frac{\partial w}{\partial z} \end{vmatrix}$$

Note: If the functions $u_1, u_2, ..., u_n$ of n independent variables $x_1, x_2, ..., x_n$ are of the following forms,

$u_1 = f_1(x_1), u_2 = f_2(x_1, x_2), ..., u_n = f_n(x_1, x_2, ..., x_n)$, then

$$\frac{\partial(u_1, u_2, ..., u_n)}{\partial(x_1, x_2, ..., x_n)} = \begin{vmatrix} \frac{\partial u_1}{\partial x_1} & 0 & 0 & \cdots & 0 \\ \frac{\partial u_2}{\partial x_1} & \frac{\partial u_2}{\partial x_2} & 0 & \cdots & 0 \\ \cdots & \cdots & \cdots & \cdots & \cdots \\ \frac{\partial u_n}{\partial x_1} & \frac{\partial u_n}{\partial x_2} & \frac{\partial u_n}{\partial x_3} & \cdots & \frac{\partial u_n}{\partial x_n} \end{vmatrix}$$

$$= \frac{\partial u_1}{\partial x_1} \frac{\partial u_2}{\partial x_2} \cdots\cdots \frac{\partial u_n}{\partial x_n},$$

i.e., in such cases the Jacobian reduces to the principal diagonal term of the determinant.

SOLVED EXAMPLES

Example 1:

If x = cos u cosh v, y = c sin u sinh v, prove that

$$\frac{\partial(x, y)}{\partial(u, v)} = \frac{1}{2}c^2(\cos 2u - \cosh 2v).$$

Solution:

We have $\frac{\partial(x, y)}{\partial(u, v)} = \begin{vmatrix} \frac{\partial x}{\partial u} & \frac{\partial x}{\partial v} \\ \frac{\partial y}{\partial u} & \frac{\partial y}{\partial v} \end{vmatrix}$

$$= \begin{vmatrix} -c \sin u \cosh v & c \cos u \sinh v \\ c \cos u \sinh v & c \sin u \cosh v \end{vmatrix}$$

$$= - c^2 \sin^2 u \cosh^2 v - c^2 \cos^2 u \sinh^2 v$$

$$= -\frac{1}{2}c^2 [(1 - \cos 2u) \cosh^2 v + (1 + \cos 2u) \sinh^2 v]$$

$$= -\frac{1}{2}c^2 [\cosh^2 v + \sinh^2 v - \cos 2u (\cosh^2 v - \sinh^2 v)]$$

$$= -\frac{1}{2}c^2 (\cosh 2v - \cos 2u) = \frac{1}{2}c^2 (\cos 2u - \cosh 2v).$$

Example 2:

If x = u (1 + v), y = v (1 + u), find the Jacobian of x, y with respect to u, v.

Solution:

We have

$$\frac{\partial(x, y)}{\partial(u, v)} = \begin{vmatrix} \frac{\partial x}{\partial y} & \frac{\partial x}{\partial v} \\ \frac{\partial y}{\partial u} & \frac{\partial y}{\partial v} \end{vmatrix} = \begin{vmatrix} 1+v & u \\ v & 1+u \end{vmatrix}$$

$$= (1 + v) (1 + u) - uv = 1 + u + v + uv - uv = 1 + u + v.$$

Example 3:

If $y_1 = r\cos\theta_1$, $y_2 = r\sin\theta_1\cos\theta_2$,
$y_3 = r\sin\theta_1\sin\theta_2\cos\theta_3, \ldots, y_{n-1}$
$= r\sin\theta_1\sin\theta_2 \ldots \sin\theta_{n-2}\cos\theta_{n-1}$
and $y_n = r\sin\theta_1\sin\theta_2 \ldots \sin\theta_{n-1}$, prove that

$$\frac{\partial(y_1, y_2, \ldots, y_n)}{\partial(r, \theta_1, \ldots, \theta_{n-1})} = r^{n-1}\sin^{n-2}\theta_1\sin^{n-3}\theta_2 \ldots \sin\theta_{n-2}.$$

Solution:

If we square and add, we get $y_1^2 + y_2^2 + \ldots + y_n^2 = r^2$.

$$\therefore\ y_1\frac{\partial y_1}{\partial r} + y_2\frac{\partial y_2}{\partial r} + \ldots + y_n\frac{\partial y_n}{\partial r} = r$$

and $y_1\dfrac{\partial y_1}{\partial\theta_k} + y_2\dfrac{\partial y_2}{\partial\theta_k} + \ldots + y_n\dfrac{\partial y_n}{\partial\theta_k} = 0$, k = 1, 2, ... n – 1. ...(1)

$$\text{Now } \frac{\partial(y_1, y_2, \ldots, y_n)}{\partial(r, \theta_1, \ldots, \theta_{n-1})} = \begin{vmatrix} \frac{\partial y_1}{\partial r} & \frac{\partial y_1}{\partial\theta_1} & \frac{\partial y_1}{\partial\theta_2} & \cdots & \frac{\partial y_1}{\partial\theta_{n-1}} \\ \frac{\partial y_2}{\partial r} & \frac{\partial y_2}{\partial\theta_1} & \frac{\partial y_2}{\partial\theta_2} & \cdots & \frac{\partial y_2}{\partial\theta_{n-1}} \\ \cdots & \cdots & \cdots & \cdots & \cdots \\ \frac{\partial y_n}{\partial r} & \frac{\partial y_n}{\partial\theta_1} & \frac{\partial y_n}{\partial\theta_1} & \cdots & \frac{\partial y_n}{\partial\theta_{n-1}} \end{vmatrix}$$

$$= \frac{1}{y_n}\begin{vmatrix} \frac{\partial y_1}{\partial r} & \frac{\partial y_1}{\partial\theta_1} & \frac{\partial y_1}{\partial\theta_2} & \cdots & \frac{\partial y_1}{\partial\theta_{n-1}} \\ \frac{\partial y_2}{\partial r} & \frac{\partial y_2}{\partial\theta_1} & \frac{\partial y_2}{\partial\theta_2} & \cdots & \frac{\partial y_2}{\partial\theta_{n-1}} \\ \cdots & \cdots & \cdots & \cdots & \cdots \\ \frac{\partial y_{n-1}}{\partial r} & \frac{\partial y_{n-1}}{\partial\theta_1} & \frac{\partial y_{n-1}}{\partial\theta_2} & \cdots & \frac{\partial y_{n-1}}{\partial\theta_{n-1}} \\ r & 0 & 0 & \cdots & 0 \end{vmatrix}$$

operating $y_nR_n + (y_1R_1 + y_2R_2 + \ldots + y_{n-1}R_{n-1})$ and using the results (1)

$$= (-1)^{n-1}\cdot\frac{r}{y_n}\begin{vmatrix} \frac{\partial y_1}{\partial\theta_1} & \frac{\partial y_1}{\partial\theta_2} & \cdots & \frac{\partial y_1}{\partial\theta_{n-1}} \\ \frac{\partial y_2}{\partial\theta_1} & \frac{\partial y_2}{\partial\theta_2} & \cdots & \frac{\partial y_2}{\partial\theta_{n-1}} \\ \cdots & \cdots & \cdots & \cdots \\ \frac{\partial y_{n-1}}{\partial\theta_1} & \frac{\partial y_{n-1}}{\partial\theta_2} & \cdots & \frac{\partial y_{n-1}}{\partial\theta_{n-1}} \end{vmatrix},$$

expanding the determinant along the nth row

$$=(-1)^{n-1}\cdot\frac{r}{y_n}\begin{vmatrix}\frac{\partial y_1}{\partial\theta_1} & 0 & 0 & \cdots & 0\\ \frac{\partial y_2}{\partial\theta_1} & \frac{\partial y_2}{\partial\theta_2} & 0 & \cdots & 0\\ \cdots & \cdots & \cdots & \cdots & \cdots\\ \frac{\partial y_{n-1}}{\partial\theta_1} & \frac{\partial y_{n-1}}{\partial\theta_2} & \frac{\partial y_{n-1}}{\partial\theta_3} & \cdots & \frac{\partial y_{n-1}}{\partial\theta_{n-1}}\end{vmatrix}$$

$$=(-1)^{n-1}\cdot\frac{r}{y_n}\cdot\frac{\partial y_1}{\partial\theta_1}\cdot\frac{\partial y_2}{\partial\theta_2}\cdot\frac{\partial y_3}{\partial\theta_3}\cdot\ldots\frac{\partial y_{n-1}}{\partial\theta_{n-1}}$$

$$=(-1)^{n-1}\cdot\frac{r}{y_n}\cdot(-r\sin\theta_1)(-r\sin\theta_1\sin\theta_2)\cdot\ldots\cdot(-r\sin\theta_1\sin\theta_2\ldots\sin\theta_{n-1})$$

$$=(-1)^{n-1}\cdot\frac{r}{y_n}(-1)^{n-1}r^{n-1}\sin^{n-1}\theta_1\sin^{n-2}\theta_2\cdot\ldots\sin^2\theta_{n-2}\sin\theta_{n-1}$$

$$=\frac{r}{r\sin\theta_1\sin\theta_2\ldots\sin\theta_{n-2}\sin\theta_{1-1}}\cdot r^{n-1}\sin^{n-1}\theta_1\sin^{n-2}\theta_2\ldots\sin^2\theta_{n-2}\sin\theta_{n-1}$$

$$=r^{n-1}\sin^{n-2}\theta_1\sin^{n-3}\theta_2\ldots\sin\theta_{n-2}.$$

Example 4:

If $u_1=\frac{x_2x_3}{x_1}$, $u_2=\frac{x_3x_1}{x_2}$, $u_3=\frac{x_1x_2}{x_3}$, *prove that* $J(u_1, u_2, u_3)=4$.

Solution:

We have $J(u_1, u_2, u_3)$

$$\begin{vmatrix}\frac{\partial u_1}{\partial x_1} & \frac{\partial u_1}{\partial x_2} & \frac{\partial u_1}{\partial x_3}\\ \frac{\partial u_2}{\partial x_1} & \frac{\partial u_2}{\partial x_2} & \frac{\partial u_2}{\partial x_3}\\ \frac{\partial u_3}{\partial x_1} & \frac{\partial u_3}{\partial x_2} & \frac{\partial u_3}{\partial x_3}\end{vmatrix}=\begin{vmatrix}-\frac{x_2x_3}{x_1^2} & \frac{x_3}{x_1} & \frac{x_2}{x_1}\\ \frac{x_3}{x_2} & -\frac{x_3x_1}{x_2^2} & \frac{x_1}{x_2}\\ \frac{x_2}{x_3} & \frac{x_1}{x_3} & -\frac{x_1x_2}{x_3^2}\end{vmatrix}$$

$$= \frac{1}{x_1^2 x_2^2 x_3^2} \begin{vmatrix} -x_2x_3 & x_3x_1 & x_1x_2 \\ x_2x_3 & -x_3x_1 & x_1x_2 \\ x_2x_3 & x_3x_1 & -x_1x_2 \end{vmatrix}$$

$$= \frac{1}{x_1^2 x_2^2 x_3^2} \begin{vmatrix} 0 & 0 & 2x_1x_2 \\ 2x_2x_3 & 0 & 0 \\ x_2x_3 & x_3x_1 & -x_1x_2 \end{vmatrix}, \text{ adding } R_2 \text{ to } R_1 \text{ and then } R_3 \text{ to } R_2$$

$$= \frac{1}{x_1^2 x_2^2 x_3^2} \cdot \left(2x_1x_2 . 2x_2x_3^2x_1\right) = 4.$$

Example 5(a):

If $x = r \sin \theta \cos \phi$, $y = r \sin \theta \sin \phi$, $z = r \cos \theta$, *show that*

$$\frac{\partial(x, y, z)}{\partial(r, \theta, \phi)} = r^2 \sin \theta.$$

Solution:

We have

$$\frac{\partial(x, y, z)}{\partial(r, \theta, \phi)} = \begin{vmatrix} \frac{\partial x}{\partial r} & \frac{\partial x}{\partial \theta} & \frac{\partial x}{\partial \phi} \\ \frac{\partial y}{\partial r} & \frac{\partial y}{\partial \theta} & \frac{\partial y}{\partial \phi} \\ \frac{\partial z}{\partial r} & \frac{\partial z}{\partial \theta} & \frac{\partial}{\partial \phi} \end{vmatrix}$$

$$= \begin{vmatrix} \sin\theta\cos\phi & r\cos\theta\cos\phi & -r\sin\theta\sin\phi \\ \sin\theta\sin\phi & r\cos\theta\sin\phi & r\sin\theta\cos\phi \\ \cos\theta & -r\sin\theta & 0 \end{vmatrix}$$

$= \cos \theta(r^2 \sin \theta \cos \theta \cos^2 \phi + r^2 \sin \theta \cos \theta \sin^2 \phi)$

$+ r \sin \theta(r \sin^2 \theta \cos^2 \phi + r \sin^2 \theta \sin^2 \phi)$, expanding the determinant along the third row

$= r^2 \sin \theta \cos^2 \theta + r^2 \sin^3 \theta = r^2 \sin \theta (\cos^2 \theta + \sin^2 \theta) = r^2 \sin \theta.$

Example 5(b):

If $y_1 = r \sin \theta_1 \sin \theta_2$, $y_2 = r \sin \theta_1 \cos \theta_2$,

$y_3 = r \cos \theta_1 \sin \theta_3$, $y_4 = r \cos \theta_1 \cos \theta_3$, *find the value of the Jacobian*

$$\frac{\partial(y_1, y_2, y_3, y_4)}{\partial(r, \theta_1, \theta_2, \theta_3)}.$$

Solution:

Squaring and adding the given relations, we have

$$y_1^2 + y_2^2 + y_3^2 + y_4^2 = r^2.$$

$$\therefore \quad y_1 \frac{\partial y_1}{\partial r} + y_2 \frac{\partial y_2}{\partial r} + y_3 \frac{\partial y_3}{\partial r} + y_4 \frac{\partial y_4}{\partial r} = r$$

$$\text{and} \quad y_1 \frac{\partial y_1}{\partial \theta_r} + y_2 \frac{\partial y_2}{\partial \theta_r} + y_3 \frac{\partial y_3}{\partial \theta_r} + y_4 \frac{\partial y_4}{\partial \theta_r} = 0, \; r = 1, 2, 3 \qquad \text{...(1)}$$

Also $\quad y_3^2 + y_4^2 = r^2 \cos^2 \theta_1$, so that

$$y_3 \frac{\partial y_3}{\partial \theta_1} + y_4 \frac{\partial y_4}{\partial \theta_1} = -r^2 \cos \theta_1 \sin \theta_1;$$

$$y_3 \frac{\partial y_3}{\partial \theta_1} + y_4 \frac{\partial y_4}{\partial \theta_r} = 0, \; r = 2, 3. \qquad \text{...(2)}$$

Now the required Jacobian

$$J = \begin{vmatrix} \frac{\partial y_1}{\partial r} & \frac{\partial y_1}{\partial \theta_1} & \frac{\partial y_1}{\partial \theta_2} & \frac{\partial y_1}{\partial \theta_3} \\ \frac{\partial y_2}{\partial r} & \frac{\partial y_2}{\partial \theta_1} & \frac{\partial y_2}{\partial \theta_2} & \frac{\partial y_2}{\partial \theta_3} \\ \frac{\partial y_3}{\partial r} & \frac{\partial y_3}{\partial \theta_1} & \frac{\partial y_3}{\partial \theta_2} & \frac{\partial y_3}{\partial \theta_3} \\ \frac{\partial y_4}{\partial r} & \frac{\partial y_4}{\partial \theta_1} & \frac{\partial y_4}{\partial \theta_2} & \frac{\partial y_4}{\partial \theta_3} \end{vmatrix}.$$

Operating $y_1R_1 + (y_2R_2 + y_3R_3 + y_4R_4)$, and using the results (1), we get

$$J = \frac{1}{y_1} \begin{vmatrix} r & 0 & 0 & 0 \\ \frac{\partial y_2}{\partial r} & \frac{\partial y_2}{\partial \theta_1} & \frac{\partial y_2}{\partial \theta_2} & \frac{\partial y_2}{\partial \theta_3} \\ \frac{\partial y_3}{\partial r} & \frac{\partial y_3}{\partial \theta_1} & \frac{\partial y_3}{\partial \theta_2} & \frac{\partial y_3}{\partial \theta_3} \\ \frac{\partial y_4}{\partial r} & \frac{\partial y_4}{\partial \theta_1} & \frac{\partial y_4}{\partial \theta_2} & \frac{\partial y_4}{\partial \theta_3} \end{vmatrix}$$

$$= \frac{r}{y_1}\begin{vmatrix} \frac{\partial y_2}{\partial \theta_1} & \frac{\partial y_2}{\partial \theta_2} & \frac{\partial y_2}{\partial \theta_3} \\ \frac{\partial y_3}{\partial \theta_1} & \frac{\partial y_3}{\partial \theta_2} & \frac{\partial y_3}{\partial \theta_3} \\ \frac{\partial y_4}{\partial \theta_1} & \frac{\partial y_4}{\partial \theta_2} & \frac{\partial y_4}{\partial \theta_3} \end{vmatrix}$$

$$= \frac{r}{y_1 y_3}\begin{vmatrix} \frac{\partial y_2}{\partial \theta_1} & \frac{\partial y_2}{\partial \theta_2} & \frac{\partial y_2}{\partial \theta_3} \\ -r^2 \cos\theta_1 \sin\theta_1 & 0 & 0 \\ \frac{\partial y_4}{\partial \theta_1} & \frac{\partial y_4}{\partial \theta_2} & \frac{\partial y_4}{\partial \theta_3} \end{vmatrix}$$

adding y_4R_3 to y_3R_2 and using the results (2)

$$= \frac{r}{y_1 y_3}\cdot r^2 \cos\theta_1 \sin\theta_1 \left[\frac{\partial y_2}{\partial \theta_2}\cdot\frac{\partial y_4}{\partial \theta_3} - \frac{\partial y_4}{\partial \theta_2}\cdot\frac{\partial y_2}{\partial \theta_3}\right]$$

$$= \frac{r^3 \cos\theta_1 \sin\theta_1}{y_1 y_3}\ [(-r\sin\theta_1 \sin\theta_2)(-r\cos\theta_1 \sin\theta_3) - 0]$$

$$= \frac{r^5 \sin^2\theta_1 \cos^2\theta_1 \sin\theta_2 \sin\theta_3}{r^2 \sin\theta_1 \cos\theta_1 \sin\theta_2 \sin\theta_3} = r^3 \sin\theta_1 \cos\theta_1.$$

Example 5(c):

If $y_1 = \cos x_1$, $y_2 = \sin x_1 \cos x_2$, $y_3 = \sin x_1 \sin x_2 \cos x_3$, ..., $y_n = \sin x_1 \sin x_2 \sin x_3 \ldots \sin x_{n-1}$, find the Jacobian of $y_1, y_2, \ldots, y_n$ with respect to $x_1, x_2, \ldots, x_n$.

Solution:

Here y_1 is a function of x_1 only, y_2 is a function of x_1, x_2 only, y_3 is a function of x_1, x_2, x_3 only, ... and y_n is a function of $x_1, x_2, \ldots, x_n$.

$$\therefore \frac{\partial(y_1, y_2, \ldots, y_n)}{\partial(x_1, x_2, \ldots, x_n)} = \text{the principal diagonal term of the determinant}$$

$$= \frac{\partial x_1}{\partial x_1}\cdot\frac{\partial y_2}{\partial x_2}\cdot\frac{\partial y_3}{\partial x_3}\cdot \ldots \cdot\frac{\partial y_n}{\partial x_n}$$

$$= (-\sin x_1)\cdot(-\sin x_1 \sin x_2)\cdot(-\sin x_1 \sin x_2 \sin x_3)\ \ldots \cdot(-\sin x_1 \sin x_2 \sin x_3 \ldots \sin x_n)$$

$$= (-1)^n \sin^n x_1 \sin^{n-1} x_2 \sin^{n-2} x_3 \ldots \sin x_n.$$

Example 5(d):

If $u_1 = x_2^2 + 2a_1x_2x_3 + x_3^2$, $u_2 = x_1^2 + 2a_2x_3x_1 + x_3^2$,

$u_3 = x_1^2 + 2a_3x_1x_2 + x_2^2$,

find the Jacobian of u_1, u_2, u_3 *w.r.t.* x_1, x_2, x_3.

Solution:

We have $J(u_1, u_2, u_3)$

$$= \begin{vmatrix} 0 & 2(x_2 + a_1x_3) & 2(a_1x_2 + x_3) \\ 2(x_1 + a_2x_3) & 0 & 2(a_2x_1 + x_3) \\ 2(x_1 + a_3x_2) & 2(a_3x_1 + x_2) & 0 \end{vmatrix}$$

$$= 8[-(a_2x_3 + x_1)(a_1x_2 + x_3)(a_3x_1 + x_2) + (x_1 + a_3x_2)(x_2 + a_1x_3)(a_2x_1 + x_3)],$$

expanding the determinant along the first column

$$= 8[(a_2x_3 + x_1)(a_1x_2 + x_3)(a_3x_1 + x_2) + (x_1 + a_3x_2)(x_2 + a_1x_3)(x_3 + a_2x_1)].$$

Example 6:

If $x = \sin\theta\sqrt{(1 - c^2\sin^2\phi)}$, $y = \cos\theta\cos\phi$, *then show that*

$$\frac{\partial(x, y)}{\partial(\theta, \phi)} = -\sin\phi\frac{\left[(1-c^2)\cos^2\theta + c^2\cos^2\phi\right]}{\sqrt{(1 - c^2\sin^2\phi)}}.$$

Solution:

$$\text{We have } \frac{\partial(x, y)}{\partial(\theta, \phi)} = \begin{vmatrix} \frac{\partial x}{\partial \theta} & \frac{\partial x}{\partial \phi} \\ \frac{\partial y}{\partial \theta} & \frac{\partial y}{\partial \phi} \end{vmatrix} = \frac{\partial x}{\partial \theta} \cdot \frac{\partial y}{\partial \phi} - \frac{\partial x}{\partial \phi} \cdot \frac{\partial y}{\partial \theta}$$

$$= \cos\theta\sqrt{(1 - c^2\sin^2\theta)} \cdot (-\cos\theta\sin\phi)$$

$$-\sin\theta \cdot \frac{1}{2}\left(1 - c^2\sin^2\phi\right)^{-1/2} . \left(-2c^2\sin\phi\cos\phi\right) \cdot (-\sin\theta\cos\phi)$$

$$= -\frac{\sin\phi}{\sqrt{(1 - c^2\sin^2\phi)}}\left[\cos^2\theta\,(1 - c^2\sin^2\phi) + c^2\sin^2\theta\cos^2\phi\right]$$

$$= -\frac{\sin \phi}{\sqrt{(1-c^2 \sin^2 \phi)}} [\cos^2 \theta - c^2 \cos^2 \theta \sin^2 \phi + c^2 (1 - \cos^2 \theta) \cos^2 \phi]$$

$$= -\frac{\sin \phi}{\sqrt{(1-c^2 \sin^2 \phi)}} [\cos^2 \theta - c^2 \cos^2 \theta (\sin^2 \phi + \cos^2 \phi) + c^2 \cos^2 \phi]$$

$$= -\frac{\sin \phi}{\sqrt{(1-c^2 \sin^2 \phi)}} [\cos^2 \theta - c^2 \cos^2 \theta + c^2 \cos^2 \phi]$$

$$= -\frac{\sin \phi}{\sqrt{(1-c^2 \sin^2 \phi)}} [(1 - c^2) \cos^2 \theta + c^2 \cos^2 \phi].$$

Example 7:

If $u^3 + v^3 = x + y$, $u^2 + v^2 = x^3 + y^3$, show that

$$\frac{\partial(u, v)}{\partial(x, y)} = \frac{1}{2} \frac{y^2 - x^2}{uv(u-v)}.$$

Solution:

The given relations can be written as

$$F_1 \equiv u^3 + v^3 - x - y = 0$$

and $$F_2 \equiv u^2 + v^2 - x^3 - y^3 = 0.$$

Now $$\frac{\partial(u, v)}{\partial(x, y)} = (-1)^2 \frac{\frac{\partial(F_1, F_2)}{\partial(x, y)}}{\frac{\partial(F_1, F_2)}{\partial(u, v)}}. \qquad ...(1)$$

We have $$\frac{\partial(F_1, F_2)}{\partial(x, y)} = \begin{vmatrix} \frac{\partial F_1}{\partial x} & \frac{\partial F_1}{\partial y} \\ \frac{\partial F_2}{\partial x} & \frac{\partial F_2}{\partial y} \end{vmatrix} = \begin{vmatrix} -1 & -1 \\ -3x^2 & -3y^2 \end{vmatrix}$$

$$= 3y^2 - 3x^2 = 3 (y^2 - x^2)$$

and $$\frac{\partial(F_1, F_2)}{\partial(u, v)} = \begin{vmatrix} 3u^2 & 3v^2 \\ 2u & 2v \end{vmatrix} = 6u^2v - 6uv^2 = 6uv (u - v).$$

$\therefore$ From (1), $$\frac{\partial(u, v)}{\partial(x, y)} = \frac{3(y^2 - x^2)}{\partial uv(u-v)} = \frac{1}{2} \frac{y^2 - x^2}{uv(u-v)}.$$

Example 8:

If $x + y + z = u,\ y + z = uv,\ z = uvw$, *show that*

$$\frac{\partial(x, y, z)}{\partial(u, v, w)} = u^2 v.$$

Solution:

The given relations can be written as

$$F_1 \equiv x + y + z - u = 0$$

$$F_2 \equiv y + z - uv = 0$$

and $\quad F_3 \equiv z - uvw = 0.$

Now $$\frac{\partial(x, y, z)}{\partial(u, v, w)} = (-1)^3 \frac{\dfrac{\partial(F_1, F_2, F_3)}{\partial(u, v, w)}}{\dfrac{\partial(F_1, F_2, F_3)}{\partial(x, y, z)}} \qquad ...(1)$$

We have $$\frac{\partial(F_1, F_2, F_3)}{\partial(u, v, w)} = \begin{vmatrix} -1 & 0 & 0 \\ -v & -u & 0 \\ -uw & -uw & -uv \end{vmatrix}$$

$$= (-1) \cdot (-u) \cdot (-uv) = -u^2v$$

and $$\frac{\partial(F_1, F_2, F_3)}{\partial(x, y, z)} = \begin{vmatrix} 1 & 1 & 1 \\ 0 & 1 & 1 \\ 0 & 0 & 1 \end{vmatrix} = 1.$$

$\therefore$ From (1), $$\frac{\partial(x, y, z)}{\partial(u, v, w)} = -\frac{-u^2 v}{1} = u^2 v.$$

Example 9:

Prove that $$\frac{\partial(u, v)}{\partial(x, y)} \times \frac{\partial(x, y)}{\partial(u, v)} = 1.$$

Solution:

Let $u = f_1(x, y),\ v = f_2(x, y)$. ...(1)

Obviously x and y can also be expressed as functions of u and v. Differentiating relations (1) partially with respect to u and v, we get

$$\left.\begin{aligned} 1 &= \frac{\partial u}{\partial x}\frac{\partial x}{\partial u} + \frac{\partial u}{\partial y}\frac{\partial y}{\partial y}, \quad & 0 &= \frac{\partial u}{\partial x}\frac{\partial x}{\partial v} + \frac{\partial u}{\partial y}\frac{\partial y}{\partial v}, \\ 0 &= \frac{\partial v}{\partial x}\frac{\partial x}{\partial u} + \frac{\partial v}{\partial y}\frac{\partial y}{\partial u}, \quad & 1 &= \frac{\partial v}{\partial x}\frac{\partial x}{\partial v} + \frac{\partial v}{\partial y}\frac{\partial y}{\partial v}. \end{aligned}\right\} \qquad ...(2)$$

Now $\dfrac{\partial(u, v)}{\partial(x, y)} \times \dfrac{\partial(x, y)}{\partial(u, v)} = \begin{vmatrix} \dfrac{\partial u}{\partial x} & \dfrac{\partial u}{\partial y} \\ \dfrac{\partial v}{\partial x} & \dfrac{\partial v}{\partial y} \end{vmatrix} \times \begin{vmatrix} \dfrac{\partial x}{\partial u} & \dfrac{\partial x}{\partial v} \\ \dfrac{\partial y}{\partial u} & \dfrac{\partial y}{\partial v} \end{vmatrix}$

$$= \begin{vmatrix} \dfrac{\partial u}{\partial x}\dfrac{\partial x}{\partial u} + \dfrac{\partial u}{\partial y}\dfrac{\partial y}{\partial u} & \dfrac{\partial u}{\partial x}\dfrac{\partial x}{\partial v} + \dfrac{\partial u}{\partial y}\dfrac{\partial y}{\partial v} \\ \dfrac{\partial v}{\partial x}\dfrac{\partial x}{\partial u} + \dfrac{\partial v}{\partial y}\dfrac{\partial y}{\partial u} & \dfrac{\partial v}{\partial x}\dfrac{\partial x}{\partial v} + \dfrac{\partial v}{\partial y}\dfrac{\partial y}{\partial v} \end{vmatrix},$$

applying row-by-column multiplication

$= \begin{vmatrix} 1 & 0 \\ 0 & 1 \end{vmatrix}$, using the relations (2) = 1.

Example 10:

If $u^3 + v + w = x + y^2 + z^2$,

$u + v^3 + w = x^2 + y + z^2$,

$u + v + w^3 = x^2 + y^2 + z$,

prove that $\dfrac{\partial(u, v, w)}{\partial(x, y, z)} = \dfrac{1 - 4(xy + yz + zx)}{2 - 3(u^2 + v^2 + w^2) + 27u^2v^2w^2}$.

Solution:

The given relations can be written as

$$F_1 \equiv u^3 + v + w - x - y^2 - z^2 = 0,$$

$$F_2 \equiv u + v^3 + w - x^{-2} - y - z^2 = 0,$$

$$F_3 \equiv u + v + w^3 - x^2 - y^2 - z = 0.$$

Now $\dfrac{\partial(u, v, w)}{\partial(x, y, z)} = (-1)^3 \dfrac{\partial(F_1, F_2, F_3)}{\partial(x, y, z)} \Big/ \dfrac{\partial(F_1, F_2, F_3)}{\partial(u, v, w)}$

Here $\dfrac{\partial(F_1, F_2, F_3)}{\partial(x, y, z)} = \begin{vmatrix} -1 & -2y & -2z \\ -2x & -1 & -2z \\ -2x & -2y & -1 \end{vmatrix}$

$$= -1(1 - 4yz) + 2x(2y - 4yz) - 2x(4yz - 2z)$$

$$= -1 + 4(yz + zx + xy) - 16xyz.$$

And $\dfrac{\partial(F_1, F_2, F_3)}{\partial(u, v, w)} = \begin{vmatrix} 3u^2 & 1 & 1 \\ 1 & 3v^2 & 1 \\ 1 & 1 & 3w^2 \end{vmatrix}$

$$= 3u^2 (9v^2w^2 - 1) - 1 (3w^2 - 1) + 1 . (1 - 3v^2)$$

$$= 2 - 3 (u^2 + v^2 + w^2) + 27u^2v^2w^2.$$

$$\therefore \text{ From (1), } \frac{\partial(u, v, w)}{\partial(x, y, z)} = \frac{1-4(yz+zx+xy)+16xyz}{2-3(u^2+v^2+w^2)+27u^2v^2w^2}.$$

Example 11:

Prove that

$$\frac{\partial(y_1, y_2, ..., y_n)}{\partial(x_1, x_2, ..., x_n)} \cdot \frac{\partial(x_1, x_2, ..., x_n)}{\partial(y_1, y_2, ..., y_n)} = 1.$$

Solution:

Let $y_1 = f_1 (x_1, x_2, ..., x_n)$, $y_2 = f_2 (x_1, x_2, ..., x_n)$, ...,

$$y_n = f_n(x_1, x_2, ..., x_n)\} \quad ...(1)$$

These relations may be put in the form

$$x_1 = F_1(y_1, y_2, ..., y_n),$$

$$x_2 = F_2(y_1, y_2, ..., y_n), ...,$$

$$x_n = F_n(y_1, y_2, ..., y_n).$$

Differentiating the relations (1) partially w.r.t. $y_1, y_2, ..., y_n$, we have

$$(A)\begin{cases} 1 = \frac{\partial y_1}{\partial x_1} \cdot \frac{\partial x_1}{\partial y_1} + \frac{\partial y_1}{\partial x_2} \cdot \frac{\partial x_2}{\partial y_1} + \ldots + \frac{\partial y_1}{\partial x_n} \cdot \frac{\partial x_n}{\partial y_1} = \Sigma \frac{\partial y_1}{\partial x_r} \cdot \frac{\partial x_r}{\partial y_1} \\ 0 = \frac{\partial y_1}{\partial x_1} \cdot \frac{\partial x_1}{\partial y_1} + \frac{\partial y_1}{\partial x_2} \cdot \frac{\partial x_2}{\partial y_1} + \ldots + \frac{\partial y_1}{\partial x_n} \cdot \frac{\partial x_n}{\partial y_1} = \Sigma \frac{\partial y_1}{\partial x_r} \cdot \frac{\partial x_r}{\partial y_1} \\ \ldots \quad \ldots \quad \ldots \quad \ldots \quad \ldots \quad \ldots \quad \ldots \quad \ldots \\ \ldots \quad \ldots \quad \ldots \quad \ldots \quad \ldots \quad \ldots \quad \ldots \quad \ldots \\ 0 = \frac{\partial y_1}{\partial x_1} \cdot \frac{\partial x_1}{\partial y_n} + \frac{\partial y_1}{\partial x_2} \cdot \frac{\partial x_2}{\partial y_n} + \ldots + \frac{\partial y_1}{\partial x_n} \cdot \frac{\partial x_n}{\partial y_n} = \Sigma \frac{\partial y_1}{\partial x_r} \cdot \frac{\partial x_r}{\partial y_n} \end{cases}$$

and similar other relations.

$$\text{Now} \quad \frac{\partial(y_1, y_2, ..., y_n)}{\partial(x_1, x_2, ..., x_n)} \cdot \frac{\partial(x_1, x_2, ..., x_n)}{\partial(y_1, y_2, ..., y_n)}$$

$$= \begin{vmatrix} \frac{\partial y_1}{\partial x_1} & \frac{\partial y_1}{\partial x_2} & \cdots & \frac{\partial y_1}{\partial x_n} \\ \frac{\partial y_2}{\partial x_1} & \frac{\partial y_2}{\partial x_2} & \cdots & \frac{\partial y_2}{\partial x_n} \\ \cdots & \cdots & \cdots & \cdots \\ \frac{\partial y_n}{\partial x_1} & \frac{\partial y_n}{\partial x_2} & \cdots & \frac{\partial y_n}{\partial x_n} \end{vmatrix} \times \begin{vmatrix} \frac{\partial x_1}{\partial y_1} & \frac{\partial x_1}{\partial y_2} & \cdots & \frac{\partial x_1}{\partial y_n} \\ \frac{\partial x_2}{\partial y_1} & \frac{\partial x_2}{\partial y_2} & \cdots & \frac{\partial x_2}{\partial y_n} \\ \cdots & \cdots & \cdots & \cdots \\ \frac{\partial x_n}{\partial y_1} & \frac{\partial x_n}{\partial y_2} & \cdots & \frac{\partial x_n}{\partial y_n} \end{vmatrix}$$

$$= \begin{vmatrix} \Sigma \frac{\partial y_1}{\partial x_r} \cdot \frac{\partial x_r}{\partial y_1} & \Sigma \frac{\partial y_1}{\partial x_r} \cdot \frac{\partial x_r}{\partial y_2} & \cdots & \Sigma \frac{\partial y_1}{\partial x_r} \cdot \frac{\partial x_r}{\partial y_n} \\ \Sigma \frac{\partial y_2}{\partial x_r} \cdot \frac{\partial x_r}{\partial y_1} & \Sigma \frac{\partial y_1}{\partial x_r} \cdot \frac{\partial x_r}{\partial y_2} & \cdots & \Sigma \frac{\partial y_2}{\partial x_r} \cdot \frac{\partial x_r}{\partial y_n} \\ \cdots & \cdots & \cdots & \cdots \\ \Sigma \frac{\partial y_n}{\partial x_r} \cdot \frac{\partial x_r}{\partial y_1} & \Sigma \frac{\partial y_n}{\partial x_r} \cdot \frac{\partial x_r}{\partial y_2} & \cdots & \Sigma \frac{\partial y_n}{\partial x_r} \cdot \frac{\partial x_r}{\partial y_n} \end{vmatrix},$$

applying row-by-column multiplication

$$= \begin{vmatrix} 1 & 0 & \cdots & 0 \\ 0 & 1 & \cdots & 0 \\ \cdots & \cdots & \cdots & \cdots \\ 0 & 0 & \cdots & 1 \end{vmatrix}, \text{ using the relations (A)}$$

$= 1.$

Example 12:

If $u^3 = xyz, \ \frac{1}{v} = \frac{1}{x} + \frac{1}{y} + \frac{1}{z}, \ w^2 = x^2 + y^2 + z^2$, *prove that*

$$\frac{\partial(u, v, w)}{\partial(x, y, z)} = \frac{-v(y-z)(z-x)(x-y)(x+y+z)}{3u^2 w(yz+zx+xy)}.$$

Solution:

The given relations can be written as

$$F_1 \equiv u^3 - xyz = 0$$

$$F_2 \equiv \frac{1}{v} - \frac{1}{x} - \frac{1}{y} - \frac{1}{z} = 0$$

and $\quad F_3 \equiv w^2 - x^2 - y^2 - z^2 = 0.$

Now $$\frac{\partial(u, v, w)}{\partial(x, y, z)} = (-1)^3 \frac{\dfrac{\partial(F_1, F_2, F_3)}{\partial(x, y, z)}}{\dfrac{\partial(F_1, F_2, F_3)}{\partial(u, v, w)}} \qquad ...(1)$$

We have $$\frac{\partial(F_1, F_2, F_3)}{\partial(x, y, z)} = \begin{vmatrix} \dfrac{\partial F_1}{\partial x} & \dfrac{\partial F_1}{\partial y} & \dfrac{\partial F_1}{\partial z} \\ \dfrac{\partial F_2}{\partial x} & \dfrac{\partial F_2}{\partial y} & \dfrac{\partial F_2}{\partial z} \\ \dfrac{\partial F_3}{\partial x} & \dfrac{\partial F_3}{\partial y} & \dfrac{\partial F_3}{\partial z} \end{vmatrix}$$

$$= \begin{vmatrix} -yz & -zx & -xy \\ \dfrac{1}{x^2} & \dfrac{1}{y^2} & \dfrac{1}{z^2} \\ -2x & -2y & -2z \end{vmatrix}$$

$$= \frac{2}{x^2y^2z^2} \begin{vmatrix} x^2yz & y^2zx & z^2xy \\ 1 & 1 & 1 \\ x^3 & y^3 & z^3 \end{vmatrix}$$

$$= \frac{2}{xyz} \begin{vmatrix} x & y & z \\ 1 & 1 & 1 \\ x^3 & y^3 & z^3 \end{vmatrix} = -\frac{2}{xyz} \begin{vmatrix} 1 & 1 & 1 \\ x & y & z \\ x^3 & y^3 & z^3 \end{vmatrix}$$

$$= -\frac{2}{xyz} \begin{vmatrix} 1 & 0 & 0 \\ x & y-x & z-x \\ x^3 & y^3-x^3 & z^3-x^3 \end{vmatrix}, \text{ by } C_2 - C_1 \text{ and } C_3 - C_1$$

$$= -\frac{2}{xyz} (y-x)(z-x) \begin{vmatrix} 1 & 1 \\ y^2+xy+x^2 & z^2+zx+x^2 \end{vmatrix}$$

$$= -\frac{2}{xyz} (y-x)(z-x) \begin{vmatrix} 1 & 0 \\ y^2+xy+x^2 & (z-y)(x+y+z) \end{vmatrix},$$

by $C_2 - C_1$

$$= -\frac{2}{xyz} (y-x)(z-x)(z-y)(x+y+z)$$

$$= -\frac{2}{xyz} (x-y)(y-z)(z-x)(x+y+z).$$

Also $\dfrac{\partial(F_1, F_2, F_3)}{\partial(u, v, w)} = \begin{vmatrix} 3u^2 & 0 & 0 \\ 0 & -\dfrac{1}{v^2} & 0 \\ 0 & 0 & 2w \end{vmatrix} = -\dfrac{6u^2w}{v^2}.$

Hence from (1), $\dfrac{\partial(u, v, w)}{\partial(x, y, z)}$

$$= -\frac{-2(x-y)(y-z)(z-x)(x+y+z)}{xyz} \cdot \frac{-v^2}{6u^2w}$$

$$= -\frac{(x-y)(y-z)(z-x)(x+y+z).v}{3u^2wxyz} \cdot \frac{xyz}{yz+zx+xy}$$

$$\left[\because \frac{1}{v} = \frac{1}{x} + \frac{1}{y} + \frac{1}{z}, \text{ so that } v = \frac{xyz}{yz+zx+xy}\right]$$

$$= -\frac{v(y-z)(z-x)(x-y)(x+y+z)}{3u^2w(yz+zx+xy)}.$$

Example 13:

If x, y, z are connected by a functional relation f(x, y, z) = 0, show that

$$\frac{\partial(y, z)}{\partial(x, z)} = \left(\frac{\partial y}{\partial x}\right)_{z=const.}$$

Solution:

We have f(x, y, z) = 0 $\Rightarrow$ y is a function of x and z.

Also from the equation, z = z, z may be regarded as a function of x and z.

$$\therefore \frac{\partial(y, z)}{\partial(x, z)} = \begin{vmatrix} \left(\dfrac{\partial y}{\partial x}\right)_{z=const.} & \left(\dfrac{\partial y}{\partial z}\right)_{x=const.} \\ \dfrac{\partial z}{\partial x} & \dfrac{\partial z}{\partial z} \end{vmatrix}$$

$$= \begin{vmatrix} \left(\dfrac{\partial y}{\partial x}\right)_{z=const.} & \left(\dfrac{\partial y}{\partial z}\right)_{x=const.} \\ 0 & 1 \end{vmatrix} \qquad \left[\because \frac{\partial z}{\partial x} = 0, \frac{\partial z}{\partial z} = 1\right]$$

$$= \left(\frac{\partial y}{\partial x}\right)_z = \text{constant.}$$

Example 14:

If $u^3 + v^3 + w^3 = x + y + z,$

$u^2 + v^2 + w^2 = x^3 + y^3 + z^3,$

$$u + v + w = x^2 + y^2 + z^2,$$

then prove that

$$\frac{\partial(u, v, w)}{\partial(x, y, z)} = \frac{(y-z)(z-x)(x-y)}{(u-v)(v-w)(w-u)}.$$

Solution:

The given relations can be written as

$$F_1 \equiv u^3 + v^3 + w^3 - x - y - z = 0,$$

$$F_2 \equiv u^2 + v^2 + w^2 - x^3 - y^3 - z^3 = 0,$$

and $$F_3 \equiv u + v + w - x^2 - y^2 - z^2 = 0.$$

Now $$\frac{\partial(u, v, w)}{\partial(x, y, z)} = (-1)^3 \frac{\dfrac{\partial(F_1, F_2, F_3)}{\partial(x, y, z)}}{\dfrac{\partial(F_1, F_2, F_3)}{\partial(u, v, w)}}. \qquad ...(1)$$

We have $$\frac{\partial(F_1, F_2, F_3)}{\partial(x, y, z)} = \begin{vmatrix} -1 & -1 & -1 \\ -3x^2 & -3y^2 & -3z^2 \\ -2x & -2y & -2z \end{vmatrix}$$

$$= -6\begin{vmatrix} 1 & 1 & 1 \\ x^2 & y^2 & z^2 \\ x & y & z \end{vmatrix} = 6\begin{vmatrix} 1 & 1 & 1 \\ x & y & z \\ x^2 & y^2 & z^2 \end{vmatrix}$$

$$= 6\begin{vmatrix} 1 & 0 & 0 \\ x & y-x & z-x \\ x^2 & y^2-x^2 & z^2-x^2 \end{vmatrix}$$

$$= 6(y - x)(z - x)\begin{vmatrix} 1 & 1 \\ y+x & z+x \end{vmatrix}$$

$$= 6(y - x)(z - x)(z - y) = 6(x - y)(y - z)(z - x).$$

Also $$\frac{\partial(F_1, F_2, F_3)}{\partial(u, v, w)} = \begin{vmatrix} 3u^2 & 3v^2 & 3w^2 \\ 2u & 2v & 2w \\ 1 & 1 & 1 \end{vmatrix}$$

$$= 6\begin{vmatrix} 1 & 1 & 1 \\ u & v & w \\ u^2 & v^2 & w^2 \end{vmatrix}$$

$$= -6(u-v)(v-w)(w-u).$$

Hence from (1), $\dfrac{\partial(u, v, w)}{\partial(x, y, z)} = -\dfrac{6(x-y)(y-z)(z-x)}{-6(u-v)(v-w)(w-u)}$

$$= \frac{(y-z)(z-x)(x-y)}{(u-v)(v-w)(w-u)}.$$

Example 15:

If λ, μ, ν are the roots of the equation in k,

$$\frac{x}{a+k} + \frac{y}{b+k} + \frac{z}{c+k} = 1,$$

prove that

$$\frac{\partial(x, y, z)}{\partial(\lambda, \mu, \nu)} = -\frac{(\mu-\nu)(\nu-\lambda)(\lambda-\mu)}{(b-c)(c-a)(a-b)}.$$

Solution:

The given equation in k can be written as

$$(a + k)(b + k)(c + k) - x(b + k)(c + k) - y(c + k)(a + k) - z(a + k)(b + k) = 0$$

or $\quad k^3 + k^2(a + b + c - x - y - z) + k\{ab + bc + ca - x(b + c) - y(c + a) - z(a + b)\} + abc - xbc - yca - zab = 0.$

Since λ, μ, ν are the roots of this equation, therefore from theory of equations, we have

$$\lambda + \mu + \nu = x + y + z - a - b - c$$

$\lambda\mu + \mu\nu + \nu\lambda = ab + bc + ca - x(b + c) - y(c + a) - z(a + b)$ and $\lambda\mu\nu = xbcc + yca + zab - abc.$

The above relations can be written as

$F_1 \equiv \lambda + \mu + \nu - x - y + z + a + b + c = 0,$

$F_2 \equiv \lambda\mu + \mu\nu + \nu\lambda + x(b + c) + y(c + a) + z(a + b) - ab - bc - ca = 0$

and $F_3 = \lambda\mu\nu - xbc - yca - zab + abc = 0.$

Now $\dfrac{\partial(x, y, z)}{\partial(\lambda, \mu, \nu)} = (-1)^3 \dfrac{\dfrac{\partial(F_1, F_2, F_3)}{\partial(\lambda, \mu, \nu)}}{\dfrac{\partial(F_1, F_2, F_3)}{\partial(x, y, z)}}.$ ...(1)

We have $\dfrac{\partial(F_1, F_2, F_3)}{\partial(\lambda, \mu, \nu)} = \begin{vmatrix} 1 & 1 & 1 \\ \mu+\nu & \lambda+\nu & \mu+\lambda \\ \mu\nu & \nu\lambda & \lambda\mu \end{vmatrix}$

$$= \begin{vmatrix} 1 & 0 & 0 \\ \mu+\nu & \lambda-\nu & \lambda-\nu \\ \mu\nu & \nu(\lambda-\nu) & \mu(\lambda-\nu) \end{vmatrix}$$

$$= (l - m)\ (l - n) \begin{vmatrix} 1 & 1 \\ \nu & \mu \end{vmatrix} = (\lambda - \mu)\ (\lambda - \nu)\ (\mu - \nu)$$

$$= -\ (\lambda - \mu)\ (\mu - \nu)\ (\nu - \lambda).$$

Also $\dfrac{\partial(F_1, F_2, F_3)}{\partial(x, y, z)} = \begin{vmatrix} -1 & -1 & -1 \\ b+c & c+a & a+b \\ -bc & -ca & -ab \end{vmatrix}$

$$= \begin{vmatrix} 1 & 1 & 1 \\ b+c & c+a & a+b \\ bc & ca & ab \end{vmatrix} = -\ (b - c)\ (c - a)\ (a - b).$$

Hence from (1), $\dfrac{\partial(x, y, z)}{\partial(\lambda, \mu, \nu)} = -\dfrac{-(\lambda-\mu)(\mu-\nu)(\nu-\lambda)}{-(b-c)(c-a)(a-b)}$

$$= -\frac{(\mu-\nu)(\nu-\lambda)(\lambda-\mu)}{(b-c)(c-a)(a-b)}.$$

Example 16:

Compute the Jacobian $\dfrac{\partial(u, v)}{\partial(r, \theta)}$ *where*

$u = 2xy,\ v = x^2 - y^2,\ x = r\cos\theta,\ y = r\sin\theta.$

Solution:

Here we have case of functions of functions. Using the formula for finding the Jacobian in the case of functions of functions, we have

$$\frac{\partial(u, v)}{\partial(r, \theta)} = \frac{\partial(u, v)}{\partial(x, y)} \cdot \frac{\partial(x, y)}{\partial(r, \theta)}. \qquad \text{...(1)}$$

Now $\dfrac{\partial(u, v)}{\partial(x, y)} = \begin{vmatrix} \dfrac{\partial u}{\partial x} & \dfrac{\partial u}{\partial y} \\ \dfrac{\partial v}{\partial x} & \dfrac{\partial v}{\partial y} \end{vmatrix}$

$$= \begin{vmatrix} 2y & 2x \\ 2x & -2y \end{vmatrix} = -4\left(x^2 + y^2\right).$$

Also $$\frac{\partial(x, y)}{\partial(r, \theta)} = \begin{vmatrix} \frac{\partial x}{\partial r} & \frac{\partial x}{\partial \theta} \\ \frac{\partial y}{\partial r} & \frac{\partial y}{\partial \theta} \end{vmatrix} = \begin{vmatrix} \cos\theta & -r\sin\theta \\ \sin\theta & r\cos\theta \end{vmatrix}$$

$$= r\,(\cos^2\theta + \sin^2\theta) = r.$$

Hence from (1), $\dfrac{\partial(u, v)}{\partial(r, \theta)} = -4\,(x^2 + y^2)\,r = -4r^2.r = -4r^3.$

[$\because$ from $x = r\cos\theta$, $y = r\sin\theta$, we have $x^2 + y^2 = r^2$]

Example 17:

If $u_1 = x_1 + x_2 + x_3 + x_4$, $u_1u_2 = x_2 + x_3 + x_4$, $u_1u_2u_3 = x_3 + x_4$, $u_1u_2u_3u_4 = x_4$, *show that*

$$\frac{\partial(x_1, x_2, x_3, x_4)}{\partial(u_1, u_2, u_3, u_4)} = u_1^3 u_2^2 u_3.$$

Solution:

The given relations can be written as

$$F_1 \equiv u_1 - x_1 - x_2 - x_3 - x_4 = 0,$$
$$F_2 \equiv u_1u_2 - x_2 - x_3 - x_4 = 0,$$
$$F_3 \equiv u_1u_2u_3 - x_3 - x_4 = 0$$

and $$F_4 \equiv u_1u_2u_3u_4 - x_4 = 0.$$

Now

$$\frac{\partial(x_1, x_2, x_3, x_4)}{\partial(u_1, u_2, u_3, u_4)} = (-1)^4 \frac{\dfrac{\partial(F_1, F_2, F_3, F_4)}{\partial(u_1, u_2, u_3, u_4)}}{\dfrac{\partial(F_1, F_2, F_3, F_4)}{\partial(x_1, x_2, x_3, x_4)}} \quad ...(1)$$

We have $\dfrac{\partial(F_1, F_2, F_3, F_4)}{\partial(u_1, u_2, u_3, u_4)}$ = the principal diagonal term of the Jacobian determinant

$$= \frac{\partial F_1}{\partial u_1} \cdot \frac{\partial F_2}{\partial u_2} \cdot \frac{\partial F_3}{\partial u_3} \cdot \frac{\partial F_4}{\partial u_4} = 1.u_1.u_1u_2.u_1u_2u_3 = u_1^3u_2^2u_3.$$

$$\text{Also } \frac{\partial(F_1, F_2, F_3, F_4)}{\partial(x_1, x_2, x_3, x_4)} = \begin{vmatrix} -1 & -1 & -1 & -1 \\ 0 & -1 & -1 & -1 \\ 0 & 0 & -1 & -1 \\ 0 & 0 & 0 & -1 \end{vmatrix} = 0.$$

$$\text{Hence from (1), } \frac{\partial(x_1, x_2, x_3, x_4)}{\partial(u_1, u_2, u_3, u_4)} = \frac{u_1^3 u_2^2 u_3}{1} = u_1^3 u_2^2 u_3.$$

Example 18:

The roots of the equation in λ,

$$(\lambda - x)^3 + (\lambda - y)^3 + (\lambda - z)^3 = 0$$

are u, v, w. Prove that

$$\frac{\partial(u, v, w)}{\partial(x, y, z)} = -\frac{(y-z)(z-x)(x-y)}{(v-w)(w-u)(u-v)}.$$

Solution:

The given equation in l can be written as

$$3\lambda^3 - 3\lambda^2 (x + y + z) + 3\lambda(x^2 + y^2 + z^2) - (x^3 + y^3 + z^3) = 0.$$

Since u, v are the roots of this equation, therefore from theory of equations, we have

$$u + v + w = x + y + z,$$

$$uv + vw + wu = x^2 + y^2 + z^2,$$

and $\quad uvw = 1/3\ (x^3 + y^3 + z^3).$

The above relations can be written as

$$F_1 \equiv u + v + w - x - y - z = 0,$$

$$F_2 \equiv uv + vw + wu - x^2 - y^2 - z^2 = 0$$

and $\quad F_3 \equiv uvw - 1/3\ (x^3 + y^3 + z^3) = 0.$

$$\text{Now } \frac{\partial(u, v, w)}{\partial(x, y, z)} = (-1)^3 \frac{\dfrac{\partial(F_1, F_2, F_3)}{\partial(x, y, z)}}{\dfrac{\partial(F_1, F_2, F_3)}{\partial(u, v, w)}} \qquad ...(1)$$

$$\text{We have } \frac{\partial(F_1, F_2, F_3)}{\partial(x, y, z)} = \begin{vmatrix} -1 & -1 & -1 \\ -2x & -2y & -2z \\ -x^2 & -y^2 & -z^2 \end{vmatrix}$$

$$= -2\begin{vmatrix} 1 & 1 & 1 \\ x & y & z \\ x^2 & y^2 & z^2 \end{vmatrix} = -2(y-z)(z-x)(x-y)$$

Also $\dfrac{\partial(F_1, F_2, F_3)}{\partial(u, v, w)} = \begin{vmatrix} 1 & 1 & 1 \\ v+w & u+w & u+v \\ vw & uw & uv \end{vmatrix}$

$= -(v-w)(w-u)(u-v).$

Hence from (1), $\dfrac{\partial(u, v, w)}{\partial(x, y, z)} = -\dfrac{-2(y-z)(z-x)(x-y)}{-(v-w)(w-u)(u-v)}$

$$= -2\frac{(y-z)(z-x)(x-y)}{(v-w)(w-u)(u-v)}.$$

Example 19:

If $u = x(1-r^2)^{-1/2}$, $v = y(1-r^2)^{-1/2}$

$w = z(1-r^2)^{-1/2}$, *where* $r^2 = x^2 + y^2 + z^2$,

show that $\dfrac{\partial(u, v, w)}{\partial(x, y, z)} = (1-r^2)^{-5/2}$

Solution:

From the given relations, we have

$$x^2 = u^2(1-r^2) = u^2(1-x^2-y^2-z^2),$$

$$y^2 = v^2(1-x^2-y^2-z^2)$$

and $\quad z^2 = w^2(1-x^2-y^2-z^2).$

The above relations can be written as

$$F_1 \equiv x^2 - u^2(1-x^2-y^2-z^2) = 0,$$

$$F_2 \equiv y^2 - v^2(1-x^2-y^2-z^2) = 0,$$

and $\quad F_3 \equiv z^2 - w^2(1-x^2-y^2-z^2) = 0.$

Now

$$\frac{(u, v, w)}{\partial(x, y, z)} = (-1)^3 \frac{\dfrac{\partial(F_1, F_2, F_3)}{\partial(x, y, z)}}{\dfrac{\partial(F_1, F_2, F_3)}{\partial(u, v, w)}}. \qquad ...(1)$$

We have $\dfrac{\partial(F_1, F_2, F_3)}{\partial(x, y, z)} = \begin{vmatrix} 2x(1+u^2) & 2yu^2 & 2zu^2 \\ 2xv & 2y(1+v^2) & 2zv^2 \\ 2xw^2 & 2yw^2 & 2z(1+w^2) \end{vmatrix}$

$$= 8xyz \begin{vmatrix} 1+u^2 & u^2 & u^2 \\ v^2 & 1+v^2 & v^2 \\ w^2 & w^2 & 1+w^2 \end{vmatrix}$$

$$= 8xyz\ (1 + u^2 + v^2 + w^2) \begin{vmatrix} 1 & 1 & 1 \\ v^2 & 1+v^2 & v^2 \\ w^2 & w^2 & 1+w^2 \end{vmatrix}$$

$$= 8xyz\ (1 + u^2 + v^2 + w^2) \begin{vmatrix} 1 & 0 & 0 \\ v^2 & 1 & 0 \\ w^2 & 0 & 1 \end{vmatrix}$$

$$= 8xyz\ (1 + u^2 + v^2 + w^2)$$

$$= 8xyz\left[1 + \frac{x^2}{1-r^2} + \frac{y^2}{1-r^2} + \frac{z^2}{1-r^2}\right] = 8xyz\left[1 + \frac{x^2+y^2+z^2}{1-r^2}\right]$$

$$= 8xyz\left[1 + \frac{r^2}{1-r^2}\right] = \frac{8xyz}{1-r^2}.$$

Also $\dfrac{\partial(F_1, F_2, F_3)}{\partial(u, v, w)}$

$$= \begin{vmatrix} -2u(1-r^2) & 0 & 0 \\ 0 & -2v(1-r^2) & 0 \\ 0 & 0 & -2w(1-r^2) \end{vmatrix}$$

$$= -\ 8uvw\ (1 - r^2)^3.$$

Hence from (1), $\dfrac{\partial(u, v, w)}{\partial(x, y, z)} = -\dfrac{\partial xyz}{1-r^2} \cdot \dfrac{1}{-8uvw\,(1-r^2)^3}$

$$= \frac{xyz}{uvw(1-r^2)^4} = \frac{xyz}{xyz(1-r^2)^{-3/2}(1-r^2)^4}$$

$$= (1 - r^2)^{-5/2}.$$

Example 20:

Given $y_1(x_1 - x_2) = 0$, $y_2(x_1^2 + x_1x_2 + x_2^2) = 0$, *show that*

$$\frac{\partial(y_1, y_2)}{\partial(x_1, x_2)} = 3y_1y_2\frac{x_1 + x_2}{x_1^3 - x_2^3}.$$

Solution:

The given relations can be written as

$$F_1 \equiv y_1 (x_1 - x_2) = 0$$

and $$F_2 \equiv y_2 \left(x_1^2 + x_1x_2 + x_2^2\right) = 0.$$

Now $$\frac{\partial(y_1, y_2)}{\partial(x_1, x_2)} = (-1)^2 \frac{\dfrac{\partial(F_1, F_2)}{\partial(x_1, x_2)}}{\dfrac{\partial(F_1, F_2)}{\partial(y_1, y_2)}}. \quad ...(1)$$

We have $$\frac{\partial(F_1, F_2)}{\partial(x_1, x_2)} = \begin{vmatrix} y_1 & -y_2 \\ y_2(2x_1 + x_2) & y_2(x_1 + 2x_2) \end{vmatrix}$$

$$= y_1y_2(x_1 + 2x_2 + 2x_1 + x_2) = 3y_1y_2(x_1 + x_2).$$

Also $$\frac{\partial(F_1, F_2)}{\partial(x_1, y_2)} = \begin{vmatrix} x_1 - x_2 & 0 \\ 0 & x_1^2 + x_1x_2 + x_2^2 \end{vmatrix} = x_1^3 - x_2^3.$$

Hence from (1), $$\frac{\partial(y_1, y_2)}{\partial(x_1, x_2)} = \frac{3y_1y_2(x_1 + x_2)}{x_1^3 - x_2^3}.$$

Example 21:

If $u = x^2 + y^2 + z^2$, $v = x + y + z$, $w = xy + yz + zx$, *show that the Jacobian* $\dfrac{\partial(u, v, w)}{\partial(x, y, z)}$ *vanishes identically. Also find the relation between u, v and w.*

Solution:

We have $$\frac{\partial(u, v, w)}{\partial(x, y, z)}$$

$$= \begin{vmatrix} 2x & 2y & 2z \\ 1 & 1 & 1 \\ y+z & z+x & x+y \end{vmatrix} = 2\begin{vmatrix} x & y & z \\ 1 & 1 & 1 \\ y+z & z+x & x+y \end{vmatrix}.$$

$$= 2\begin{vmatrix} x+y+z & x+y+z & x+y+z \\ 1 & 1 & 1 \\ y+z & z+x & x+y \end{vmatrix}, \text{ by } R_1 + R_3$$

$$= 2(x+y+z)\begin{vmatrix} 1 & 1 & 1 \\ 1 & 1 & 1 \\ y+z & z+x & x+y \end{vmatrix} = 0, \text{ the first two rows being identical}$$

Since the Jacobian of the functions u, v, w, is zero, therefore these functions are not independent and there must exist a relation between them.

We have $v^2 = (x + y + z)^2 = x^2 + y^2 + z^2 + 2(xy + yz + zx)$

$= u + 2w.$

Thus, $v^2 = u + 2w$ is the required relation between u, v and w.

Example 22:

Show that the functions

$$u = x + y - z,\ v = x - y + z,\ w = x^2 + y^2 + z^2 - 2yz$$

are not independent of one another. Also find the relation between them.

Solution:

Here $\dfrac{\partial(u, v, w)}{\partial(x, y, z)} = \begin{vmatrix} 1 & 1 & -1 \\ 1 & -1 & 1 \\ 2x & 2(y-z) & 2(x-y) \end{vmatrix}$

$$= \begin{vmatrix} 1 & 1 & 0 \\ 1 & -1 & 0 \\ 2x & 2(y-z) & 0 \end{vmatrix}, \text{ adding } C_2 \text{ to } C_3$$

Since the Jacobian is zero, the functions are not independent.

Now $u + v = 2x$ and $u - v = 2(y - z)$.

Therefore $(u + v)^2 + (u - v)^2 = 4(x^2 + y^2 + z^2 - 2yz) = 4w.$

This is the required relation between u, v, w.

Example 23:

Show that $ax^2 + 2hxy + by^2$ and $Ax^2 + 2Hxy + By^2$. If the functions u, v are not independent, then

$$\frac{\partial(u, v)}{\partial(x, y)} = 0$$

or $$\begin{vmatrix} \frac{\partial u}{\partial x} & \frac{\partial u}{\partial y} \\ \frac{\partial v}{\partial x} & \frac{\partial v}{\partial y} \end{vmatrix} = 0$$

or $$\begin{vmatrix} 2(ax+hy) & 2(hx+by) \\ 2(Ax+Hy) & 2(Hx+By) \end{vmatrix} = 0$$

or $(ax + hy)(Hx + By) - (hx + by)(Ax + Hy) = 0$

or $(aH - Ah)x^2 + (aB - Ab)xy + (Bh - bH)y^2 = 0.$

Since the variables x, y are independent, the coefficients of x^2 and y^2 in the above equation must be separately zero. Hence, we have

$$aH - Ah = 0 \text{ and } Bh - bH = 0$$

whence $\dfrac{a}{A} = \dfrac{h}{H} = \dfrac{b}{B}.$

Example 24:

If $u = x + 2y + z$, $v = x - 2y + 3z$ and $w = 2xy - xz + 4yz - 2z^2$, show that they are not independent. Find the relation between u, v and w.

Solution:

We have $\dfrac{\partial(u, v, w)}{\partial(x, y, z)}$

$$= \begin{vmatrix} 1 & 2 & 1 \\ 1 & -2 & 3 \\ 2y-z & 2x+4z & -x+4y-4z \end{vmatrix}$$

$$= \begin{vmatrix} 1 & 0 & 0 \\ 1 & -4 & 2 \\ 2y-z & 2x-4y+6z & -x+2y-3z \end{vmatrix}, \text{ by } C_2 - 2C_1 \text{ and } C_3 - C_1$$

$$= -2\begin{vmatrix} 1 & 0 & 0 \\ 1 & 2 & 2 \\ 2y-z & -x+2y-3z & -x+2y-3z \end{vmatrix} = 0,$$

the last two columns being identical.

Since the Jacobians of the functions u, v, w is zero, therefore these functions are not independent and so there must exist a relation between them.

We have

$$u^2 - v^2 = (x + 2y + z)^2 - (x - 2y + 3z)^2$$
$$= (x + 2y + z + x - 2y + 3z)(x + 2y + z - x + 2y - 3z)$$
$$= (2x + 4z)(4y - 2z) = 4(x + 2z)(2y - z)$$
$$= 4(2xy - xz + 4yz - 2z^2) = 4w.$$

Therefore $u^2 - v^2 = 4w$ is the required relation between u, v and w.

Example 25:

Find the Jacobian of $y_1, y_2, y_3, \ldots, y_n$, being given

$y_1 = x_1(1 - x_2),\ y_2 = x_1x_2(1 - x_3),\ \ldots,\ y_{n-1} = x_1x_2 \ldots x_{n-1}(1 - x_n)$

$y_n = x_1x_2x_3 \ldots x_n$

Solution:

Adding all the given relations, we get

$$y_1 + y_2 + \ldots + y_n = x_1.$$

$$\therefore \quad \frac{\partial y_1}{\partial x_1} + \frac{\partial y_2}{\partial x_1} + \ldots + \frac{\partial y_n}{\partial x_1} = 1$$

and $$\frac{\partial y_1}{\partial x_r} + \frac{\partial y_2}{\partial x_r} + \ldots + \frac{\partial y_n}{\partial x_r} = 0,\ r = 2, 3, \ldots, n. \qquad \ldots(1)$$

Now $$\frac{\partial(y_1, y_2, \ldots, y_n)}{\partial(x_1, x_2, \ldots, x_n)} = \begin{vmatrix} \frac{\partial y_1}{\partial x_1} & \frac{\partial y_1}{\partial x_2} & \frac{\partial y_1}{\partial x_3} & \cdots & \frac{\partial y_1}{\partial x_n} \\ \frac{\partial y_2}{\partial x_1} & \frac{\partial y_2}{\partial x_2} & \frac{\partial y_2}{\partial x_3} & \cdots & \frac{\partial y_2}{\partial x_n} \\ \cdots & \cdots & \cdots & \cdots & \cdots \\ \frac{\partial y_n}{\partial x_1} & \frac{\partial y_n}{\partial x_2} & \frac{\partial y_n}{\partial x_3} & \cdots & \frac{\partial y_n}{\partial x_n} \end{vmatrix}$$

$$= \begin{vmatrix} \frac{\partial y_1}{\partial x_1} & \frac{\partial y_1}{\partial x_2} & \frac{\partial y_1}{\partial x_3} & \cdots & \frac{\partial y_1}{\partial x_n} \\ \frac{\partial y_2}{\partial x_1} & \frac{\partial y_2}{\partial x_2} & \frac{\partial y_2}{\partial x_3} & \cdots & \frac{\partial y_2}{\partial x_n} \\ \cdots & \cdots & \cdots & \cdots & \cdots \\ \frac{\partial y_{n-1}}{\partial x_1} & \frac{\partial y_{n-1}}{\partial x_2} & \frac{\partial y_{n-1}}{\partial x_3} & \cdots & \frac{\partial y_{n-1}}{\partial x_n} \\ 1 & 0 & 0 & \cdots & 0 \end{vmatrix}$$

adding $R_1, R_2, \ldots, R_{n-1}$ to R_n and using the relations (1)

$$=(-1)^{n-1}\begin{vmatrix} \frac{\partial y_1}{\partial x_2} & \frac{\partial y_1}{\partial x_3} & \frac{\partial y_1}{\partial x_4} & \cdots & \frac{\partial y_1}{\partial x_n} \\ \frac{\partial y_2}{\partial x_2} & \frac{\partial y_2}{\partial x_3} & \frac{\partial y_2}{\partial x_4} & \cdots & \frac{\partial y_2}{\partial x_n} \\ \cdots & \cdots & \cdots & \cdots & \cdots \\ \frac{\partial y_{n-1}}{\partial x_2} & \frac{\partial y_{n-1}}{\partial x_3} & \frac{\partial y_{n-1}}{\partial x_4} & \cdots & \frac{\partial y_{n-1}}{\partial x_n} \end{vmatrix},$$

expanding the determinant along the nth row

$$=(-1)^{n-1}\begin{vmatrix} \frac{\partial y_1}{\partial x_2} & 0 & 0 & \cdots & 0 \\ \frac{\partial y_2}{\partial x2} & \frac{\partial y_2}{\partial x_3} & 0 & \cdots & 0 \\ \cdots & \cdots & \cdots & \cdots & \cdots \\ \frac{\partial y_{n-1}}{\partial x_2} & \frac{\partial y_{n-1}}{\partial x_3} & \frac{\partial y_{n-1}}{\partial x_4} & \cdots & \frac{\partial y_{n-1}}{\partial x_n} \end{vmatrix}$$

$$=(-1)^{n-1}\cdot\frac{\partial y_1}{\partial x_2}\cdot\frac{\partial y_2}{\partial x_3}\cdot\frac{\partial y_3}{\partial x_4}\cdot\ldots\cdot\frac{\partial y_{n-1}}{\partial x_n}$$

$$= (-1)^{n-1}\cdot(-x_1)\cdot(-x_1x_2)\cdot(-x_1x_2x_3)\cdot\ldots\cdot(-x_1x_2\ldots x_{n-1})$$

$$=(-1)^{n-1}\cdot(-1)^{n-1}x_1^{n-1}x_2^{n-2}x_3^{n-3}\ldots x_{n-1}$$

$$=(-1)^{2n-2}x_1^{n-1}x_2^{n-2}x_3^{n-2}x_2^{n-2}x_3^{n-3}\ldots x_{n-1}$$

$$=x_1^{n-1}x_2^{n-2}\ldots x_{n-1}.$$

Example 26:

Find the Jacobian $\frac{\partial(x,y,z)}{\partial(r,\theta,\phi)}$ *being given*

$x = r\cos\theta\cos\phi,\ y = r\sin\theta\sqrt{(1-m^2\sin^2\phi)},$

$z = r\sin\phi\sqrt{(1-n^2\sin^2\theta)}.$

Solution:

Here $x^2 + y^2 + z^2$

$= r^2\cos^2\theta\cos^2 f + r^2\sin^2\theta - r^2m^2\sin^2\theta\sin^2 f + r^2\sin^2 f$

$- r^2n^2\sin^2 f\sin^2\theta$

$= r^2(\cos^2\theta \cos^2 f + \sin^2\theta + \sin^2 f - \sin^2\theta \sin^2 f)$ $[Q\ m^2 + n^2 = 1]$

$= r^2 (\cos^2\theta \cos^2 f + \sin^2\theta + \sin^2 f \cos^2\theta)$

$= r^2 (\sin^2\theta + \cos^2\theta) = r^2.$

$$\therefore\ x\frac{\partial x}{\partial r}+y\frac{\partial y}{\partial r}+z\frac{\partial z}{\partial r}=r;\ x\frac{\partial x}{\partial \theta}+y\frac{\partial y}{\partial \theta}+z\frac{\partial z}{\partial \theta}=0 \qquad ...(1)$$

and $x\frac{\partial x}{\partial \phi}+y\frac{\partial y}{\partial \phi}+z\frac{\partial z}{\partial \phi}=0$

$$\text{Now } J(x, y, z)=\begin{vmatrix} \frac{\partial x}{\partial r} & \frac{\partial x}{\partial \theta} & \frac{\partial x}{\partial \phi} \\ \frac{\partial y}{\partial r} & \frac{\partial y}{\partial \theta} & \frac{\partial y}{\partial \phi} \\ \frac{\partial z}{\partial r} & \frac{\partial z}{\partial \theta} & \frac{\partial z}{\partial \phi} \end{vmatrix}$$

$$=\frac{1}{x}\begin{vmatrix} x\frac{\partial x}{\partial r} & x\frac{\partial x}{\partial \theta} & x\frac{\partial x}{\partial \phi} \\ \frac{\partial y}{\partial r} & \frac{\partial y}{\partial \theta} & \frac{\partial y}{\partial \phi} \\ \frac{\partial z}{\partial r} & \frac{\partial z}{\partial \theta} & \frac{\partial z}{\partial \phi} \end{vmatrix}$$

$$=\frac{1}{x}\begin{vmatrix} r & 0 & 0 \\ \frac{\partial y}{\partial r} & \frac{\partial y}{\partial \theta} & \frac{\partial y}{\partial \phi} \\ \frac{\partial z}{\partial r} & \frac{\partial z}{\partial \theta} & \frac{\partial z}{\partial f} \end{vmatrix}, \quad \text{by adding } R_2 + zR_3 \text{ to } R_1 \text{ and using the relations (1)}$$

$$=\frac{r}{x}\begin{vmatrix} \frac{\partial y}{\partial \theta} & \frac{\partial y}{\partial \phi} \\ \frac{\partial z}{\partial \theta} & \frac{\partial z}{\partial \phi} \end{vmatrix}=\frac{r}{x}\left\{\frac{\partial y}{\partial \theta}\frac{\partial z}{\partial \phi}-\frac{\partial z}{\partial \theta}\frac{\partial y}{\partial \phi}\right\}$$

$$=\frac{r}{x}\left\{r\cos\theta\sqrt{\left(1-m^2\sin^2\phi\right)}.r\cos\phi\sqrt{\left(1-n^2\sin^2\theta\right)}-\frac{r\sin\phi.n^2\sin\theta\cos\theta}{\sqrt{\left(1-n^2\sin^2\theta\right)}}\right.$$

$$\left.\cdot\frac{r\sin\theta.m^2\sin\phi\cos\phi}{\sqrt{\left(1-m^2\sin^2\phi\right)}}\right\}$$

$$= \frac{r^3 \cos\theta \cos\phi}{x}\left[\frac{\left(1-m^2\sin^2\phi\right)\left(1-n^2\sin^2\theta\right)-n^2m^2\sin^2\theta\sin^2\phi}{\sqrt{\left[\left(1-n^2\sin^2\theta\right)\left(1-m^2\sin^2\phi\right)\right]}}\right]$$

$$= \frac{r^3 \cos\theta \cos\phi}{r\cos\theta\cos\phi}\left[\frac{\begin{array}{r}1-m^2\sin^2\phi-n^2\sin^2\theta+m^2n^2\sin^2\phi\sin^2\theta\\ -m^2n^2\sin^2\theta\sin^2\phi\end{array}}{\sqrt{\left[\left(1-n^2\sin^2\theta\right)\left(1-m^2\sin^2\phi\right)\right]}}\right]$$

$$= \frac{r^2\left(m^2\cos^2\phi+n^2\cos^2\theta\right)}{\sqrt{\left[\left(1-n^2\sin^2\theta\right)\left(1-m^2\cos^2\phi\right)\right]}}. \qquad [\because m^2+n^2=1]$$

Example 27:

If $y_1 = 1 - x_1$, $y_2 = x_1(1 - x_2)$, $y_3 = x_1x_2(1 - x_3)$, ..., $y_n = x_1x_2 \ldots x_{n-1}(1 - x_n)$, *prove that*

$$J(y_1, y_2, \ldots, y_n) = (-1)^n x_1^{n-1}x_2^{n-2}\ldots x_{n-1}.$$

Solution:

Here y_1 is a function of x_1 only, y_2 is a function of x_1, x_2 only, y_3 is a function of x_1, x_2, x_3 only, ..., and y_n is a function of x_1, x_2, ..., x_n.

$$\therefore \frac{\partial(y_1, y_2, \ldots, y_n)}{\partial(x_1, x_2, \ldots, x_n)} = \begin{vmatrix} \frac{\partial y_1}{\partial x_1} & 0 & 0 & \cdots & 0 \\ \frac{\partial y_2}{\partial x_1} & \frac{\partial y_2}{\partial x_2} & 0 & \cdots & 0 \\ \cdots & \cdots & \cdots & \cdots & \cdots \\ \frac{\partial y_n}{\partial x_1} & \frac{\partial y_n}{\partial x_2} & \frac{\partial y_n}{\partial x_3} & \cdots & \frac{\partial y_n}{\partial x_n} \end{vmatrix}$$

$$= \frac{\partial y_1}{\partial x_1}\cdot\frac{\partial y_2}{\partial x_2}\cdot\frac{\partial y_3}{\partial x_3}\cdot \ldots \cdot\frac{\partial y_n}{\partial x_n}$$

$$= (-1)\cdot(-x_1)(-x_1x_2)\ldots(-x_1x_2x_3\ldots x_{n-1})$$

$$= (-1)^n x_1^{n-1}x_2^{n-2}\ldots x_{n-1}.$$

5

Partial Differentiation

HOMOGENEOUS FUNCTIONS

An expression in which every term is of the same degree is called a homogeneous function. Thus

$$a_0x^n + a_1x^{n-1}y + a_2x^{n-2}y^2 + ... + a_{n-1}xy^{n-1} + a_ny^n$$

is a homogeneous function of x and y of degree n. This can also be written as

$$x^n\left\{a_0 + a_1\left(\frac{y}{x}\right) + a_2\left(\frac{y}{x}\right)^2 + ... + a_{n-1}\left(\frac{y}{x}\right)^{n-1} + a_n\left(\frac{y}{x}\right)^n\right\}$$

or $x^n f\left(\frac{y}{x}\right)$, where $f\left(\frac{y}{x}\right)$ is some function of $\left(\frac{y}{x}\right)$.

Note 1: *To test whether a given function f(x, y) is homogeneous or not we put tx for x and ty for y in it.*

If we get $f(tx, ty) = t^n f(x, y)$,

the function f(x, y) is homogeneous of degree n; otherwise f(x, y) is not a homogeneous function.

Note 2: *If u is a homogeneous function of x and y of degee n then* $\frac{\partial u}{\partial x}$ *and* $\frac{\partial u}{\partial y}$ *are also homogeneous functions of x and y each being of degree n – 1.*

Let $u = x^n f\left(\frac{y}{x}\right)$.

[∵ u is a homogeneous function of x and y of degree n]

Then $\frac{\partial u}{\partial x} = nx^{n-1} f\left(\frac{y}{x}\right) + x^n\left\{f'\left(\frac{y}{x}\right)\right\}.\left(-\frac{y}{x^2}\right)$

$$= x^{n-1}\left[n\, f\left(\frac{y}{x}\right) - \left(\frac{y}{x}\right) f'\left(\frac{y}{x}\right)\right]$$

$= x^{n-1}. \left[\text{some function of } \frac{y}{x}\right]$

= a homogeneous function of x and y of degree (n – 1).

Similarly, $\frac{\partial u}{\partial y} = x^n \left\{f'\left(\frac{y}{x}\right)\right\}.\frac{1}{x} = x^{n-1} f'\left(\frac{y}{x}\right)$

$= x^{n-1}. \left(\text{some function of } \frac{y}{x}\right)$

= a homogeneous function of x and y of degree (n – 1).

TOTAL DERIVATIVES

If $u = f(x, y)$, *where* $x = \phi_1(t)$ *and* $y = \phi_2(t)$, *then*

$$\frac{du}{dt} = \frac{\partial u}{\partial x}.\frac{dx}{dt} + \frac{\partial u}{\partial y}.\frac{dy}{dt}.$$

Here $\frac{du}{dt}$ is called the **Total Differential Coefficient** of u with respect to t while $\frac{\partial u}{\partial x}$ and $\frac{\partial u}{\partial y}$ are partial derivatives of u.

Proof:

Let t be given a small increment δt, and let the corresponding changes in u, x and y be δu, δx and δy rexpectively. We have then

$u = f(x, y),$ ...(1)

and $u = \delta u = f(x + \delta x, y + \delta y).$ ...(2)

$\therefore\ \delta u = f(x + \delta x, y + \delta y) - f(x, y)$

$= [f(x + \delta x, y + \delta y) - f(x, y + \delta y)] + [f(x, y + \delta y) - f(x, y)]$. **(Note)**

$$\therefore\ \frac{\delta u}{\delta t} = \left\{\frac{f(x+\delta x, y+\delta y) - f(x, y+\delta y)}{\delta t}\right\} + \left\{\frac{f(x, y+\delta y) - f(x, y)}{\delta t}\right\}$$

$$= \left\{\frac{f(x+\delta x, y+\delta y) - f(x, y+\delta y)}{\delta x}\right\}.\frac{\delta x}{\delta t}$$

$$+ \left\{\frac{f(x, y+\delta y) - f(x, y)}{\delta y}\right\}.\frac{\delta y}{\delta t} \quad ...(3)$$

Let $\delta t \to 0$ so that $\delta x \to 0$ and $\delta y \to 0$.

Now $\lim\limits_{\delta t \to 0} \dfrac{\delta u}{\delta t} = \dfrac{du}{dt}$, $\lim\limits_{\delta t \to 0} \dfrac{\delta x}{\delta t} = \dfrac{dx}{dt}$ and $\lim\limits_{\delta t \to 0} \dfrac{\delta y}{\delta t} = \dfrac{dy}{dt}$.

Also $\lim\limits_{\delta y \to 0} \dfrac{f(x, y + \delta y) - f(x, y)}{\delta y} = \dfrac{\partial f}{\partial y} = \dfrac{\partial u}{\partial y}$,

and $\lim\limits_{\delta y \to 0} \dfrac{f(x + \delta x, y + \delta y) - f(x, y + \delta y)}{\delta x} = \dfrac{\partial f}{\partial x} = \dfrac{\partial u}{\partial x}$.

Since δx and δy tend to zero with δt and the functions involved are all supposed to be continuous, therefore the limit (3) becomes

$$\frac{du}{dt} = \frac{\partial u}{\partial x} \cdot \frac{dx}{dt} + \frac{\partial u}{\partial y} \cdot \frac{dy}{dt}.$$

In the same way if $u = f(x, y, z)$, where x, y, z are all functions of some variable t, then

$$\frac{du}{dt} = \frac{\partial u}{\partial x} \cdot \frac{dx}{dt} + \frac{\partial u}{\partial y} \cdot \frac{dy}{dt} + \frac{\partial u}{\partial z} \cdot \frac{dz}{dt}.$$

This result can be extended to any number of variables.

Cor. 1: *If u be a function of x and y, where y is a function of x, then*

$$\frac{du}{dx} = \frac{\partial u}{\partial x} + \frac{\partial u}{\partial y} \cdot \frac{dy}{dx}.$$

This result follows immediately by taking t = x in the formula of 6.6.

Cor. 2: *If $u = f(x, y)$ and $x = f_1(t_1, t_2)$ and $y = f_2(t_1, t_2)$, then*

$$\frac{\partial u}{\partial t_1} = \frac{\partial u}{\partial x} \cdot \frac{\partial x}{\partial t_1} + \frac{\partial u}{\partial y} \cdot \frac{\partial y}{\partial t_1}$$

and $$\frac{\partial u}{\partial t_2} = \frac{\partial u}{\partial x} \cdot \frac{\partial x}{\partial t_2} + \frac{\partial u}{\partial y} \cdot \frac{\partial y}{\partial t_2}.$$

Cor 3: *If x and y are connected by an equation of the form $f(x, y) = 0$ then*

$$\frac{dy}{dx} = -\frac{\partial f / \partial x}{\partial f / \partial y}.$$

Since $f(x, y) = 0$, therefore by cor. 1, we get

$$0 = \frac{\partial f}{\partial x} + \frac{\partial f}{\partial y} \cdot \frac{dy}{dx},$$

from which the required result follows.

EULER'S THEOREM ON HOMOGENEOUS FUNCTIONS

If u is a homogeneous function of x and y of degree n, then

$$x\frac{\partial u}{\partial x}+y\frac{\partial u}{\partial y}=nu.$$

Proof:

Since u is a homogeneous function of x and y of degree n, therefore u may be put in the form

$$u=x^n f\left(\frac{y}{x}\right). \quad ...(1)$$

Differentiating (1) partially w.r.t. 'x', we hae

$$\frac{\partial u}{\partial x}=\frac{\partial}{\partial x}\left[x^n f\left(\frac{y}{x}\right)\right]$$

$$=\left[f\left(\frac{y}{x}\right)\right]nx^{n-1}+x^n\left[f'\left(\frac{y}{x}\right)\right]\left(-\frac{y}{x^2}\right).$$

$$\therefore\ x\frac{\partial u}{\partial x}=nx^n f\left(\frac{y}{x}\right)-x^{n-1}y.f'\left(\frac{y}{x}\right). \quad ...(2)$$

Again differentiating (1) partially w.r.t. 'y', we have

$$\frac{\partial u}{\partial y}=\frac{\partial}{\partial y}\left[x^n f\left(\frac{y}{x}\right)\right]=x^n\left[f'\left(\frac{y}{x}\right)\right].\frac{1}{x}=x^{n-1}f'\left(\frac{y}{x}\right).$$

$$\therefore\ y\frac{\partial u}{\partial y}=y.x^{n-1}f'\left(\frac{y}{x}\right). \quad ...(3)$$

Adding (2) and (3), we have

$= 2xy\ u^3 + 3y^2u^5\ [(x - y)^2 - (1 - 2xy + y^2)],$

$[\because u^{-2} = 1 - 2xy + y^2]$

$= 2xy\ u^3 + 3y^2u^5\ [x^2 - 1] = 2xy\ u^3 - 3y^2u^5\ (1 - x^2).$ ...(2)

Adding (1) and (2), we have

$$\frac{\partial}{\partial x}\left\{\left(1-x^2\right)\frac{\partial u}{\partial x}\right\}+\frac{\partial}{\partial y}\left\{y^2\frac{\partial u}{\partial y}\right\}=0.$$

Example:

If $\theta = t^n e^{-r^2-4t}$*, what value of n will make*

$$\frac{1}{r^2}\frac{\partial}{\partial r}\left(r^2\frac{\partial\theta}{\partial r}\right)=\frac{\partial\theta}{\partial t}?$$

Solution:

We have $\frac{\partial\theta}{\partial r} = t^n.e^{-r^2/4}.\left(-\frac{2r}{4t}\right) = -\frac{r}{2}t^{n-1}e^{-r^2/4t}$.

$\therefore \quad r^2\frac{\partial\theta}{\partial r} = -\frac{1}{2}r^3\,t^{n-1}\,e^{-r^2/4t}$.

$\therefore \quad \frac{\partial}{\partial r}\left(r^2\frac{\partial\theta}{\partial r}\right) = -\frac{3r^2}{2}t^{n-1}e^{-r^2/4t} - \frac{1}{2}r^3t^{n-1}\,e^{-r^2/4t}.\left(-\frac{2r}{4t}\right)$

$= -\frac{3}{2}r^2t^{n-1}\,e^{-r^2/4t} + \frac{1}{4}r^4t^{n-2}\,e^{-r^2/4t}$.

$\therefore \quad \frac{1}{r^2}\frac{\partial}{\partial r}\left(r^2\frac{\partial\theta}{\partial r}\right) = -\frac{3}{2}t^{n-1}e^{-r^2/4t} + \frac{1}{2}r^2t^{n-2}\,e^{-r^2/4t}$

Also $\quad \frac{\partial\theta}{\partial r}\ nt^{n-1}e^{-r^2/4t} + t^n\,e^{-r^2/4t}\ .\ \frac{r^2}{4t^2}$

$= n\ t^{n-1}\ e^{-r2} + 1/4\ r^2\ t^{n-2}\ e^{-r2/4t}$.

Now $\quad \frac{1}{r^2}\frac{\partial}{\partial r}\left(r^2\frac{\partial\theta}{\partial r}\right) = \frac{\partial\theta}{\partial t}$

$\Rightarrow -\frac{3}{2}t^{n-1}e^{-r^2/4t} + \frac{1}{4}r^2t^{n-2}e^{-r^2/4t} = nt^{n-1}\,e^{-r^2/4t} + \frac{1}{4}r^2\,t^{n-2}\,e^{-r^2/4t}$

$\Rightarrow -\frac{3}{2}t^{n-1}\,e^{-r^2/4t} = nt^{n-1}\,e^{-r^2/4t}$, for all posible values of r and t

$\Rightarrow n = -\frac{3}{2}$.

FUNCTIONS OF TWO OR MORE INDEPENDENT VARIABLES

We after come across which depend two or more variables. For example, volume of a right circular cylinder of radius r and height h is given by $v = \pi r^2 h$ for a given value of r and h V has a definite values.

A function of x, y and z may be written as f(x, y, z) or ϕ(x, y, z), etc. Similarly, a function of x and y is generally denoted by the symbols f(x, y) or ϕ(x, y) etc.

If a derivative of a function of several independent variables be found with respect to any of them, keeping the others as constants, it is said to be a Partial Derivative. The operation of finding the partial derivatives of a function of more than one independent variables is called *Partial*

Differentiation. The symbols $\partial/\partial x$, $\partial/\partial y$, etc., areused to denote such differentiations and the represents $\partial u/\partial x$, $\partial u/\partial y$, etc., are respectively called the *Partial Differential Coefficients* of u w.r.t. x, y, etc. Thus, if $u = f(x, y, z)$, then the partial differential coefficient of u w.r.t. x *i.e.* $\partial u/\partial x$ is obtained by differentiating u w.r.t. x keeping y and z as constants. Sometimes $\partial f/\partial x$ is also denoted by f_x.

Second Order Partial Differential Coefficiens: If $u = f(x, y)$, then $\partial u/\partial x$ or f_x and $\partial u/\partial y$ or f_y are themselves functions of x and y and can be again differentiated partially.

We call $\frac{\partial}{\partial x}\left(\frac{\partial u}{\partial y}\right), \frac{\partial}{\partial y}\left(\frac{\partial u}{\partial x}\right), \frac{\partial}{\partial x}\left(\frac{\partial u}{\partial x}\right)$ and $\frac{\partial}{\partial y}\left(\frac{\partial u}{\partial y}\right)$

as the second order partial derivatives are u and these are respectively as denoted by $\frac{\partial^2 u}{\partial x\,\partial y}, \frac{\partial^2 u}{\partial y\,\partial x}, \frac{\partial^2 u}{\partial x^2}$ and $\frac{\partial^2 u}{\partial y^2}$

Note: If $u = f(x, y)$ and its partial derivatives are continuous (as is true in all ordinary cases), the order of differentiation is immaterial *i.e.*, $\frac{\partial^2 u}{\partial x\,\partial y} = \frac{\partial^2 u}{\partial y\,\partial x}$. The partial differentiation co-efficient of fx and fy are, fxu, fxy, fyx, fyy or

$\frac{\partial^2 f}{\partial x^2}, \frac{\partial^2 f}{\partial y\,\partial x}, \frac{\partial^2 f}{\partial x\,\partial y}, \frac{\partial^2 f}{\partial y^2}$ respectively. It should be specially noted that

$$\frac{\partial^2 f}{\partial y\,\partial x} \text{ means } \frac{\partial}{\partial y}\left(\frac{\partial f}{\partial x}\right) \text{ and } \frac{\partial^2 f}{\partial x\,\partial y} \text{ means } \frac{\partial}{\partial x}\left(\frac{\partial f}{\partial y}\right).$$

The student will be able to convince himself that in all ordinary cases.

$$\frac{\partial^2 f}{\partial y\,\partial x} = \frac{\partial^2 f}{\partial x\,\partial y}.$$

SOLVED EXAMPLES

Example 1:

If $z = \sin\left(\frac{y}{x}\right)$, *then prove that* $\frac{\partial^2 z}{\partial x\,\partial y} = \frac{\partial^2 z}{\partial y\,\partial x}$.

Solution:

We have

$$\frac{\partial z}{\partial x} = \cos\left(\frac{y}{x}\right)\left(\frac{-y}{x^2}\right), \frac{\partial z}{\partial y} = \cos\left(\frac{y}{x}\right)\left(\frac{1}{x}\right)$$

$$\frac{\partial^2 z}{\partial x\,\partial y} = \frac{\partial}{\partial x}\left[\frac{\partial z}{\partial y}\right] = \frac{\partial}{\partial x}\left[\frac{1}{z}\cos\left(\frac{y}{x}\right)\right]$$

$$= -\frac{1}{x^2}\cos\left(\frac{y}{x}\right) - \frac{1}{x}\sin\left(\frac{y}{x}\right)\left(\frac{-y}{x^2}\right)$$

$$= -\frac{1}{x^2}\cos\left(\frac{y}{x}\right) + \frac{y}{x^2}\sin\left(\frac{y}{x}\right) \qquad \text{...(1)}$$

Now $$\frac{\partial^2 z}{\partial y\,\partial x} = \frac{\partial}{\partial y}\left(\frac{\partial z}{\partial x}\right) = \frac{\partial}{\partial y}\left(-\frac{y}{x^2}\cos\frac{y}{x}\right)$$

$$= -\frac{1}{x^2}\left[1.\cos\left(\frac{y}{x}\right) - y\sin\left(\frac{y}{x}\right).\frac{1}{x}\right]$$

$$= -\frac{1}{x^2}\cos\left(\frac{y}{x}\right) + \frac{y}{x^2}\sin\left(\frac{y}{x}\right) \qquad \text{...(2)}$$

from eqaution (1) and (2) we get

$$\frac{\partial^2 z}{\partial x\,\partial y} = \frac{\partial^2 z}{\partial y\,\partial x}.$$

Example 2:

If $z = f(y/x)$, show that $x\,(\partial z/\partial x) = +\,y(\partial z/\partial y) = 0$.

Solution:

$$\frac{\partial z}{\partial x} = \left[f'\left(\frac{y}{x}\right)\right]\left(-\frac{y}{x^2}\right), \text{ (diff. particularly w.r.t. x).}$$

$$\therefore \quad x\left(\frac{\partial z}{\partial x}\right) = -\left(\frac{y}{x}\right)f'\left(\frac{y}{x}\right). \qquad \text{...(1)}$$

Again $$\frac{\partial z}{\partial y} = \left[f'\left(\frac{y}{x}\right)\right].\left(\frac{1}{x}\right), \text{ (diff. partially w.r.t. y)}$$

$$\therefore \quad y\left(\frac{\partial z}{\partial y}\right) = \left(\frac{y}{x}\right)f'\left(\frac{y}{x}\right). \qquad \text{...(2)}$$

Adding equations (1) and (2), we get $x\left(\frac{\partial z}{\partial x}\right) + y\left(\frac{\partial z}{\partial y}\right) = 0$.

Example 3:

If $u = \sin^{-1}\frac{x}{y} + \tan^{-1}\frac{y}{x}$, show that $\frac{\partial u}{\partial x} + y\,\frac{\partial u}{\partial y} = 0$.

Solution:

Here we have

$$\frac{\partial u}{\partial x}=\frac{1}{\sqrt{\left\{1-\left(\frac{x}{y}\right)^2\right\}}}.\frac{1}{y}+\frac{1}{1+\left(\frac{y}{x}\right)^2}.\left(-\frac{y}{x^2}\right), \text{ (treating as constant)}$$

$$=\frac{1}{\sqrt{(y^2-x^2)}}-\frac{1}{(x^2+y^2)}.$$

$$\therefore \qquad x\frac{\partial u}{\partial x}=\frac{x}{\sqrt{(y^2-x^2)}}-\frac{xy}{x^2+y^2}. \qquad ...(1)$$

Again

$$\frac{\partial u}{\partial y}=\frac{1}{\sqrt{\left\{1-\left(\frac{x}{y}\right)^2\right\}}}\left(1-\frac{x}{y^2}\right)+\frac{1}{1+\left(\frac{y}{x}\right)^2}.\frac{1}{x}, \text{ (treating x as constant)}$$

$$=-\frac{1}{y\sqrt{(y^2-x^2)}}+\frac{x}{x^2+y^2}.$$

$$\therefore \qquad y\frac{\partial u}{\partial y}=-\frac{r}{\sqrt{(y^2-x^2)}}+\frac{xy}{x^2+y^2}. \qquad ...(2)$$

Adding (1) and (2), we have $x\left(\frac{\partial u}{\partial x}\right)+y\left(\frac{\partial u}{\partial y}\right)=0$.

Example 4(a):

If $z = f(x + ay) + \phi(x - ay)$, prove that

$$\frac{\partial^2 z}{\partial y^2}=a^2\left(\frac{\partial^2 z}{\partial x^2}\right).$$

Solution:

We have z = f(x + ay) + f(x – ay).

$$\therefore \quad \frac{\partial z}{\partial x} = f'(x + ay) + \phi'(x - ay), \text{ (diff. partially w.r.t. ‘x’)}$$

$$\text{and } \frac{\partial^2 z}{\partial x^2} = f''(x - y) + \phi''(x - y) \qquad ...(1)$$

Again $\frac{\partial z}{\partial y}$ = af ' (x + ay) – aϕ ' (x – ay).

$\therefore \quad \frac{\partial^2 z}{\partial y^2}$ = a^2f " (x + ay) + a^2ϕ " (x – ay). ...(2)

From (1) and (2), we get $\frac{\partial^2 z}{\partial y^2} = a^2\left(\frac{\partial^2 z}{\partial x^2}\right)$.

Example 4(b):

If $z = x f\left(\frac{y}{x}\right)$ *prove that*

$$x\frac{\partial z}{\partial x} + y\frac{\partial z}{\partial y} = 2z.$$

Solution:

We have $\frac{\partial z}{\partial x} = y\left\{f\left(\frac{y}{x}\right) + xf'\left(\frac{y}{x}\right)\left(-\frac{y}{x^2}\right)\right\}$

$$\frac{\partial z}{\partial y} = x\left\{f\left(\frac{y}{x}\right) + yf'\left(\frac{y}{x}\right)\frac{1}{x}\right\}.$$

Hence $x\frac{\partial z}{\partial x} + y\frac{\partial z}{\partial y} = 2xy\ f\left(\frac{y}{x}\right) = 22$.

Example 4(c):

If $u = \sin^{-1}\left(\frac{x^2 + y^2}{x + y}\right)$, *show that*

$$x\left(\frac{\partial u}{\partial x}\right) + y\left(\frac{\partial u}{\partial y}\right) = \tan u.$$

Solution:

We have sin u = $(x^2 + y^2)/(x + y)$.

$\therefore \quad$ log sin u = log $(x^2 + y^2)$ – log (x + y). ...(1)

Differentiating (1) partially w.r.t. x, we get

$$\frac{1}{\sin u}\cos u.\frac{\partial u}{\partial x} = \frac{2x}{x^2 + y^2} - \frac{1}{x + y},$$

$\therefore \quad (\cot u)\ x\frac{\partial u}{\partial x} = \frac{2x^2}{x^2 + y^2} - \frac{x}{x + y}.$ ...(2)

Again differentiating (1) partially w.r.t. y, we get

$$\frac{\cos u}{\sin u}.\frac{\partial u}{\partial y}=\frac{2y}{x^2+y^2}-\frac{1}{x+y}.$$

$$\therefore \qquad (\cot u).y\frac{\partial u}{\partial y}=\frac{2y^2}{x^2+y^2}-\frac{y}{x+y}. \qquad ...(3)$$

Adding (2) and (3), we get

$$(\cot u)\left(x\frac{\partial u}{\partial x}+y\frac{\partial u}{\partial y}\right)=\frac{2x^2+2y^2}{x^2+y^2}-\frac{x+y}{x+y}=2-1=1.$$

$$\therefore \qquad x\left(\frac{\partial u}{\partial x}\right)+y\left(\frac{\partial u}{\partial y}\right) = 1/\cot u = \tan u.$$

Example 5:

If $z = xy\ f(y/x)$

prove that $x\ \frac{\partial z}{\partial x} + y\ \frac{\partial z}{\partial y} = 2z$

Solution:

We have, $\frac{\partial z}{\partial x} = y\ [f(y/x) + xf'(y/x)\ (-y/x^2)]$

$$\frac{\partial z}{\partial y} = x\ [f(y/x) + yf'(y/x)\ 1/x]$$

Hence, $x\ \frac{\partial z}{\partial x} + y\ \frac{\partial z}{\partial y} = xy\ f(y/x) = 22$

Example 6:

If $z = tan^{-1} \frac{x^2+y^2}{x-y}$. *Prove that*

$$x\frac{\partial z}{\partial u}+y\frac{\partial z}{\partial y}=\sin 2z.$$

Solution:

We have

$$z = \tan^{-1} \frac{x^2+y^2}{x-y}$$

$$\Rightarrow \qquad \tan z = \frac{x^2+y^2}{x-y}$$

Let $u = \tan z$ ...(1)

$$\therefore \quad u = \frac{x^2+y^2}{x-y} = \frac{x^2\left\{1+\left(\frac{y}{x}\right)^2\right\}}{x\left\{1-\left(\frac{y}{x}\right)\right\}} = x^2 f\left(\frac{y}{x}\right).$$

Thus, u is a homogeneous function of degree z.

Hence, the Euler's theorem;

$$x\frac{\partial u}{\partial x} + y\frac{\partial u}{\partial y} = 2u \quad ...(2)$$

From (1) $\frac{\partial u}{\partial u} = \sec^2 z\frac{\partial z}{\partial u}, \frac{\partial u}{\partial y} = \sec^2 z\frac{\partial z}{\partial y}$ ...(3)

Putting (3) and (1) in (2) we get

$$\sec^2 z\left(x\frac{\partial z}{\partial x} + y\frac{\partial z}{\partial y}\right) = 2\tan z$$

or
$$x\frac{\partial z}{\partial x} + y\frac{\partial z}{\partial y} = 2\frac{\sin z}{\cos z}.\cos^2 z$$

$$2\sin z \cos z = \sin 2z.$$

Example 7:

If $u = x^2 \tan^{-1}\frac{y}{x} - y^2 \tan^{-1}\frac{x}{y}$, *find* $\frac{\partial^2 u}{\partial x\, \partial y}$.

Solution:

We have

$$\frac{\partial u}{\partial y} = x^2.\frac{1}{1+\left(\frac{y}{x}\right)^2}.\frac{1}{x} - 2y\tan^{-1}\frac{x}{y} - y^2\frac{1}{1+\left(\frac{x}{y}\right)^2}.\left(-\frac{x}{y^2}\right)$$

$$= \frac{x^3}{x^2+y^2} - 2y\tan^{-1}\frac{x}{y} + \frac{xy^2}{x^2+y^2}$$

$$= \frac{x(x^2+y^2)}{x^2+y^2} - 2y\tan^{-1}\frac{x}{y} = x - 2y\tan^{-1}\frac{x}{y}.$$

Now
$$\frac{\partial^2 u}{\partial x\, \partial y} = \frac{\partial}{\partial x}\left(\frac{\partial u}{\partial y}\right) = 1 - 2y\frac{1}{1+\left(\frac{x}{y}\right)^2}.\frac{1}{y} = 1 - \frac{2y^2}{x^2+y^2}$$

$$= \frac{x^2 + y^2 - 2y^2}{x^2 + y^2} = \frac{x^2 - y^2}{x^2 + y^2}.$$

Hence $$\frac{\partial^2 u}{\partial x\, \partial y} = \frac{x^2 - y^2}{x^2 + y^2}.$$

Example 8:

Find the value of

$$\frac{1}{a^2}\frac{\partial^2 z}{\partial x^2} + \frac{1}{b^2}\frac{\partial^2 z}{\partial y^2} \text{ when } a^2x^2 + b^2y^2 - c^2z^2 = 0.$$

Solution:

We have $$z^2 = \frac{a^2}{c^2}x^2 + \frac{b^2}{c^2}y^2. \qquad ...(1)$$

Differentiating (1) partially with respect to x taking y as constant, we have

$$2z\frac{\partial z}{\partial x} = 2\frac{a^2}{c^2}x \text{ or } \frac{\partial z}{\partial x} = \frac{a^2}{c^2}.\frac{x}{z}.$$

$$\therefore \quad \frac{\partial^2 z}{\partial x^2} = \frac{\partial}{\partial x}\left(\frac{a^2}{c^2}x.\frac{1}{z}\right) = \frac{a^2}{c^2}.\frac{1}{z} + \frac{c^2}{c^2}x.\left(-\frac{1}{z^2}\right)\frac{\partial z}{\partial x}$$

$$= \frac{a^2}{c^2 z} - \frac{a^2 x}{c^2 z^2}.\left(\frac{a^2 x}{c^2 z}\right) = \frac{a^2}{c^2 z} - \frac{a^4 x^2}{c^4 z^3}.$$

$$\therefore \quad \frac{1}{a^2}\frac{\partial^2 z}{\partial x^2} = \frac{1}{c^2 z} - \frac{a^2 x^2}{c^4 z^3} \qquad ...(2)$$

Again differentaiting (1) partially with respect to y taking x as constant, we have

$$2z\frac{\partial z}{\partial y} = 2\frac{b^2}{c^2}y \text{ or } \frac{\partial z}{\partial y} = \frac{b^2}{c^2}.\frac{y}{z}.$$

$$\therefore \quad \frac{\partial^2 z}{\partial y^2} = \frac{b^2}{c^2}.\frac{1}{z} + \frac{b^2}{c^2}y.\left(-\frac{1}{z^2}\right)\frac{\partial z}{\partial y}$$

$$= \frac{b^2}{c^2 z} - \frac{b^2 y}{c^2 z^2}.\left(\frac{b^2 y}{c^2 z}\right) = \frac{b^2}{c^2 z} - \frac{b^4 y^2}{c^4 z^3}.$$

$$\therefore \quad \frac{1}{b^2}\frac{\partial^2 z}{\partial y^2} = \frac{1}{c^2 z} - \frac{b^2 y^2}{c^4 z^3}. \qquad ...(3)$$

Adding (2) and (3), we get

$$\frac{1}{a^2}\frac{\partial^2 z}{\partial x^2}+\frac{1}{b^2}\frac{\partial^2 z}{\partial y^2}=\frac{2}{c^2 z}-\frac{a^2x^2+b^2y^2}{c^4z^3}$$

$$=\frac{2}{c^2 z}-\frac{c^2z^2}{c^4z^3}-\frac{1}{c^2z}-\frac{1}{c^2z}=\frac{1}{c^2z}.$$

Example 9:

If $z = xf(x + y) + y\phi(x + y)$, *prove that*

$$\left(\frac{\partial^2 z}{\partial x^2}\right)+\left(\frac{\partial^2 z}{\partial y^2}\right)=2\left(\frac{\partial^2 z}{\partial x\,\partial y}\right).$$

Solution:

We have $z = xf(x + y) + y\phi(x + y)$.

$$\therefore \frac{\partial z}{\partial x} - f(x + y) + xf'(x + y) + y\phi(x + y),$$

and $$\frac{\partial^2 z}{\partial x^2} = f'(x + y) + f'(x + y) + xf''(x + y) + y\phi''(x + y)$$

$$= 2f'(x + y) + xf''(x + y) + y\phi''(x + y).$$

Also $$\frac{\partial y}{\partial z} = xf'(x + y) + \phi(x + y) + y\phi'(x + y)$$

and $$\frac{\partial^2 z}{\partial y^2} = xf''(x + y) + \phi'(x + y) + \phi'(x + y) + y\phi''(x + y)$$

$$= 2f'(x + y) + xf''(x + y) + yf''(x + y).$$

Again $$\frac{\partial^2 z}{\partial x\,\partial y}=\frac{\partial}{\partial x}\left(\frac{\partial z}{\partial y}\right)$$

$$=\frac{\partial}{\partial x}\{xf'(x + y) + \phi(x + y) + y\phi'(x + y)\}$$

$$= f'(x + y) + xf''(x + y) + \phi'(x + y) + y\phi''(x + y).$$

Now $$\left(\frac{\partial^2 z}{\partial x^2}\right)+\left(\frac{\partial^2 x}{\partial y^2}\right)$$

$$= 2[f'(x + y) + f''(x + y) + xf'(x + y) + y\phi''(x + y)]$$

$$=2\left(\frac{\partial^2 z}{\partial x\,\partial y}\right).$$

Example 10:

If $u = tan^{-1}\frac{xy}{\sqrt{(1+x^2+y^2)}}$, *show that*

$$\frac{\partial^2 u}{\partial x\, \partial y} = \frac{1}{(1+x^2+y^2)^{3/2}}.$$

Solution:

We have

$$\frac{\partial u}{\partial x} = \frac{1}{1+\frac{x^2y^2}{1+x^2+y^2}}$$

$$y.\frac{1.\sqrt{(1+x^2+y^2)} - x.\frac{1}{2}(1+x^2+y^2)^{-1/2}(-2x)}{1+x^2+y^2}$$

$$= \frac{1+x^2+y^2}{1+x^2+y^2+x^2y^2}.\frac{(1+x^2+y^2)-x^2}{(1+x^2+y^2)(1+x^2+y^2)^{1/2}}$$

$$= \frac{y(1+y^2)}{(1+x^2)(1+y^2)(1+x^2+y^2)^{1/2}} = \frac{1}{1+x^2}.\frac{y}{(1+x^2+y^2)^{1/2}}.$$

$$\therefore \quad \frac{\partial^2 u}{\partial x\, \partial y} = \frac{\partial}{\partial y}\left(\frac{\partial u}{\partial x}\right)$$

$$= \frac{1}{1+x^2}.\frac{1.(1+x^2+y^2)^{1/2} - y.\frac{1}{2}(1+x^2+y^2)^{-1/2}.2y}{1+x^2+y^2}$$

$$= \frac{1}{1+x^2}.\frac{1+x^2+y^2-y^2}{(1+x^2+y^2)(1+x^2+y^2)^{1/2}}$$

$$= \frac{1}{1+x^2}.\frac{1+x^2}{(1+x^2+y^2)^{3/2}} = \frac{1}{(1+x^2+y^2)^{3/2}}.$$

Example 11:

If $\frac{x^2}{a^2+u} + \frac{y^2}{b^2+u} + \frac{z^2}{c^2+u} = 1$, *prove that*

$$\left(\frac{\partial u}{\partial x}\right)^2+\left(\frac{\partial u}{\partial y}\right)^2+\left(\frac{\partial u}{\partial z}\right)^2=2\left(x\frac{\partial u}{\partial x}+y\frac{\partial u}{\partial y}+z\frac{\partial u}{\partial z}\right).$$

Solution:

From the given equation we observe that u is a function of three independent variables x, y and z. Differentiating the given equation partially w.r.t. 'x', we get

$$\frac{2x}{a^2+u}-\left\{\frac{x^2}{(a^2+u)^2}+\frac{y^2}{(b^2+u)^2}+\frac{z^2}{(c^2+u)^2}\right\}\frac{\partial u}{\partial x}=0.$$

$$\therefore \frac{\partial u}{\partial x}=\frac{2x/(a^2+u)}{\Sigma\left\{x^2/(a^2+u)^2\right\}}.$$

Similarly, by symmetry, we can write the values of $\frac{\partial u}{\partial y}$ and $\frac{\partial u}{\partial z}$.

Now $\left(\frac{\partial u}{\partial x}\right)^2=\frac{4x^2/(a^2+u)^2}{\left[\Sigma\left\{x^2/(a^2+u)^2\right\}\right]}.$

$$\therefore \Sigma\left(\frac{\partial u}{\partial x}\right)^2=\frac{4\Sigma\left\{x^2/(a^2+u)^2\right\}}{\left[\left\{x^2/(a^2+u)^2\right\}\right]}$$

$$=\frac{4}{\Sigma\left\{x^2/(a^2+u)^2\right\}} \qquad \text{...(1)}$$

Again $2x\frac{\partial u}{\partial x}=\frac{4x^2/(a^2+u)}{\Sigma\left\{x^2/(a^2+u)^2\right\}}.$

$$\therefore 2\Sigma x\left(\frac{\partial u}{\partial x}\right)=\frac{4\Sigma\left\{x^2/(a^2+u)\right\}}{\Sigma\left\{x^2/(a^2+u)^2\right\}}=\frac{4}{\Sigma\left\{x^2/(a^2+u)^2\right\}}. \qquad \text{...(2)}$$

[$\because \Sigma\ \{x^2/(a^2+u)\}=1$, from the given relation]

Now from (1) and (2), we have

$$\Sigma\left(\frac{\partial u}{\partial x}\right)^2=2\Sigma\left\{x\left(\frac{\partial u}{\partial x}\right)\right\}$$

Example 12:

If $z = (x^2 + y^2)/(x + y)$, *show that*

$$\left(\frac{\partial z}{\partial x}-\frac{\partial z}{\partial y}\right)^2 = 4\left(1-\frac{\partial z}{\partial x}-\frac{\partial z}{\partial y}\right).$$

Solution:

We have $z = (x^2 + y^2)/(x + y)$.

$$\therefore \frac{\partial z}{\partial x}=\frac{(x+y).2x-(x^2+y^2).1}{(x+y)^2}=\frac{x^2-y^2+2xy}{(x+y)^2}$$

$$\text{and } \frac{\partial z}{\partial y}=\frac{(x+y)2y-(x^2+y^2).1}{(x+y)^2}=\frac{y^2-x^2+2xy}{(x+y)^2}.$$

$$\therefore \left(\frac{\partial z}{\partial x}-\frac{\partial z}{\partial y}\right)^2=\left[\frac{(x^2-y^2+2xy)-(y^2-x^2+2xy)}{(x+y)^2}\right]^2$$

$$=\left[\frac{2(x^2-y^2)^2}{(x^2+y^2)}\right]=4\left[\frac{x-y}{x+y}\right]^2.$$

$$\text{Also } 1-\frac{\partial z}{\partial x}-\frac{\partial z}{\partial y}=1-\frac{x^2-y^2+2xy}{(x+y^2)}-\frac{y^2-x^2+2xy}{(x+y)^2}$$

$$=\frac{(x^2+y^2+2xy)-x^2+y^2-2xy-y^2+x^2-2xy}{(x+y)^2}=\frac{x^2-2xy+y^2}{(x+y)^2}$$

$$=\left(\frac{x-y}{x+y}\right)^2.$$

$$\text{Hence } \left(\frac{\partial z}{\partial x}-\frac{\partial z}{\partial y}\right)^2=4\left(1-\frac{\partial z}{\partial x}-\frac{\partial z}{\partial y}\right).$$

Example 13:

If $u = \log r$, *where* $r^2 = (x - a)^2 + (y - b^2) + (z - c)^2$, *show that*

$$\frac{\partial^2 u}{\partial x^2}+\frac{\partial^2 u}{\partial y^2}+\frac{\partial^2 u}{\partial z^2}=\frac{1}{r^2}.$$

Solution:

We have $r^2 = (x - a)^2 + (y - b)^2 + (z - c)^2$. ...(1)

Differentiating (1) partially wrt x we have

$$2r\frac{\partial r}{\partial x} = 2(x-a)$$

or $\dfrac{\partial r}{\partial x} = \dfrac{1}{r}\left(\dfrac{x-a}{r}\right)$ ...(2)

Now u = log r. $\therefore \dfrac{\partial u}{\partial x} = \dfrac{1}{r}$

$\therefore \dfrac{\partial r}{\partial x} = \dfrac{1}{r}\left(\dfrac{x-a}{r}\right).$ ...[from (2)]

Thus, $\dfrac{\partial u}{\partial x} = \dfrac{(x-a)}{r^2}$

$$\therefore \frac{\partial^2 u}{\partial x^2} = \frac{\partial}{\partial x}\left(\frac{\partial u}{\partial x}\right) = \frac{\partial}{\partial x}\left(\frac{x-a}{r^2}\right) = \frac{r^2(1)-(x-a).2r\left(\frac{\partial r}{\partial x}\right)}{r^4}$$

$= \dfrac{r^2-2(x-a)^2}{r^4},$ $\left[\because \text{from } (2), \dfrac{\partial r}{\partial x} = (x-a)/r\right]$

Similarly, by symmetry

$$\frac{\partial^2 u}{\partial y^2} = \frac{r^2-2(y-b)^2}{r^4}$$

and $\dfrac{\partial^2 u}{\partial z^2} = \dfrac{r^2-2(z-c)^2}{r^4}.$

Hence $\dfrac{\partial^2 u}{\partial x^2} + \dfrac{\partial^2 u}{\partial y^2} + \dfrac{\partial^2 u}{\partial z^2} = \dfrac{3r^2-2\left\{(x-a)^2+(y-b)^2+(z-c)^2\right\}}{r^4}$

$= \dfrac{3r^2-2r^2}{r^4}$, {using (1)} $= \dfrac{r^2}{r^4} = \dfrac{1}{r^2}.$

Example 14:

If $1/u = \sqrt{(x^2+y^2+z^2)}$, *show that*

$$x\frac{\partial u}{\partial x} + y\frac{\partial u}{\partial y} + z\frac{\partial u}{\partial z} = -u,$$

and $$\frac{\partial^2 u}{\partial x^2} + \frac{\partial^2 u}{\partial y^2} + \frac{\partial^2 u}{\partial z^2} = 0.$$

Solution:

We have $u = (x^2 + y^2 + z^2)^{-1/2}$.

$$\therefore \frac{\partial u}{\partial x} = -\frac{1}{2}(x^2 + y^2 + z^2)^{-3/2}.(2x) = -x(x^2 + y^2 + z^2)^{-3/2},$$

and $$\frac{\partial^2 u}{\partial x^2} = x.\frac{3}{2}(x^2 + y^2 + z^2)^{-5/2}(2x) - 1(x^2 + y^2 + z^2)^{-3/2}$$

$$= (x^2 + y^2 + z^2)^{-5/2}[3x^2 - (x^2 + y^2 + z^2)]$$

$$= u^5(2x^2 - y^2 - z^2).$$

Similarly, by symmetry $\frac{\partial u}{\partial y} = -y(x^2 + y^2 + z^2)^{-3/2}$,

$$\frac{\partial^2 u}{\partial y^2} = u^5(2y^2 - x^2 - z^2), \left(\frac{\partial u}{\partial z}\right) = -z(x^2 + y^2 + z^2)^{-3/2},$$

Now $x\frac{\partial u}{\partial x} + y\frac{\partial u}{\partial y} + z\frac{\partial u}{\partial z}$

$$= -x^2(x^2 + y^2 + z^2)^{-3/2} - y^2(x^2 + y^2 + z^2)^{-3/2} - z^2(x^2 + y^2 + z^2)^{-3/2}$$

$$= -(x^2 + y^2 + z^2)^{-3/2}(x^2 + y^2 + z^2) = -(x^2 + y^2 + z^2)^{-1/2} = -u.$$

Also $\frac{\partial^2 u}{\partial x^2} + \frac{\partial^2 u}{\partial y^2} + \frac{\partial^2 u}{\partial z^2}$

$$= u^5(2x^2 - y^2 - z^2 + 2y^2 - x^2 - z^2 - x^2 - y^2) = u^5.0 = 0.$$

Example 15:

If $u = e^{xyz}$, show that

$$\frac{\partial^3 u}{\partial x\,\partial y\,\partial z} = (1 + 3xyz + x^2y^2z^2)e^{xyz}.$$

Solutions:

Here $u = e^{xyz}$. $\therefore \frac{\partial u}{\partial z} = xy\,e^{xyz}$.

Now $$\frac{\partial^2 u}{\partial y\,\partial z} = \frac{\partial}{\partial y}\left(\frac{\partial u}{\partial z}\right) = \frac{\partial}{\partial y}\left(xy\,e^{xyz}\right) = x\frac{\partial}{\partial y}\left(y\,e^{xyz}\right)$$

$$= x[y.xz\,e^{xyz} + e^{xyz}] = e^{xyz}(x^2yz + z).$$

Again $$\frac{\partial^3 u}{\partial x\,\partial y\,\partial z} = \frac{\partial}{\partial x}\left(\frac{\partial^2 u}{\partial y\,\partial z}\right) = \frac{\partial}{\partial x}\left[e^{xyz}\left(x^2yz + x\right)\right]$$

$= e^{xyz}(2xyz + 1) + yz\, e^{xyz}(x^2yz + x)$

$= e^{xyz}[2xyz + 1 + x^2y^2z^2 + xyz]$

$= e^{xyz}[1 + 3xyz + x^2y^2z^2].$

Example 16:

If $u = \log(x^3 + y^3 + z^3 - 3xyz)$, *show that*

$$\frac{\partial u}{\partial x} + \frac{\partial u}{\partial y} + \frac{\partial u}{\partial z} = \frac{3}{x+y+z}$$

and $$\left(\frac{\partial}{\partial x} + \frac{\partial}{\partial y} + \frac{\partial}{\partial z}\right)^2 u = \frac{-9}{(x+y+z)^2}.$$

Solution:

We have $u = \log(x^3 + y^3 + z^3 - 3xyz)$.

$$\therefore \quad \frac{\partial u}{\partial x} = \frac{3x^2 - 3yz}{x^3 + y^3 + z^3 - 3xyz}, \quad \frac{\partial u}{\partial y} = \frac{3y^2 - 3zx}{x^3 + y^3 + z^3 - 3xyz}$$

and $$\frac{\partial u}{\partial z} = \frac{3z^2 - 3xy}{x^3 + y^3 + z^3 - 3xyz}$$

$$\therefore \quad \frac{\partial u}{\partial x} + \frac{\partial u}{\partial y} + \frac{\partial u}{\partial z} = \frac{3(x^2 + y^2 + z^2 - yz - zx - xy)}{x^3 + y^3 + z^3 - 3xyz}$$

$$= \frac{3(x^2 + y^2 + z^2 - yz - zx - xy)}{(x+y+z)(x^2 + y^2 + z^2 - yz - zx - xy)} = \frac{3}{x+y+z}. \qquad ...(1)$$

Now $$\left(\frac{\partial}{\partial x} + \frac{\partial}{\partial y} + \frac{\partial}{\partial z}\right)^2 u = \left(\frac{\partial}{\partial x} + \frac{\partial}{\partial y} + \frac{\partial}{\partial z}\right)\left(\frac{\partial u}{\partial x} + \frac{\partial u}{\partial y} + \frac{\partial u}{\partial z}\right)$$

$$= \left(\frac{\partial}{\partial x} + \frac{\partial}{\partial y} + \frac{\partial}{\partial z}\right)\left(\frac{3}{x+y+z}\right), \qquad \text{from (1)}$$

$$= 3\left[\frac{\partial}{\partial x}\left(\frac{1}{x+y+z}\right) + \frac{\partial}{\partial y}\left(\frac{1}{x+y+z}\right) + \frac{\partial}{\partial z}\left(\frac{1}{x+y+z}\right)\right]$$

$$= 3\left[\frac{-1}{(x+y+z)^2} + \frac{-1}{(x+y+z)^2} + \frac{-1}{(x+y+z)^2}\right] = \frac{-9}{(x+y+z)^2}.$$

Example 17:

If u log($x^2 + y^2 + z^2$), show that

$$x\frac{\partial^2 u}{\partial y\,\partial z} = y\frac{\partial^2 u}{\partial z\,\partial x} = z\frac{\partial^2 u}{\partial x\,\partial y}.$$

Solution:

We have $u = \log(x^2 + y^2 + z^2)$.

$$\therefore\ \frac{\partial u}{\partial z} = \frac{2z}{x^2 + y^2 + z^2},\ \text{[treating x and y as constants]}$$

$$\text{and } \frac{\partial^2 u}{\partial y\,\partial z} = \frac{\partial}{\partial y}\left(\frac{\partial u}{\partial z}\right) = 2z.\left[-\left(x^2 + y^2 + z^2\right)^{-2}.2y\right]$$

$$= -4yz/(x^2 + y^2 + z^2)^2.$$

$$\text{Now } x\frac{\partial^2 u}{\partial y\,\partial z} = -4xyz/(x^2 + y^2 + z^2)^2.$$

$$\text{By symmetry, } y\frac{\partial^2 u}{\partial z\,\partial x} = z\frac{\partial^2 u}{\partial x\,\partial y} = -\frac{4xyz}{\left(x^2 + y^2 + z^2\right)^2}.$$

$$\text{Hence } x\frac{\partial^2 u}{\partial y\,\partial z} = y\frac{\partial^2 u}{\partial z\,\partial x} = z\frac{\partial^2 u}{\partial x\,\partial y}.$$

Example 18:

If $x^x\ y^y\ z^z = c$ show that $x = y - z$,

$$\frac{\partial^2 z}{\partial x\,\partial y} = -\{x \log(ex)\}^{-1}.$$

Solution:

We have $x^x.y^y.z^z = c$. From this equation we observe that we can regard z as a function of two independent variables x and y. Taking logarithms of both sides of the given equation, we get

$$x \log x + y \log y + z \log z = \log c. \qquad ...(1)$$

Now differentiating (1) partially w.r.t. x taking y as constant, we have

$$x.\frac{1}{x} + 1.\log x + \left[z.\frac{1}{z} + 1.\log z\right]\frac{\partial z}{\partial x} = 0.$$

[Note that z is not a constant but is a function of x and y]

$$\therefore \quad \frac{\partial z}{\partial x} = -\frac{(1+\log x)}{(1+\log z)}. \qquad \text{...(2)}$$

Similarly differentiating (1) partially w.r.t. y, we have

$$\frac{\partial z}{\partial y} = -\frac{(1+\log y)}{(1+\log z)}. \qquad \text{...(3)}$$

Now $\dfrac{\partial^2 z}{\partial x\,\partial y} = \dfrac{\partial}{\partial x}\left(\dfrac{\partial z}{\partial y}\right) = \dfrac{\partial}{\partial x}\left[-\left(\dfrac{1+\log y}{1+\log z}\right)\right]$, [from (3)]

$$= -(1+\log y).\frac{\partial}{\partial x}\left[(1+\log z)^{-1}\right]$$

$$= -(1+\log y).\left[-(1+\log z)^{-2}.\frac{1}{z}.\frac{\partial z}{\partial x}\right]$$

$$= \frac{(1+\log y)}{z(1+\log z)^2}.\left[-\left(\frac{1+\log x}{1+\log z}\right)\right], \qquad \text{[from (2)].}$$

Hence, when x = y = z, we have

$$\frac{\partial^2 z}{\partial x\,\partial y} = -\frac{(1+\log x)^2}{x(1+\log x)^3}, \quad \left[\text{putting } y = z = x \text{ in the value of } \left(\frac{\partial^2 z}{\partial x\,\partial y}\right)\right]$$

$$= -\frac{1}{x(1+\log x)} = -\frac{1}{x(\log e+\log x)}, \qquad [\because \log e = 1]$$

$$= -\frac{1}{x\log(ex)} = -\{x\log(ex)\}^{-1}.$$

Example 19:

If $z = \tan(y + ax) + (y - ax)^{3/2}$, find the value of

$$\left(\frac{\partial^2 z}{\partial x^2}\right) - a^2\left(\frac{\partial^2 z}{\partial y^2}\right).$$

Solution:

Here $z = \tan(y + ax) + (y - ax)^{3/2}$.

$$\therefore \quad \left(\frac{\partial z}{\partial x}\right) = \{\sec^2(y+ax)\}.a + \frac{3}{2}(y-ax)^{1/2}.(-a),$$

and $$\left(\frac{\partial^2 z}{\partial x^2}\right) = 2a^2\tan(y+ax)\sec^2(y+ax) + \frac{3}{4}a^2(y-ax)^{-1/2}.$$

Again $\left(\frac{\partial z}{\partial y}\right) = \sec^2(y+ax) + \frac{3}{2}(y-ax)^{1/2}$

and $\left(\frac{\partial^2 z}{\partial y^2}\right) = 2\sec^2(y+ax)\tan(y+ax) + \frac{3}{4}(y-ax)^{-1/2}$

Thus, $\left(\frac{\partial^2 z}{\partial x^2}\right) - a^2\left(\frac{\partial^2 z}{\partial y^2}\right) = 0.$

Example 20:

If u = xφ (y/x) = ψ(y/x), prove that

$$x^2\frac{\partial^2 u}{\partial x^2} + 2xy\frac{\partial^2 u}{\partial x\,\partial y} + y^2\frac{\partial^2 u}{\partial y^2} = 0.$$

Solution:

We have u = xϕ (y/x) + ψ (y/x). ...(1)

Differentiating (1) partially w.r.t. x and y, we get

$$\left(\frac{\partial u}{\partial x}\right) = x\left\{\phi'\left(\frac{y}{x}\right)\right\}.\left(-\frac{y}{x^2}\right) + \phi\left(\frac{y}{x}\right) + \left\{\psi'\left(\frac{y}{x}\right)\right\}.\left(-\frac{y}{x^2}\right),$$

and $\left(\frac{\partial u}{\partial x}\right) = x\left\{\phi'\left(\frac{y}{x}\right)\right\}.\left(\frac{1}{x}\right) + \phi\left(\frac{y}{x}\right) + \left\{\psi'\left(\frac{y}{x}\right)\right\}.\left(\frac{1}{x}\right).$

$$\therefore\ x\left(\frac{\partial u}{\partial x}\right) + y\left(\frac{\partial u}{\partial y}\right) = x\phi\left(\frac{y}{x}\right) \qquad ...(2)$$

Now differentiating (2) partially w.r.t. x and y respectively, we get

$$x\frac{\partial^2 u}{\partial x^2} + \frac{\partial u}{\partial y} + y\frac{\partial^2 u}{\partial x\,\partial x} = x\left\{\phi'\left(\frac{y}{x}\right)\right\}\left(-\frac{y}{x^2}\right) + \phi\left(\frac{y}{x}\right),$$

and $x\frac{\partial^2 u}{\partial x\,\partial y} + y\frac{\partial^2 u}{\partial y^2} + \frac{\partial u}{\partial y} = x\left\{\phi'\left(\frac{y}{x}\right)\right\}.\frac{1}{x}.$

Multiplying these equations by x and y respectively and adding, we get

$$x^2\frac{\partial^2 u}{\partial x^2} + 2xy\frac{\partial^2 u}{\partial x\,\partial y} + y^2\frac{\partial^2 u}{\partial y^2} + x\frac{\partial u}{\partial x} + y\frac{\partial u}{\partial y} = x\phi\left(\frac{y}{x}\right)$$

or $x^2\frac{\partial^2 u}{\partial x^2} + 2xy\frac{\partial^2 u}{\partial x\,\partial y} + y^2\frac{\partial^2 u}{\partial y^2} = 0,$ [from (2)]

Example 21:

If $u = f(r)$ where $r^2 = x^2 + y^2$, show that

$$\frac{\partial^2 u}{\partial x^2}+\frac{\partial^2 u}{\partial y^2}=f''(r)+\frac{1}{r}f'(r).$$

Solution:

Differentiating $r^2 = x^2 + y^2$ partially w.r.t. x and y, we get

$$2r\frac{\partial r}{\partial x}=2x \text{ or } \frac{\partial r}{\partial x}=\frac{x}{r};\ 2r\frac{\partial r}{\partial y}=2y \text{ or } \frac{\partial r}{\partial y}=\frac{y}{r}. \quad ...(1)$$

Now $u = f(r)$. Therefore $\frac{\partial u}{\partial x}=\{f'(r)\}=\frac{x}{r}f'(r)$ [from (1)]

and $$\frac{\partial^2 u}{\partial x^2}=\frac{\partial}{\partial x}\left(\frac{\partial u}{\partial x}\right)=\frac{\partial}{\partial x}\left[x.\frac{1}{r}.f'(r)\right]$$

$$=1.\frac{1}{r}.f'(r)+\{xf'(r)\}\left(-\frac{1}{r^2}\frac{\partial r}{\partial x}\right)+\frac{x}{r}\{f''(r)\}\frac{\partial r}{\partial x}$$

$$=\frac{1}{r}f'(r)-\frac{x}{r^2}.\frac{x}{r}f'(r)+\frac{x^2}{r^2}f''(r), \quad \left[\because \text{ from (1), } \frac{\partial x}{\partial r}=\frac{x}{r}\right]$$

$$=\left(\frac{1}{r}\right)f'(r)-\left(\frac{x^2}{r^3}\right)f'(r)+\left(\frac{x^2}{r^2}\right)f''(r). \quad ...(2)$$

Similarly, by symmetry, we have

$$\frac{\partial^2 u}{\partial y^2}=\frac{1}{r}f'(r)-\frac{y^2}{r^3}f'(r)+\frac{y^2}{r^2}f''(r). \quad ...(3)$$

Adding (2) and (3), we get

$$\frac{\partial^2 u}{\partial x^2}+\frac{\partial^2 u}{\partial y^2}=\frac{2}{r}f'(r)-\frac{x^2+y^2}{r^3}f'(r)+\frac{x^2+y^2}{r^2}f''(r)$$

$$=\left(\frac{2}{r}\right)f'(r)-\left(\frac{r^2}{r^3}\right)f'(r)+\left(\frac{r^2}{r^2}\right)f''(r), \quad [\because r^2 = x^2 + y^2]$$

$$=\left(\frac{2}{r}\right)f'(r)-\left(\frac{1}{r}\right)f'(r)+f''(r)=f''(r)+\left(\frac{1}{r}\right)f'(r).$$

Example 22:

If $x = r\cos\theta$, $y = r\sin\theta$, prove that

(a) $\frac{\partial^2 r}{\partial x^2}+\frac{\partial^2 r}{\partial y^2}=\frac{1}{r}\left[\left(\frac{\partial r}{\partial x}\right)^2+\left(\frac{\partial r}{\partial y}\right)^2\right]$,

(b) $\frac{\partial^2 r}{\partial x^2}.\frac{\partial^2 r}{\partial y^2}=\left(\frac{\partial^2 r}{\partial x\,\partial y}\right)^2$,

(c) $\left(\frac{\partial r}{\partial x}\right)^2+\left(\frac{\partial r}{\partial y}\right)^2=1$.

Solution:

(a) We have $x = r\cos\theta,\ y = r\sin\theta.$

Therefore $r^2 = x^2 + y^2$ ...(1)

Now $2r\left(\frac{\partial r}{\partial x}\right)=2x;$ [diff. (1) partially w.r.t 'x']

$\therefore\ \frac{\partial r}{\partial x}=\frac{x}{r}.$

$$\frac{\partial^2 r}{\partial x^2}=\frac{r.1-x.\frac{\partial r}{\partial x}}{r^2}=\frac{r-x.\frac{x}{r}}{r^2},$$ ...[using (2)]

$$=\frac{r^2-x^2}{r^3}=\frac{\left(x^2+y^2\right)-x^2}{r^3}=\frac{y^2}{r^3}.$$ ...(3)

Again differentiating (1) partially w.r.t. 'y', we get

$$2r\frac{\partial r}{\partial y}=2y;\qquad \therefore\ \frac{\partial r}{\partial y}=\frac{y}{r}$$ ...(4)

Differentiating (4) partially w.r.t. y, we get

$$\frac{\partial^2 r}{\partial y}=\frac{r.1-y.\frac{\partial r}{\partial y}}{r^2}=\frac{r-y.\frac{y}{r}}{r^2}$$ $\left[\because \text{from (4)}, \frac{\partial r}{\partial r}=\frac{y}{r}\right]$

$$=\frac{r^2-y^2}{r^3}=\frac{\left(x^2+y^2\right)-y^2}{r^3}=\frac{x^2}{r^3}.$$ ...(5)

Adding (3) and (5), we get

$$\frac{\partial^2 r}{\partial x^2}+\frac{\partial^2 r}{\partial y^2}=\frac{y^2}{r^3}+\frac{x^2}{r^3}=\frac{x^2+y^2}{r^3}=\frac{1}{r}.$$

Also $\frac{1}{r}\left[\left(\frac{\partial r}{\partial x}\right)^2+\left(\frac{\partial r}{\partial y}\right)^2\right]=\frac{1}{r}\left[\frac{x^3}{r^2}+\frac{y^2}{r^2}\right]=\frac{x^2+y^2}{r^3}=\frac{r^2}{r^3}=\frac{1}{r}.$

$$\therefore \quad \frac{\partial^2 r}{\partial x^2}+\frac{\partial^2 r}{\partial y^2}=\frac{1}{r}\left[\left(\frac{\partial r}{\partial x}\right)^2+\left(\frac{\partial r}{\partial y}\right)^2\right].$$

(b) Differentiating (4) partially w.r.t. 'x', we get

$$\frac{\partial^2 r}{\partial x\,\partial y}=-\frac{y}{r^2}\cdot\frac{\partial r}{\partial x}=-\frac{xy}{r^3}, \qquad \left[\because \text{from (2), } \frac{\partial r}{\partial x}=\frac{x}{r}\right]$$

$$\text{Now } \frac{\partial^2 r}{\partial x^2}\cdot\frac{\partial^2 r}{\partial y^2}=\frac{y^2}{r^3}\cdot\frac{x^2}{r^3}, \qquad \text{[from (3) and (5)]}$$

$$=\frac{x^2y^2}{r^6}=\left(\frac{-xy}{r^3}\right)^2=\left(\frac{\partial^2 r}{\partial x\,\partial y}\right)^2.$$

(c) From (2) and (4), on squaring and adding, we get

$$\left(\frac{\partial r}{\partial x}\right)^2+\left(\frac{\partial r}{\partial y}\right)^2=\frac{x^2}{r^2}+\frac{y^2}{r^2}=\frac{x^2+y^2}{r^2}=\frac{r^2}{r^2}=1.$$

Example 23:

If $u = (1 - 2xy + y^2)^{-1/2}$, prove that

$$\frac{\partial}{\partial x}\left\{(1-x^2)\frac{\partial u}{\partial x}\right\}+\frac{\partial}{\partial y}\left\{y^2\frac{\partial u}{\partial y}\right\}=0.$$

Solution:

Here $u = (1 - 2xy + y^2)^{-1/2}$.

$$\therefore \quad \frac{\partial u}{\partial x}=-\frac{1}{2}(1-2xy+y^2)^{-3/2}(-2y)=yu^3$$

$$\text{and} \quad \frac{\partial u}{\partial y}=-1/2\,(1-2xy+y^2)^{-3/2}\,(-2y+2y)=(x-y)\,u^3.$$

$$\text{Now} \quad \frac{\partial}{\partial x}\left\{(1-x^2)\frac{\partial u}{\partial x}\right\}=\frac{\partial}{\partial x}\left\{(1-x^2).yu^3\right\}$$

$$= y(-2x)u^3 + y(1-x^2).3u^2\frac{\partial u}{\partial x} = -2xy\,u^3 + 3y\,(1-x^2)u^2.yu^3$$

$$= -2xy\,u^3 + 3y^2u^5\,(1-x)^2. \qquad ...(1)$$

$$\text{Also } \frac{\partial}{\partial y}\left\{y^2\frac{\partial u}{\partial y}\right\}=\frac{\partial}{\partial y}\{y^2(x-y)u^3\}=\frac{\partial}{\partial y}\{(y^2x-y^3)u^3\}$$

$$= (2xy - 3y^2)\,u^3 + (y^2x - y^3).3u^2\frac{\partial u}{\partial y}$$

$= (2xy - 3y^2)\, u^3 + y^2(x - y).3u^2.(x - y)u^3$

$= (2xy - 3y^2)u^3 + y^2(x - y)^2.3u^5$

$= 2xy\, u^3 + 3y^2u^5\, [(x - y)^2 - u^{-2}]$

Example 24:

If $z = \tan^{-1}\left(\dfrac{x+y}{\sqrt{x}+\sqrt{y}}\right)$ *then*

$$x\frac{\partial z}{\partial x}+y\frac{\partial z}{\partial y}=\frac{1}{4}\sin 2z.$$

Solution:

$$z = \tan^{-1}\left(\frac{x+y}{\sqrt{x}+\sqrt{y}}\right) \Rightarrow \tan z = \frac{x+y}{\sqrt{x}+\sqrt{y}}.$$

Let u = tan z ...(1)

u is a homogeneous functions in x and y of degree $\frac{1}{2}$.

By Euler's Theorem $x\dfrac{\partial y}{\partial x}+y\dfrac{\partial y}{\partial y}=\dfrac{1}{2}$.

or $$x\left(\sec^2 z\frac{\partial z}{\partial x}\right)+y\left(\sec^2 z\frac{\partial z}{\partial y}\right)=\frac{1}{2}\tan x \text{ by (1)}$$

or $$x\frac{\partial z}{\partial x}+y\frac{\partial z}{\partial y}=\frac{1}{2}\frac{\tan z}{\sec^2 z}=\frac{1}{2}\sin z\cos z$$

$$=\frac{1}{4}\sin 2z.$$

Example 25:

If f(x, y, z) is a homogeneous function of the nth *degree in x, y, z, prove that*

$$x^2\frac{\partial^2 f}{\partial x^2}+y^2\frac{\partial^2 f}{\partial y^2}+z^2\frac{\partial^2 f}{\partial z^2}+2yz\frac{\partial^2 f}{\partial y\,\partial z}+2zx\frac{\partial^2 f}{\partial z\,\partial x}+2xy\frac{\partial^2 f}{\partial x\,\partial y}$$

$$= n(n-1)\, f(x, y, z).$$

Solution:

Here f(x, y, z) is a homogeneous function of the nth degree n, x, y, z.

Therefore $\dfrac{\partial f}{\partial x}, \dfrac{\partial f}{\partial y}$ and $\dfrac{\partial f}{\partial z}$ are homogeneous functions f the (n – 1)th degree in x, y, z. So using Euler's theorem for $\dfrac{\partial f}{\partial x}$,

$$x\frac{\partial}{\partial x}\left(\frac{\partial f}{\partial x}\right)+y\frac{\partial}{\partial y}\left(\frac{\partial f}{\partial x}\right)+z\frac{\partial}{\partial z}\left(\frac{\partial f}{\partial x}\right)=(n-1)\frac{\partial f}{\partial x}$$

or $$x\frac{\partial^2 f}{\partial x^2}+y\frac{\partial^2 f}{\partial y\,\partial x}+z\frac{\partial^2 f}{\partial z\,\partial x}=(n-1)\frac{\partial f}{\partial x}. \quad ...(1)$$

Similarly, $$x\frac{\partial^2 f}{\partial x\,\partial y}+y\frac{\partial^2 f}{\partial y^2}+z\frac{\partial^2 f}{\partial y\,\partial z}=(n-1)\frac{\partial f}{\partial y} \quad ...(2)$$

and $$x\frac{\partial^2 f}{\partial x\,\partial y}+y\frac{\partial^2 f}{\partial z^2}=(n-1)\frac{\partial f}{\partial z} \quad ...(3)$$

Multiplying (1) by x, (2) by y and (3) by z and adding, we get

$$x^2\frac{\partial^2 f}{\partial x^2}+y^2\frac{\partial^2 f}{\partial y^2}+z^2\frac{\partial^2 f}{\partial z^2}+2yz\frac{\partial^2 f}{\partial y\,\partial z}+2zx\frac{\partial^2 f}{\partial z\,\partial x}+2xy\frac{\partial^2 f}{\partial x\,\partial y}$$

$$=(n-1)\left(x\frac{\partial f}{\partial y}+\frac{\partial f}{\partial y}z\frac{\partial f}{\partial z}\right)$$

$= (n - 1)\, nf(x, y, z) = n (n - 1) f(x, y, z).$

Example 26:

Verify Euler's theorem in the following cases:

(i) $u = x^{-4}\ 3x^3y + 5x^2y^2 + 4xy^3 - 2y^4.$

(ii) $u = \dfrac{x(x^3 - y^3)}{x^3 + y^3},$

(iii) $u = \dfrac{x^{1/4} + y^{1/4}}{x^{1/5} + y^{1/5}},$

(iv) $u = axy + byz = czx,$

(v) $u = x^n \log\left(\dfrac{y}{x}\right),$

(vi) $u = 1/\sqrt{(x^2 + y^2)}.$

Solution:

(i) We have $= x^4 - 3x^3y + 5x^2y^2 + 4xy^3 - 2y^4$. Obviously u is a homogeneous function of x and y of degree 4. So by Euler's Theorem, we must have

$$x\left(\frac{\partial u}{\partial x}\right)+y\left(\frac{\partial u}{\partial y}\right) = 4u. \text{ Let us verify it.}$$

We have $\left(\dfrac{\partial u}{\partial x}\right) = 4x^3 - 9x^2y + 10xy^2 + 4y^3,$

and $\left(\dfrac{\partial u}{\partial y}\right) = -3x^3 + 10x^2y + 12xy^2 - 8y^3.$

$$\therefore \quad x\frac{\partial u}{\partial x} + y\frac{\partial u}{\partial y} = x(4x^3 - 9x^2y + 10xy^2 + 4y^3)$$

$$+ y(-3x^3 + 10x^2y + 12xy^2 - 8y^3)$$

$= 4(x^4 - 3x^3y + 5x^2y^2 + 4xy^3 - 2y^4)$

$= 4u$. This verifies Euler's Theorem.

(ii) We have $u = \dfrac{x(x^3 - y^3)}{x^3 + y^3}$ which is obviously a homogeneous function of x and y of degree 4 – 3 *i.e.*, 1. Note that each term in the numerator is of degree 4 while each term in the denominator is of degree 3. In order to verify Euler's Theorem we are to know that

$$x\frac{\partial u}{\partial x} + y\frac{\partial u}{\partial y} = 1.u = u.$$

Now $\log u = \log x + \log(x^3 - y^3) - \log(x^3 + y^3)$. ...(1)

Differentiating (1) partially w.r.t. x and y respectively, we get

$$\frac{1}{u}\frac{\partial u}{\partial x} = \frac{1}{x} + \frac{3x^2}{x^3 - y^3} - \frac{3x^2}{x^3 + y^3} \qquad ...(2)$$

and $$\frac{1}{u}\frac{\partial u}{\partial y} = 0 - \frac{3y^2}{x^3 - y^3} - \frac{3y^2}{x^3 + y^3}. \qquad ...(3)$$

Multiplying (2) by x and (3) by y and adding, we get

$$\frac{1}{u}\left(x\frac{\partial u}{\partial x} + y\frac{\partial u}{\partial y}\right) = 1 + \frac{3(x^3 - y^3)}{x^3 - y^3} - \frac{3(x^3 + y^3)}{x^3 + y^3}$$

$$= 1 + 3 - 3 = 1.$$

$\therefore \; x\dfrac{\partial u}{\partial x} + y\dfrac{\partial u}{\partial y} = u$. This verifies Euler's Theorem.

(iii) Here u is a homogeneous function of x and y of degree $\dfrac{1}{4} - \dfrac{1}{5}$ *i.e.*, $\dfrac{1}{20}$. So by Euler's Theorem we must have

$$x\left(\frac{\partial u}{\partial x}\right)+y\left(\frac{\partial u}{\partial y}\right)=\frac{1}{20}u.$$

Let us verify it. We have

$\log u = \log(x^{1/4} + y^{1/4}) - \log (x^{1/5} + y^{1/5})$.

$$\therefore \ \frac{1}{u}\frac{\partial u}{\partial x}=\frac{\frac{1}{4}x^{-3/4}}{x^{1/4}+y^{1/4}}-\frac{\frac{1}{5}x^{-4/5}}{x^{1/5}+y^{1/5}}$$

and $$\frac{1}{u}\frac{\partial u}{\partial y}=\frac{\frac{1}{4}y^{-3/4}}{x^{1/4}+y^{1/4}}-\frac{\frac{1}{5}y^{-4/5}}{x^{1/5}+y^{1/5}}.$$

$$\therefore \ \frac{1}{u}\left(x\frac{\partial u}{\partial x}+y\frac{\partial u}{\partial y}\right)=\frac{1}{4}\frac{x^{1/4}+y^{1/4}}{x^{1/4}+y^{1/4}}-\frac{1}{5}\frac{x^{1/5}+y^{1/5}}{x^{1/5}+y^{1/5}}=\frac{1}{4}-\frac{1}{5}=\frac{1}{20}.$$

$\therefore \ x\frac{\partial u}{\partial x}+y\frac{\partial u}{\partial y}=\frac{1}{20}u.$ This verifies Euler's Theorem.

(iv) We have = axy + byz + czx, which is a homogeneous function of x, y and z of degree 2. So in order to verify Euler's Theorem, we must show that $x\frac{\partial u}{\partial x}+y\frac{\partial u}{\partial y}+z\frac{\partial u}{\partial z}=2u$.

Now $\frac{\partial u}{\partial x}=ay+cz$, $\frac{\partial u}{\partial y}=ax+bz$, and $\frac{\partial u}{\partial z}=by+cx$.

$\therefore \ x\frac{\partial u}{\partial x}+y\frac{\partial u}{\partial y}+z\frac{\partial u}{\partial z}$ = x (ay + cz) + y (ax + bz) + z (by + cx)

= 2 (axy + byz + czx) = 2u. This verifies Euler's Theorem.

(v) Here u is a homogeneous function of x and y of degree n. So by Euler's Theorem we must have

$$x\left(\frac{\partial u}{\partial x}\right)+y\left(\frac{\partial u}{\partial y}\right)=nu.$$

Now do the verification yourself.

(vi) Here $u=\frac{1}{\sqrt{(x^2+y^2)}}=\frac{1}{x\sqrt{\left[1+\left(y/x^2\right)\right]}}=x^{-1}.\frac{1}{\sqrt{\left[1+\sqrt{(y/x)^2}\right]}}$ is a homogeneous function of x and y of degree – 1. Now proceed yourself.

Example 27:

If $u = \tan^{-1}\left(\frac{x^3+y^3}{x+y}\right)$, *show that*

$$x\left(\frac{\partial u}{\partial x}\right)+y\left(\frac{\partial u}{\partial y}\right)=\sin 2u.$$

Solution:

We have tan u = $(x^3 + y^3)/(x + y)$ = v, say. Then v is a homogeneous function of x and y of degree 3 – 1 *i.e.*, 2. Therefore by Euler's theorem.

We have $$x\left(\frac{\partial v}{\partial x}\right)+y\left(\frac{\partial v}{\partial y}\right)=2v. \quad ...(1)$$

Now v = tan u.

$$\therefore \quad \frac{\partial v}{\partial x}=\sec^2 u\frac{\partial u}{\partial x} \text{ and } \frac{\partial v}{\partial y}=\sec^2 u\frac{\partial u}{\partial y}.$$

Putting these values in (1), we get

$$x\sec^2 u\frac{\partial u}{\partial x}+y\sec^2 u\frac{\partial u}{\partial y}=2v$$

or $$x\frac{\partial u}{\partial x}+y\frac{\partial u}{\partial y}=\frac{2v}{\sec^2 u}=\frac{2\tan u}{\sec^2 u}=2\sin u\cos u=\sin 2u.$$

Example 28:

If $z = xy\, f\left(\frac{y}{x}\right)$, *show that* $x\left(\frac{\partial z}{\partial x}\right)+y\left(\frac{\partial z}{\partial y}\right)=2z.$

Show also that if z is a constant,

$$\frac{f'\left(\frac{y}{x}\right)}{f\left(\frac{y}{x}\right)}=\frac{x\left\{y+x\left(\frac{dy}{dx}\right)\right\}}{y\left\{y-x\left(\frac{dv}{dx}\right)\right\}}.$$

Solution:

We have, $z = x^2.\left(\frac{y}{x}\right)f\left(\frac{y}{x}\right)$, so that z is a homogeneous function of x and y of degree 2.

Hence by Euler's Theorem, we have

$$x\left(\frac{\partial z}{\partial x}\right)+y\left(\frac{\partial z}{\partial y}\right)=2z.$$

If z be a constant, then differentiating

$z = xy\, f\left(\frac{y}{x}\right)$ logarithmically, w.r.t. x, we get

$$0=\frac{1}{x}+\frac{1}{y}\frac{dy}{dx}+\frac{f'\left(\frac{y}{x}\right)}{f\left(\frac{y}{x}\right)}.\frac{x\left(\frac{dy}{dx}\right)-y}{x^2}$$

$$=\frac{y+x\left(\frac{dy}{dx}\right)}{xy}+\frac{f'\left(\frac{y}{x}\right)}{f\left(\frac{y}{x}\right)}.\frac{x\left(\frac{dy}{dx}\right)-y}{x^2}$$

Hence $\dfrac{f'\left(\frac{y}{x}\right)}{f\left(\frac{y}{x}\right)}=\dfrac{x\left\{y+x\left(\frac{dy}{dx}\right)\right\}}{y\left\{y-x\left(\frac{dy}{dx}\right)\right\}}$.

Example 29:

If $u = \log\dfrac{x^3+y^3}{x+y}$, *show that* $x\dfrac{\partial u}{\partial x}+y\dfrac{\partial u}{\partial y}=2$.

Solution:

We have $e^u = \dfrac{x^3+y^3}{x+y}=v$, say.

Obviously $v = (x^3 + y^3)/(x + y)$ is a homogeneous function of x and y of degree 3 – 1 *i.e.*, 2. Therefore by Euler's theorem, we have

$$x\frac{\partial v}{\partial x}+y\frac{\partial v}{\partial y}=2.v=2v \qquad ...(1)$$

Now $v = e^u$.

$$\therefore\ \frac{\partial v}{\partial x}=e^u\frac{\partial u}{\partial x}\ \text{ and }\ \frac{\partial v}{\partial y}=e^u\frac{\partial u}{\partial y}.$$

Putting these values in (1), we get

$$xe^u\frac{\partial u}{\partial x}+y\,e^u\frac{\partial u}{\partial y}=2e^u$$

or $e^u\left(x\dfrac{\partial u}{\partial x}+y\dfrac{\partial u}{\partial y}\right)=2e^u$ or $x\dfrac{\partial u}{\partial x}+y\dfrac{\partial u}{\partial y}=2$.

Example 30:

If $u = x\phi\left(\dfrac{y}{x}\right)+\psi\left(\dfrac{y}{x}\right)$, *prove that*

$$x^2 \frac{\partial^2 u}{\partial x^2} + 2xy \frac{\partial^2 u}{\partial x \, \partial y} + y^2 \frac{\partial^2 u}{\partial y^2} = 0.$$

Solution:

Let $u = z_1 + z_2$, where $z_1 = x\phi\left(\frac{y}{x}\right)$ and $z_2 = \psi\left(\frac{y}{x}\right)$. Obviously z_1 is a homogeneous function of x and y of degree 1 and z_2 is a homogeneous function of x and y of degree zero. Now

$$x\frac{\partial u}{\partial x} + y\frac{\partial u}{\partial y} = x\frac{\partial}{\partial x}(z_1 + z_2) + y\frac{\partial}{\partial y}(z_1 + z_2)$$

$$= \left(x\frac{\partial z_1}{\partial x} + y\frac{\partial z_1}{\partial y}\right) + \left(x\frac{\partial z_2}{\partial x} + y\frac{\partial z_2}{\partial y}\right) = 1.z_1 + 0.z_2. \text{ (by Euler's theorem).}$$

Thus, $x\left(\frac{\partial u}{\partial x}\right) + y\left(\frac{\partial u}{\partial y}\right) = z_1$ ...(1)

Differentiating equation (1) partially w.r.t. x and y respectively, we get

$$x\frac{\partial^2 u}{\partial x^2} + \frac{\partial u}{\partial x} + y\frac{\partial^2 u}{\partial x \, \partial y} = \frac{\partial z_1}{\partial x}, \quad \text{...(2)}$$

and $x\frac{\partial^2 u}{\partial x \, \partial y} + \frac{\partial u}{\partial y} + y\frac{\partial^2 u}{\partial y^2} = \frac{\partial z_1}{\partial y}$. ...(3)

Multiplying (2) by x and (3) by y and adding, we get

$$x^2 \frac{\partial^2 u}{\partial x^2} + 2xy\frac{\partial^2 u}{\partial x \, \partial y} + y^2 \frac{\partial^2 u}{\partial y^2} + x\frac{\partial u}{\partial x} + y\frac{\partial u}{\partial y} = x\frac{\partial z_1}{\partial x} + y\frac{\partial z_1}{\partial y}$$

or $x^2 \frac{\partial^2 u}{\partial x^2} + 2xy\frac{\partial^2 u}{\partial x \, \partial y} + y^2 \frac{\partial^2 u}{\partial y^2} + z_1 = 1.z_1,$

$$\left[\because x\left(\frac{\partial u}{\partial x}\right) + y\left(\frac{\partial u}{\partial y}\right) = z_1 \text{by (1), and } x\left(\frac{\partial z_1}{\partial x}\right) + y\left(\frac{\partial z_1}{\partial y}\right) = 1.z_1 \text{ by Euler's theorem}\right]$$

or $$x^2 \frac{\partial^2 u}{\partial x^2} + 2xy + y^2 \frac{\partial^2 u}{\partial y^2} = 0.$$

Example 31:

If $u = x^2y^2/(x + y)$, *show that*

$$x\left(\frac{\partial u}{\partial x}\right) + y\left(\frac{\partial u}{\partial y}\right) = 3u.$$

Solution:

We have $u = \dfrac{x^2y^2}{x+y^2} = \dfrac{x^3\left(\dfrac{y}{x}\right)^2}{\left[1+\left(\dfrac{y}{x}\right)\right]} = x^3 f\left(\dfrac{y}{x}\right)$, say.

Thus u is a homogeneous function of x and y of degree 3.

Therefore by Euler's theorem, we have $x\dfrac{\partial u}{\partial x} + y\dfrac{\partial u}{\partial y} = 3u$.

Example 32:

If $u = \sin^{-1}\left(\dfrac{x+y}{\sqrt{x}+\sqrt{y}}\right)$, *show that*

$$x\left(\frac{\partial u}{\partial x}\right) + y\left(\frac{\partial u}{\partial y}\right) = \frac{1}{2}\tan u.$$

Solution:

We have $\sin u = (x+y)/(\sqrt{x}+\sqrt{y}) = v$, say.

Then v is a homogeneous function of x and y of degree $\left(1-\dfrac{1}{2}\right)$ *i.e.,* $\dfrac{1}{2}$. Applying Euler's Theorem for v, we have

$$x\left(\frac{\partial v}{\partial x}\right) + y\left(\frac{\partial v}{\partial y}\right) = \frac{1}{2}v$$

or $x\dfrac{\partial}{\partial x}(\sin u) + y\dfrac{\partial}{\partial y}(\sin u) = \dfrac{1}{2}\sin u$, $\qquad [\because v = \sin u]$

or $x\cos u\left(\dfrac{\partial u}{\partial x}\right) + y\cos u\left(\dfrac{\partial u}{\partial y}\right) = \dfrac{1}{2}\sin u$

or $x\left(\dfrac{\partial u}{\partial x}\right) + y\left(\dfrac{\partial u}{\partial y}\right) = \dfrac{1}{2}\tan u$.

Example 33:

If u be a homogeneous function of x and y of degree n, show that

$$x\left(\frac{\partial^2 u}{\partial x^2}\right) + y\left(\frac{\partial^2 u}{\partial x\,\partial y}\right) = (n-1)\left(\frac{\partial u}{\partial x}\right),$$

and $x\left(\dfrac{\partial^2 u}{\partial x\, \partial y}\right)+y\left(\dfrac{\partial^2 u}{\partial y^2}\right)=(n-1)\left(\dfrac{\partial u}{\partial y}\right).$

Hence deduce that

$$x^2\frac{\partial^2 u}{\partial x^2}+2xy\frac{\partial^2 u}{\partial x\,\partial y}+y^2\frac{\partial^2 u}{\partial y^2}=n(n-1)\,u.$$

Solution:

By Euler's Theorem, we have

$$x\left(\frac{\partial u}{\partial x}\right)+y\left(\frac{\partial u}{\partial y}\right)=nu \qquad ...(1)$$

Differentiating (1) partially w.r.t. x, we get

$$x\frac{\partial^2 u}{\partial x^2}+\frac{\partial u}{\partial x}+y\frac{\partial^2 u}{\partial x\,\partial y}=n\frac{\partial u}{\partial x}$$

or $$x\frac{\partial^2 u}{\partial x^2}+y\frac{\partial^2 u}{\partial x\,\partial y}=n\frac{\partial u}{\partial x}-\frac{\partial u}{\partial x}=(n-1)\frac{\partial u}{\partial x} \qquad ...(2)$$

Similarly, differentiating (1) partially w.r.t. 'y', we get

$$x\frac{\partial^2 u}{\partial x\,\partial y}+y\frac{\partial^2 u}{\partial y^2}=(n-1)\frac{\partial u}{\partial y}. \qquad ...(3)$$

Now multiplying (2) by x and (3) by y and adding, we get

$$x^2\frac{\partial^2 u}{\partial x^2}+2xy\frac{\partial^2 u}{\partial x\,\partial y}+y^2\frac{\partial^2 u}{\partial y^2}$$

$$=(n-1)\left[x\frac{\partial u}{\partial x}+y\frac{\partial u}{\partial y}\right]=(n-1)\,nu=n(n-1)u.$$

Example 34:

If $u=\sin^{-1}\left\{\left(\sqrt{x}-\sqrt{y}\right)/\left(\sqrt{x}+\sqrt{y}\right)\right\}$, *show that* $\dfrac{\partial u}{\partial x}=-\dfrac{y}{x}\dfrac{\partial u}{\partial y}.$

Solution:

We have $\sin u=\left(\sqrt{x}-\sqrt{y}\right)/\left(\sqrt{x}+\sqrt{y}\right)=v$, say. Then v is a homogeneous function of x and y of degree $\dfrac{1}{2}-\dfrac{1}{2}$ *i.e.*, 0.

Therefore by Euler's Theorem, we have $x\dfrac{\partial v}{\partial x}+y\dfrac{\partial v}{\partial y}=0$. v = 0....(1)

Now v = sin u.

$$\therefore \frac{\partial v}{\partial x} = \cos u \frac{\partial u}{\partial x}, \frac{\partial v}{\partial y} = \cos u \frac{\partial u}{\partial y}.$$

Putting these values in (1), we get

$$x \cos u \frac{\partial u}{\partial x} + y \cos u \frac{\partial u}{\partial y} = 0 \text{ or } \cos u \left(x \frac{\partial u}{\partial x} + y \frac{\partial u}{\partial y} \right) = 0$$

or $x \frac{\partial u}{\partial x} + y \frac{\partial u}{\partial y} = 0$, since $\cos u \neq 0$. Hence $\frac{\partial u}{\partial x} = -\frac{y}{x} \frac{\partial u}{\partial y}$.

Example 35:

Prove that $\frac{d^2 y}{dx^2} + \frac{2a^2 x^2}{y^5} = 0$, *where* $y^3 - 3ax^2 + x^3 = 0.$

Solution:

Let $f(x, y) \equiv y^3 - 3ax^2 + x^3 = 0.$

Then $p = \frac{\partial f}{\partial x} = -6ax + 3x^2, q = \frac{\partial f}{\partial y} = 3y^2,$

$$r = \frac{\partial^2 f}{\partial x^2} = -6a + 6x, s = \frac{\partial^2 f}{\partial x \partial y} = 0, t = \frac{\partial^2 f}{\partial y^2} = 6y.$$

Now $\frac{d^2 y}{dx^2} = -\frac{q^2 r - 2pqs + p^2 t}{q^3},$

$$= -\frac{(-6a + 6x)(3y^2)^2 + 6y(-6x + 3x^2)^2}{(3y^2)^3}$$

$$= -\frac{2(-a + x)y^3 + 2(-2ax + x^2)^2}{y^5}$$

$$= -\frac{2x(y^3 + x^3 - 3ax^2) - 2ay^3 + 8a^2x^2 - 2ax^3}{y^5}$$

$$= -\frac{-2a(y^3 + x^3) + 8a^2x^2}{y^5}, \qquad [\because y^3 + x^3 - 3ax^2 = 0]$$

$$= -\frac{-2a(3ax^2) + 8a^2x^2}{y^5} = -\frac{2a^2x^2}{y^5}$$

Hence $\frac{d^2 y}{dx^2} + \frac{2a^2x^2}{y^5} = 0.$

Example 36:

If $(\tan x)^y + y^{\cot x} = a$, *find* $\frac{dy}{dx}$.

Solution:

Let $f(x, y) = (\tan x)^y + y^{\cot x} - a$. Then

$$\left(\frac{\partial f}{\partial x}\right) = y\,(\tan x)^{y-1}.\sec^2 x + y^{\cot x}.\log y.(-\operatorname{cosec}^2 x),$$

and $$\left(\frac{\partial f}{\partial y}\right) = (\tan x)^y \log \tan x + (\cot x).y^{\cot x - 1}$$

Now we are given that $f(x, y) = 0$.

$$\therefore \quad \frac{dy}{dx} = -\frac{\partial f/\partial x}{\partial f/\partial y} = -\frac{y(\tan x)^{y-1}\sec^2 x - y^{\cot x}.\log y.\operatorname{cosec}^2 x}{(\tan x)^y \log \tan x + \cot x.y^{\cot x - 1}}.$$

Example 37:

If $x^3y^3 + 3x \sin y = e^y$, *find* $\frac{dy}{dx}$.

Solution:

Let $f(x, y) = x^3y^3 + 3x \sin y - e^x$. Then we have $f(x, y) = 0$.

$$\therefore \quad \frac{dy}{dx} = -\frac{\partial f/\partial x}{\partial f/\partial y} = -\frac{\left(3x^2y^3 + 3 \sin y\right)}{3x^3y^2 + 3x \cos y - e^y}.$$

Example 38:

If $x^y + y^x = a^b$, *find* $\frac{dy}{dx}$.

Solution:

Let $f(x, y) = x^y + y^x - a^b$. Then we have $f(x, y) = 0$.

$$\therefore \quad \frac{dy}{dx} = -\frac{\partial f/\partial x}{\partial f/\partial y} = \frac{yx^{y-1} + y^x \log y}{x^y \log x + xy^{x-1}}.$$

Example 39:

If $\sqrt{\left(1 - x^2\right)} + \sqrt{\left(1 - x^2\right)} = a(x - y)$, *prove that*

$$\frac{dy}{dx}=\frac{\sqrt{\left(1-y^2\right)}}{\sqrt{\left(1-x^2\right)}}.$$

Solution:

It is given that $\dfrac{\sqrt{\left(1-x^2\right)}+\sqrt{\left(1-y^2\right)}}{x-y}=a$.

Let $f(x, y) = \dfrac{\sqrt{(1-x^2)}+\sqrt{(1-y^2)}}{x-y} - a$. Then $f(x, y) = 0$.

$$\therefore \qquad \frac{dy}{dx}=-\frac{\partial f/\partial x}{\partial f/\partial y}$$

$$=-\frac{\dfrac{\left\{\frac{1}{2}\left(1-x^2\right)^{-1/2}.(-2x)\right\}(x-y)-1.\left\{\sqrt{(1-x)^2}+\sqrt{\left(1-y^2\right)}\right\}}{(x-y)^2}}{\dfrac{\left\{\frac{1}{2}(1-y)^{-1/2}.(-2y)\right\}(x-y)-(-1).\left\{\sqrt{\left(1-x^2\right)}+\sqrt{\left(1-y^2\right)}\right\}}{(x-y)^2}}$$

$$=-\frac{\dfrac{-x(x-y)}{\sqrt{\left(1-x^2\right)}}-\sqrt{\left(1-x^2\right)}-\sqrt{\left(1-y^2\right)}}{\dfrac{-y(x-y)}{\sqrt{\left(1-y^2\right)}}+\sqrt{\left(1-x^2\right)}+\sqrt{\left(1-y^2\right)}}$$

$$=-\frac{\sqrt{\left(1-y^2\right)}}{\sqrt{\left(1-x^2\right)}}.\frac{-x^2+xy-\left(1-x^2\right)-\sqrt{\left(1-x^2\right)}\sqrt{\left(1-y^2\right)}}{-yx+y^2+\sqrt{\left(1-x^2\right)}\sqrt{\left(1-y^2\right)}+1-y^2}$$

$$=-\frac{\sqrt{\left(1-y^2\right)}}{\sqrt{\left(1-x^2\right)}}.\frac{xy-1-\sqrt{\left(1-x^2\right)}\sqrt{\left(1-y^2\right)}}{-xy+1-\sqrt{\left(1-x^2\right)}\sqrt{\left(1-y^2\right)}}=\frac{\sqrt{(1-y^2)}}{\sqrt{(1-r^2}}$$

EXERCISES

1. If $u = e^{xyz}$ show that

$$\frac{\partial^2 u}{\partial x\,\partial y\,\partial z}=(1+3xyz+x^2y^2z^2)e^{xyz}.$$

2. If $u = (x^2 + y^2 + z^2)^{-3/2}$ show that $x\dfrac{\partial u}{\partial x} + y\dfrac{\partial u}{\partial y} + z\dfrac{\partial u}{\partial z} = -4$.

3. If $z = (x + y) + (x + y)\,\phi\left(\dfrac{y}{x}\right)$ prove that

$$x\left(\frac{\partial^2 z}{\partial x^2} - \frac{\partial^2 z}{\partial y\,\partial x}\right) = y\left(\frac{\partial^2 z}{\partial y^2} - \frac{\partial^2 z}{\partial x\,\partial y}\right).$$

4. If $u = \sin^{-1}\left(\dfrac{\sqrt{x} - \sqrt{y}}{\sqrt{x} + \sqrt{y}}\right)$ show that $x\dfrac{\partial u}{\partial x} + y\dfrac{\partial u}{\partial y} = 0$.

5. Verify Euler's Theorem for

$$z = \tan^{-1}\left(\frac{y}{x}\right).$$

6. If $u = At^{-1/2}\, e^{-x^2/4a^2t}$ prove that

$$\frac{\partial u}{\partial t} = a^2\frac{\partial^2 u}{\partial x^2}.$$

7. Verify $\dfrac{\partial^2 u}{\partial x\,\partial y} = \dfrac{\partial^2 u}{\partial y\,\partial x}$ for the function

$$u = x^2 \tan^{-1}\frac{y}{x} - y^2 \tan^{-1}\frac{x}{y}.$$

8. If $u = e^{xyz}$ show that

$$\frac{\partial^2 u}{\partial x\,\partial y\,\partial z} = (1 + 3xyz + x^2y^2z^2)e^{xyz}.$$

9. It $u = \sin^{-1}\dfrac{x^2 + y^2}{x + y}$ show that

$$x\frac{\partial u}{\partial x} + y\frac{\partial u}{\partial y} = \tan u.$$

10. If $z = \tan^{-1}\left(\dfrac{x^2 + y^2}{x + y}\right)$ find $\dfrac{\partial z}{\partial x} + \dfrac{\partial z}{\partial y}$ and $\dfrac{\partial^2 z}{\partial x\,\partial y}$.

11. Show that the function $z = \sin\left(\dfrac{y}{z}\right)\ \dfrac{\partial^2 z}{\partial y\,\partial x} = \dfrac{\partial^2 z}{\partial x\,\partial y}$.

12. If $z = x\left(\dfrac{y}{x}\right)$ prove that $x\dfrac{\partial z}{\partial x} + y\dfrac{\partial z}{\partial y} = z$.

6

Theoretical Distributions

INTRODUCTION

This distribution is based on the actual data of the experiment it is also called *Observed Frequency Distribution.* In many cases predictions can also be made on the basis of theoretical considerations that in our judgement are pertinent. We may feel that the data should be distributed in a certain way because of basic assumptions about the nature of forces acting on the example at hand. If our actually observed data do not conform to the values expected on the basis of our assumptions we shall have serious doubts about our assumptions. Such test of assumptions often leads to a *theoretical frequency distribution* also known as probability distribution. This .distribution is not based on actual experimental data but on certain theoretical considerations.

HYPERGEOMETRIC DISTRIBUTION

The hypergeometric distribution occupies a place of great significance in statistical theory. It applies to sampling without replacement from a finite population whose elements can be classified into two categories—one which possesses a certain characteristic and another which does not possess that characteristic. The categories could be—male, female, employed, unemployed, etc. When n random selections are made without replacement from the population, each subsequent draw is dependent and the probability of success changes in each draw. The following conditions characterise the hypergeometric distribution:

(1) The result of each draw can be classified into one of two categories;

(2) The probability of a success changes on each draw;

(3) Successive draws are dependent; and

(4) The drawing is repeated a fixed number of times

The hypergeometric distribution which gives the probability of r successes in a random sample of n elements drawn without replacement is:

$$p(r)\frac{\binom{n-X}{b-r}(X_r)}{\binom{N}{n}} \text{ for } r = 0, 1, 2, \ldots,[N\ X].$$

Both the binomial distribution and the hypergeometric distribution are concerned with the same thing, viz., number of successes in a sample containing n observations. What differentiates these two discrete probability distributions is the manner in which data are obtained. For the binomial model, the sample data are drawn with replacement from a finite population or without replacement from an infinite population. On the other hand, for the hypergeometric model, the sample data are drawn without replacement from a finite population. Hence, while the probability of success p is constant over all observations of a binomial experiment and the outcome of any particular hypergeometric experiment; here the outcome of one observation is affected by the outcomes of the previous observations.

The symbol [n, X] means the smaller of n or X. The hypergeometric distribution bears a very interesting relationship to the binomial distribution. When N increases without limit, the hypergeometric distribution approaches the binomial distribution. Hence, the binomial probabilities may be used as approximation to hypergeometric probabilities where σ/N is small. A frequently used rule of thumb is that the population size should be at least ten times the sample size ($N > 10n$) for the approximation to be used.

Example:

A bag contains 20 balls of which 15 are of red colour and 5 of black colour. A random sample (without replacement) of 5 balls is taken. Find the probability that the sample contains 2 black balls.

Solution:

The problem can be solved with the help of hypergeometric probability model. Here N = 20, n = 5, r = 2. Hence the required probability is given by:

$$\frac{5_{C_2} \cdot 15_{C_3}}{39_{C}} = 0.29$$

NORMAL DISTRIBUTION

The binomial and the Poisson distributions described above are the most useful theoretical distributions for discrete variables, i.e., they relate to the occurrence of distinct events. In order to have mathematical distribution suitable for dealing with quantities whose magnitude is continuously variable, a continuous distribution is needed. The *normal distribution,* also called the *normal probability distribution* happens to be most useful theoretical distribution for continuous variables. Many statistical data concerning business and economic problems are displayed in the form of normal distribution. In fact normal distribution is the *cornerstone* of modern statistics.

The normal distribution is an *approximation* to binomial distribution. Whether or not p is equal to q, the binomial distribution tends to the form of the continuous curve and when n becomes large at least for the material part of the range. As a matter of fact, the correspondence between the binomial and the curve is surprisingly close even for comparatively low values of n, provided that p and q are fairly near equality. The limiting frequency curve obtained as n becomes large is called the normal frequency curve or simply the normal curve.

The normal curve is represented in several forms. The following is the basic form relating to the curve with mean μ and standard deviation σ.

The Normal Distribution

$$P(X) = \frac{1}{\sigma\sqrt{2\pi}} e^{\frac{-(x-\mu)}{2\sigma^2}}$$

X = Values of the continuous random variable

μ = Mean of the normal random variable

e = Mathematical constant approximated by 2.7183

π = Mathematical constant approximated by 3.1416

$$\left(\sqrt{2\pi} = 2.5066\right)$$

When we say that the curve has unit area, we mean that the total frequency N is equated to i for convenience in representation and calculation. To obtain ordinates for a particular distribution the ordinates given by the above formula are multiplied by N. The equation to a normal curve corresponding to a particular distribution is thus given by

$$y = \frac{N}{\sigma\sqrt{2\pi}} e^{-x^2/2\sigma^2}$$

The quantity $\frac{N}{\sigma\sqrt{2\pi}}$ in the above formula is equal to the maximum ordinate (y_0) of the normal curve corresponding to distribution of stated total frequency N and stated standard deviation σ.

FITTING A BINOMIAL DISTRIBUTION

When a binomial distribution is to be fitted to observe data, the following procedure is adopted :

(1) Determine the values of p and q. If one of these values is known the other can be found out by the simple relationship p = (1 – q), and q = (1 – p). When p and p are equal the distribution is symmetrical, for p and q may be interchanged without alternating the value of any terms, and consequently terms equidistant from the two ends of the series are equal. If p and q are unequal, the distribution is skew. If p is less than 1/2, the distribution is positively skewed and when p is more than 1/2 the distribution is negatively skewed.

(2) Expand the binomial $(q + p)^n$. The power n is equal to one less than the number of terms in the expanded binomial. Thus when two coins are tossed (n = 2) there will be three terms in the binomial, Similarly, when four coins are tossed (n = 4) there will be five terms, and so on.

(3) Multiply each term of the expanded binomial by N (the total frequency), in order to obtain the expected frequency in each category.

POISSON DISTRIBUTION

Poisson distribution is a discrete probability distribution and is very widely used in statistical work. It was developed by a French mathematician. Simeon Denis Poisson (1781-1840), in 1837. Poisson distribution may be expected in cases where the chance of any individual event being a success is small. The distribution is used to describe the behaviour of rare events such as the number of accidents on road, number of printing mistakes in a book, etc., and has been called "the law of improbable events". In recent years the statisticians have had a renewed interest in the occurrence of comparatively rare events, such as serious floods, accidental release of radiation from a nuclear rector, and the like.

The Poisson distribution is defined as :

The Poisson Distribution

$$P(r) = \frac{e^{-m} m^r}{r!}$$

where r = 0, 1, 2, 3, 4,.......

e = 2.7183 (the base of natural Logarithms)

m = the mean of the Poisson distribution, i.e., np or the average number of occurrences of an event.

All Poisson probability distributions are skewed to the right. This is the reason why the Poisson probability distribution has been called the probability distribution of rare events (the probabilities tend to be high for small numbers of occurrences).

The Poisson probability Distribution is concerned with certain processes that can be described by a discrete random variable. The probabilities of 0. 1, 2,.... successes are given by the successive terms of the expansion.

= m.

Thus σ^2 or μ_2 = m and $\sigma = \sqrt{m}$.

In a similar manner we can show that for Poisson distribution $\mu_2 = m$ and $\mu_4 = m + 3m^2$.

Constants of Poisson Distribution

$$\mu_1 = 0,\ \mu_2 = m,\ \mu_3 = m,\ \mu_4 = m + 3m^2$$

$$\beta_1 = \frac{\mu_3^2}{\mu_2^3} = \frac{m^2}{m^3} = \frac{1}{m};\quad \beta_2 = \frac{\mu_4}{\mu_2^2} = \frac{m + 3m^2}{m^2} = 3 + \frac{1}{m}$$

One great advantage of the Poisson distribution is that we need only the value of mean in order to compute the values of various constants. This shall be clear from the following illustration.

ROLE OF THE POISSON DISTRIBUTION

The Poisson distribution is used in practice in a wide variety of problems where there are infrequently occurring events with respect to time, area, volume or similar units. Some practical situations in which Poisson distribution can be used are given below :

(1) In waiting-time problems to count the number of incoming telephone calls or incoming customers.

(2) Number of traffic arrivals such as trucks at terminals, aeroplanes at airports, ships at docks and so forth.

(3) It is used in quality control statistics to count the number of defects of an item.

(4) In biology to count the number of bacteria.

(5) In physics to count the number of particles emitted from a radioactive substance.

(6) In insurance problems to count the number of casualties.

(7) To model the distribution of the number of persons joining a queue (a line) to receive a service or purchase of a product.

(8) In problems dealing with the inspection of manufactured products with the probability that any one piece is defective is very small and the lots are very large, and

(9) The number of typographical errors per page in typed material, number of deaths as a result of road accidents, etc.,

(10) In determining the number of deaths in a district in a given period, say, a year, by a rare disease.

In general, the Poisson distribution explains the behaviour of those discrete variates where the probability of occurrence of the event is small and the total number of possible cases is sufficiently large.

Example:

Suppose that manufactured product has 2 defects per unit of product inspected. Using Poisson distribution, calculate the probabilities of finding a product without any defect, 3 defects and 4 defects. (Given $e^{-2} = 0.135$)

Solution:

Average number of defects of m = 2

$$P(r) = p(r) = \frac{e^{-m}\, m^{r}}{r!},\ r = 0, 1,2,.....$$

$P(0) = e^{-2} = 0.135$ (given)^x

$P(1) = (P_0) \times m = .135 \times 2 = .27$

$P(2) = (P_1) \times m/2 = .27 \times 2/2 = .27$

$P(3) = (P_2) \times m/3 = .27 \times 2/3 = .18$

$P(4) = (P_3) \times m/4 = .18 \times .5 = .09$

Hence the probability that a product has no defect is 0.135, product has 3 defects is 0.18 and product has 4 defects is 0.09.

Poisson Distribution as an Approximation of the Binomial Distribution

The Poisson distribution can be a reasonable approximation of the binomial under certain conditions like:

(i) *number of trials, i.e., n is indefinitely large, i.e. $n \to \infty$.*

(ii) *p, i.e., the probability of success for each trial is indenfinitely small,*

(iii) *np = = m (say) is finite.*

The rule most often used by statisticians is that the Poisson is a good approximation of the binomial when n is equal to or greater than 20 and p is equal to or less than .05.

If above conditions hold good, we can substitute the mean of the binomial distribution (np) in place of the mean of the Poisson distribution (m), so that the formula becomes:

$$p(r) = \frac{e^{-np}(np)^r}{r!}$$

Proof:

In case of binomial distribution the probability of r successes is given by

$$p^{(r)} = {}^nC_r\, q^{n-r} p^r = \frac{n(n-1)\ldots\ldots(n-r+1)}{r!}\, p^r q^{n-r}$$

Put $p = \frac{m}{n}$ $\qquad \therefore q = 1 - p = 1 - \frac{m}{n}$

We now get $\qquad p(r) = \frac{n(n-1\ldots\ldots n-r+1)}{r!}\left(\frac{m}{n}\right)^r \times \left(1-\frac{m}{n}\right)^{n-r}$

$$= \frac{1\left(1-\frac{1}{n}\right)\left(1-\frac{2}{n}\right)\ldots\ldots\left(1-\frac{r-1}{n}\right) m^r}{r!} \left\{\frac{\left(1-\frac{m}{n}\right)^n}{\left(1-\frac{m}{n}\right)^r}\right\}$$

For fixed r as $n \to \infty$

$\left(1-\frac{1}{n}\right)\ldots\left(1-\frac{r-1}{n}\right)\left(1-\frac{m}{n}\right)^r$ all tend to 1 and $\left(1-\frac{m}{n}\right)^n$ to $e^{-m}m^r$.

Hence in the limiting case $p^{(r)} = \frac{e^{-m}\, m^r}{r!}$

Needless to point out that use of Poisson as an approximation to binomial probability distribution if the conditions given above are satisfied can simplify calculation work and save time with more or less similar results as one would expect from binomial distribution.

RELATION BETWEEN BINOMIAL, POISSON AND NORMAL DISTRIBUTIONS

The three distributions, namely, Binomial, Poisson and Normal, are very closely related to each other. As explained earlier when N is large while the probability P of the occurrence of an event is close to zero so that $q = (1 - p)$ the binomial distribution is very closely approximated by the Poisson distribution with $m = np$.

Since there is a relation between the binomial and normal distributions, it follows that there is also relation between the Poisson and normal distributions.

In fact it can be proved that the Poisson distribution approaches a normal distribution with standardized variable $\frac{(x-m)}{\sqrt{m}}$ as m increases to infinity.

PROPERTIES OF THE NORMAL DISTRIBUTION

The following the important properties of the normal curve and the normal distribution.

(1) The points of inflexion, i.e., the points where the change in curvature occurs are $\overline{X} \pm \sigma$.

(2) Since there is only one maximum point, the normal curve is unimodal, i.e., it has only one mode.

(3) There is one maximum point of the normal curve which occurs at the mean. The height of the curve declines as we go in either direction from the mean. The curve approaches nearer and nearer to the base but it never touches it, i.e., the curve is asymptotic to the base on either side. Hence its range is unlimited or infinite in both directions.

(4) The height of the normal curve is at its maximum at the mean. Hence the mean and mode of the normal distribution coincide. Thus for a normal distribution mean, median and mode are all equal.

(5) The normal curve is "bell-shaped" and symmetrical in its appearance. If the curves were folded along its vertical axis, the two halves would coincide. The number of cases below the mean in a normal distribution is equal to the number of cases above the mean, which make the mean and median concide. The height of the curve for a positive deviation of 3 units is the same as the height of the curve for negative deviation of 3 units.

(6) The area under the normal curve distributed as follows:

(a) Mean $\pm$ 1 σ covers 68.27% area; 34.135% area will lie on either side of the mean.

(b) Mean $\pm$ 2 σ covers 95.45% area.

(c) Mean $\pm$ 3 σ covers 99.73% area.

The following table shows the area of the normal curve between mean ordinate and ordinates at various sigma distances from the mean as percentage of the total area.

Thus the two ordinates at distance 1.96 σ from the mean on either side would enclose 47.5 + 47.5 = 95% of the total area, and two ordinates at 2.5758 σ distance from the mean on either side would enclose 49.5 + 49.5 = 99% of the total area. The area enclosed between ordinates at 3 σ distance from the mean on either side would be 49.865 + 49.865 = 99.73% of the total area.

The various hypotheses are generally tested either at 5% level or at 1% level (i.e., taking into account 95% and 99% of the total area of the normal curve).

Area Relationship

Distance from the mean Ordinate	Percentage of Total Area
0.5 σ	19.146
0.0 σ	34.134
0.5 σ	43.319
1.96 σ	47.500
2.0 σ	47.725
2.5 σ	49.379
2.5758 σ	49.500
3.0 σ	49.865

Conditions for Normality

This following four conditions must prevail among the factors affecting the individual events that make up a given population, if the distribution of observations is to be normal:

(1) The causal forces must be numerous and of approximately equal weight.

(2) These forces must be the same over the universe from which the observations are drawn (although their incidence ill vary from event to event). This is the condition of homogeneity.

(3) The forces affecting events must be independent of one another.

(4) The operation of the causal forces must be such that deviation above the population mean are balanced as to magnitude and number by deviations below the mean. This is the condition of symmetry.

Constants of the Normal Distribution

The mean of the normal distribution is $\overline{X}$

The standard deviation of the normal distribution is σ

$$\mu_2 = \sigma^2,\ \mu^3 = 0 \text{ and } \mu_4 = 3\sigma^4$$

β_1 or moment coefficient of skewness

$$\beta_1 = \frac{\mu_3^2}{\mu_2^3} = 0$$

β_2 or moment coefficient of kurtosis

$$\beta_2 = \frac{\mu_4}{\mu_2^2} = \frac{3\sigma^4}{\sigma^4} = 3$$

For a normal distribution, the value of β_2 shall be 3. If the value of β_2 is more than 3, the curve is Leptokurtic and if the value of β_2 is less than three, the curve is platykurtic.

THE MULTINOMIAL DISTRIBUTION

An important generalization of the binomial distribution arises where there are more than two possible outcomes for each trial, the probabilities of the various outcomes remain in same for each trial, and the trials are independent. Whereas in case of binomial distribution, there are only two possible outcomes on each experimental trial, in the multinomial distribution there are more than two possible outcomes on each trial. For example, when a symmetric six-side dice is rolled there are six possible outcomes. 1, 2, 3, 4, 5 or 6. The assumptions underlying the multinomial distribution are analogous to the binomial distribution. These are :

(1) An experiment is performed under the same conditions for a fixed number of trials, say, n.

(2) There are K mutually exclusive and exhaustive outcomes of the experiment which may be referred to E_1, E_2, E_3.... E_k. Thus the sample space of possible outcomes on each trial shall be :

$$S = \{E_1 , E_2 , E_3 ,...... E_k\}$$

(3) The respective probabilities of the various outcomes, i.e., E_1 , E_2 , E_3 .....E_k denoted by p_1 , p_2 , p_3, ... p_k respectively remain constant from trial to trial.

$$p_1 + p_2 + p_3 +.....p_k = 1$$

(4) The trials are independent.

Under the above assumptions, the probability that there will be r_1 occurrence of E_1, r_2 occurrence of E_2 , ... r_k occurrence of E_k in n trials is given by :

$$P\left(r_1 , r_2 ,.....r_k\right) = \frac{n\,!}{r_1\,!\; r_2\,!\; r_k\,!}\, p_1^{r1}\, p_2^{r2} .....p_k^{r3}$$

Where $r_1 + r_2 +....+ r_k = n$ and $p_1 = p_2 +...... p_k = 1$.

The general form of the multinomial distribution is

$$(p_1 + p_2 +.....p_k)^n$$

Example 1:

A dice is role five times. Find the probability of getting a1, a2 and three other numbers.

Solution:

The multinomial mode is appropriate for this problem. Applying the formula :

$$P(r_1, r_2, \ldots\ldots r_k) = \frac{n}{r_1!\, r_2!\, r_k!} p_1^{r1} p_2^{r2} \ldots\ldots p_k^{kr}$$

$r_1 = 1, r_2 = 1, r_3 = 3$. Hence, the probability of obtaining a1, a2 and three other number is :

$$p = \frac{5!}{1!\,1!\,3!}\left(\frac{1}{6}\right)\left(\frac{1}{6}\right)\left(\frac{4}{6}\right)^2 = \frac{40}{243} = 0.165.$$

Example 2:

A community consists of 50 percent Hindus, 30 percent Muslims and 20 percent Sikhs. If a sample of six individuals is selected at random, what is the probability that two are Hindus, three are Muslims and one is a Sikh?

Solution:

The answer is provided by the multinomial distribution. The probabilities of persons of these races are respectively 0.50, 0.30 and 0.20. Hence the required probability shall be given by :

$$p = \frac{6!}{2!\,3!\,1!}(0.5)^2(0.3)^2(0.2) = 0.081.$$

NEGATIVE BINOMIAL DISTRIBUTION

The negative binomial distribution is very much similar to the binomial probability model. It is applicable when the following conditions hold good:

1. An experiment is performed under the same conditions till a fixed number of successes, say C_2 is achieved.
2. The result of each experiment can be classified into one of the two categories, success or failure.
3. The probability p of success is the same for each experiment.
4. Each experiment is independent of all the others.

Consider a sequence of Bernoulli trials with p as the probability of success. In the sequences success and failure will occur randomly and in each trial the probability of success will be p and that of a failure will be–p=q. Let us investigate how much time will be taken to reach the rth success. Here r is fixed, let the number of failures preceding the rth success be x (= 0, 1, 2,...). Then the total number of trials to be performed to reach the rth success will be x + r. (Here x is a chance variable.) Then the probability that rth success occurs at (x + r)th trial equals the probability that exactly x failures (r + x – 1) trials there are (r – 1) successes and the (r + x)th trial is a success; the corresponding probabilities are $\binom{hr+x-1}{r-1} p^{r-1}$ 1^x and p respectively. The probability of the compound event will be :

$$\binom{r+x-1}{r-1} p^{r-1} q^x . p = \binom{r+x-1}{r-1} p^r q^x \qquad ...(i)$$

Since the events are independent.

The probabilities are generally denoted by f (x, r, p). The probabilities given by (i) are known as 'negative binomial probabilities' and the law is known as the negative binomial probability law.

Definition : A discrete random variable X is said to follow 'negative binomial distribution' if it takes only non-negative values and its probability mass function is given by

$$f(x, r, p) = \binom{r+x-1}{r-1} p^r q^x ; \; x = 0, 1, 2, ...$$

$$= 0 \; ; \text{ otherwise} \qquad ...(ii)$$

The fixed quantities r and p are the two paramcters of the distribution.

It can be easily provided that Σ f (x, r, p) summation extends over all the possible values of x, is equal to unity therefore (2) is a probability distribution.

By simple calculations, it can be shown that

$$\binom{r+x-1}{r-1} = (-1)^x \binom{-r}{x}$$

Then (2) can be rewritten as

$$f(x, r, p) = \binom{-r}{x} p^r (-q)^x, \; x = 0, 1, 2, ...$$

Which is the (x + 1)th term in the expansion of $p^r (1 - q)^{-r}$ and it is the binomial theorem with negative index. Therefore the terms given by (2) are the successive terms of the expansion of $p^r (1 - q)^{-r}$, hence the name 'Negative Binomial Distribution'.

Example:

Suppose that 30% of the items taken from the end of a production line are defective. If the items taken from the line are checked until 6 defective items are found, what is the probability that 12 items are examined ?

Solution:

Suppose the occurrence of a defective item is a success. Then our question reduces to find the probability that there will be $(12 - 6) = 6$ failures preceding the 6th success, the probability of success in a trial being 0.30. Then by negative binomial probability law, the required probability will be given by :

$$\binom{6+6-1}{6-1}(.30)^6(.70)^6 = 0.396$$

BINOMIAL DISTRIBUTION

Consider a situation in which a person trie? ten times to catch the fish from a pond. The outcome of each trial can be referred to as a *success* if the fish is caught and *failure* if the fish is not caught. If the probability of catching the fish is assumed to be same for each trial, and if the chance of fish being caught is not affected by the previous trials, we have an experiment which is called *Binomial experiment.*

This distribution has been used to describe a wide variety of processes in business and the social sciences as well as other areas. The type of process which gives rise to this distribution is usually referred to as Bernoulli process is series of experimental trials. These assumptions are:

1. An experiment is performed under the same conditions for a fixed number of trials, say, n.
2. In each trial, there are only two possible outcomes of the experiment. For lack of a better nomenclature they are called "success" or "failure". Stated in somewhat different language, the sample space of possible outcomes on each experimental trial is :

   S = {failure, success}
3. The probability of a success denoted by p remains constant from trial to trial. The probability of a failure denoted by q is equal to (q – p). If the probability of success is not the same in each trial, we will not have binomial distribution. For example, if 5 balls are drawn at random from an urn containing 10 white and 20 red balls, this is a binomial experiment if each ball is replaced before another is drawn. If the balls are drawn without replacement, the probability of drawing white ball changes each time a ball is taken from the urn and we no longer have a binomial experiment.
4. The trials are statistically independent, i.e., the outcomes of any trial or sequence of trials do not affect the outcomes of subsequent trials.

This model is useful to answer questions such as this: If we conduct an experiment n times under the stated conditions, what is the probability of obtaining exactly x successes? More specifically suppose 10 coins are tossed together, what is the probability of obtaining exactly two heads?

If a coin is tossed once there are two outcomes, namely, tall or head. The probability of obtaining a head or p = 1/2 and the probability of obtaining a tail or q = 1/2. Thus (q + p) = 1. These are terms of the binomial (q+p).

Similarly, if two coins are tossed simultaneously there are four possible outcomes:

AB	AB	AB	AB
TT	TH	HT	HH

Since by expanding the binomial $(q + p)^n$ we obtain probability of 0, 1, 2... n heads, the probability distribution is naturally called the *Binomial Probability Distribution* or simply the *Binomial Distribution.* The general form of the distribution is :

The Binomial Distribution

$P(r) = {}^nC_r q^{n-r} p^r$

where p = *Probability of success in a single trial*

$q = 1 - p$

n = *Number of trials*

r = *Number of successes in n trials*

If we want to obtain the probable frequencies of the various outcomes in N sets on n trials, the following expression shall be used:

$N(q + p)^n$

$N(q + p)^n = Nq^n + {}^nC_1\, q^{n-1} p + {}^nC_2 q^{n-2} p^2 + \ldots + {}^nC_r q^{n-r} p^r + \ldots + p^n$

The frequencies obtained by the above expansion are known as expected of theoretical frequencies. On the other hand, the frequencies actually obtained by making experiments are called actual or observed frequencies. Generally, there is some difference between the observed and expected frequencies but the difference becomes smaller and smaller as N increases.

Obtaining Coefficients of the Binomial

For obtaining coefficients from the binomial expansion, the following rules may be remembered. To find the terms of the expansion of $(q + p)^n$

1. The first term is q^n.
2. The second term is ${}^nC_1 q^{n-1} p$.
3. In each succeeding term the power of q is reduced by 1 and the power of p is increased by 1.

4. The coefficient of any term is found by multiplying the coefficient of the proceeding term by the power of q in that preceding term, and dividing the products so obtained by one more than the power of p in that preceding term.

When we expand $(q + p)^n$. We get

$$(q + p)^n = q^n + {}^nC_1\, q^{n-1}\, p + {}^nC_2\, q^{n-2}\, p^2 + ... + {}^nC_r\, q^{n-r}\, p^r + ...p^n.$$

where 1. ${}^nC_2$ ..... are called the binomial coefficients. Thus, in the expansion of $(q + p)^5$ we will have

$$(q + p)^5 = q^5 + 5q^4p + 10q^3\, p^2 + 10q^2p^3 + 5qp^4 + p^5$$

and the coefficients will be 1, 5, 10, 10, 5, 1.

From the above binomial expansion, the following general relationship should be noted :

(1) The number of terms in a binomial expansion is always n + 1.

(2) The exponents of p and q, for any single term, when added together, always sum to n.

(3) The exponents of q are n, (n – 1), (n – 2)....1, 0, respectively and the exponents of p are 0, 1, 2..(n – 1). n respectively (note : $p^0 = 1 : q^0 = 1$).

(4) The coefficients for the n + 1 terms of the distribution are always symmetrical ascending to the middle of the series and then descending, when n is odd number, n + 1 is even and the coefficients of the two central terms are identical.

PROPERTIES OF THE BINOMIAL DISTRIBUTION

1. The mode of the binomial distribution is equal to the value of x which has the largest probability. For example, if n = 6 and p = 0.3, the mode is equal to 2. While for n = 6 and p = 0.9 the mode is equal to 6. The mean and mode are equal if np is an integer. For example, when n = 6 and p = 0.50, the mean and mode are both equal to 3. For fixed n, both mean and mode increase as p increases.
2. If n is large and if neither p nor q is too close to zero, the binomial distribution can be closely approximated by a normal distribution with standardized variable given by $z = \frac{X - np}{\sqrt{npq}}$. The approximation becomes better with increasing n.
3. The shape and location of binomial distribution changes as p changes for a given n or as n changes for a given p. As p increases for a fixed n, the binomial distribution shifts to the right.

4. As n increases for a fixed p, the binomial distribution moves to the right, flattens, and spreads out. The mean of the binomial distribution, np obviously increases as n increases with p held constant. For larger n there are more possible outcomes of a binomial experiment and the probability associated with any particular outcome becomes smaller.

Constants of the Binomial Distribution

The mean of the binomial distribution is np and standard deviation $\sqrt{npq}$.

Proof:

If p is the probability of success and p the probability of failure in one trial then in n independent trials the probabilities of 0, 1, 2, 3,....n successes are given by the 1st, 2nd 3rd.......n + 1th term of the binomial expansion $(q + p)^n$. Thus we have :

x	p(x)	xp(x)
0	q^n	$0 \times q^n$
1	$^nC_1\, q^{n-1}\, p$	$1 \times nq^{n-1}\, p$
2	$^nC_2\, q^{n-2}\, p^2$	$\frac{2n(n-1)}{2\times 1} q^{n-2}\, p^2$
:	:	:
:	:	:
:	:	:
n	p^n	np^n
	$\Sigma\, p(x) = 1$	

The arithmetic mean $= \dfrac{\Sigma x.p(x)}{\Sigma p(x)}$

$$\Sigma x.p(x) = 0.q^n + nq^{n-1} p + \frac{2n(n-1)}{2} q^{n-2} p^2 + ... np^n$$

$= nq^{n-1} p + n(n-1) q^{n-2} p^2 + ...np^n.$

Taking np common we have

$$= np\left[q^{n-1} + (n-1)\, q^{n-2}\, p\right] + \frac{n(n-1)\,(n-2)}{2!} q^{n-3} + ... + p^{n-1}$$

$= np\,(q + p)^{n-1}$

[since the expansion in brackets is the expansion of the binomial $(q + p)^{n-1}$]

$= np\ (1)^{n-1} = np$ $\qquad (\because q + p = 1)$

Thus $\Sigma\ x\ .\ p(x) = np$ $\qquad (\because$ the sum probabilities $= 1)$

Thus the mean of binomial distribution is np.

The standard deviation of binomial distribution is $\sqrt{npq}$.

Proof:

σ^2 or $\mu_2 = \nu_2 - \nu^2_1$ (where ν_1 and ν_2 are moments about origin, zero)

$\nu_2 = \Sigma\ (x^2\ .\ p\ (x))$

$\nu_1 = np$

$$\Sigma\ \{x^2\ .\ p\ (x)\} = 0^2\ .\ q^n + 1^2\ nq^{n-1}\ p + \frac{2^2\ .n(n-1)}{2\,!}\ q^{n-2}\ p^2$$

$$+ \frac{3^2 n\ (n-1)\ (n-2)}{3\,!}\ q^{n-2}\ p^3 + \ldots\ldots + n^2\ p^n$$

$$= nq^{n-1} + 2.n\left(n-1\right) q^{n-2} p^2 + \frac{3n\left(n-1\right)\left(n-2\right)}{2 \times 1}\ q^{n-3}\ p^3 + \ldots\ldots + n^2 p^n$$

$$= np\ q^{n-1} + 2\left(n-1\right) q^{n-2} p + \frac{3\left(n-1\right)\left(n-2\right)}{2 \times 1}\ q^{n-3}\ p^2 + \ldots\ldots + p^{n-1}$$

Breaking second, third and following terms into pars, we get

$$\Sigma x^2 . p(x) = p\left[\left(q^{n-1} + (n-1) q^{n-2}\ p + \frac{(n-1)(n-2)}{2 \times 1} q^{n-2}\ p^2 + \ldots\ldots + p^{n-1}\right)\right]$$

$$+ \{(n-1)\ q^{n-2} + \frac{(n-1)\ (n-2)}{2 \times 1} q^{n-3}\ p^2 \ldots. + (n-1)\ p^{n-1}\}$$

$= np\ (q + p)^{n-1} + (n - 1)\ p\ \{q^{n-2} + (n - 2)\ q^{n-2}\ p + \ldots\ldots + p^{n-1}\}$

$= np\ [1 + (n - 1)\ p\ (q + p)^{n-2}]$

[$\therefore$ $(q + p)^{n-1} = 1$ and the expression to the right is expansion of $(q + p)^{n-2} = 1$]

$= np\ [1 + (n - 1)\ p.1]$

$= np\ [1 + np - p]$

$= np + n^2\ p^2 - np^2$

$\mu_2 = \nu_2 - \nu^2_1.$

$= np + n^2p^2 - np^2 - (np)^2$

$= np + n^2p^2 - np^2 - n^2p^2.$

$= np - np^2$

$= np\ (1 - p)$

$= npq \qquad [\because (1 - p) = q)$

$\therefore \sigma$ or $\sqrt{\mu_2} = \sqrt{npq}$

Thus the standard deviations of binomial distribution is $\sqrt{npq}$ or the variance is npq.

Similarly,

$\mu_3 = npq\ (q - p)$, and

$\mu_4 = 3n^2\ p^2\ q^2 + npq\ (1 - 6pq)$

From the various moments the value of β_1 and β_2 can also be computed as

$$\beta_1 = \frac{\mu_3^2}{\mu_2^3} = \frac{n^2p^2q^2(q-p)^2}{n^3p^2q^3} = \frac{(q-p)^2}{npq}$$

If β_1 = the distribution is symmetrical, i.e., there is no skewness. Prof. Fisher gave the following measure of skewness :

$$\gamma_1 = \sqrt{\beta_1}$$

If $\gamma_1 = 0$ the distribution is symmetrical, if γ_1 is more than zero the distribution is positively skewed and if it is less than zero the distribution is negatively skewed. It should be noted that γ_1 is a better measure of skewness compared to β_1 because β_1 will always be positive whereas γ_1 can be both positive as well as negative.

$$\beta_2 = \frac{\mu_4}{\mu_2^2} = \frac{3n^2p^2 + npq\,(1-6pq)}{n^2p^2q^2} = 3 + \frac{1-6pq}{npq}$$

Prof. Fisher gave the following measure of kurtosis :

$$\gamma_2 = \beta_2 - 3$$

If the distribution is normal, γ_1 would be zero and γ_2 would also be zero. But the converse is not true, i.e., even if both γ_1 and γ_2 are zero the distribution is not necessarily normal. If γ_2 is positive, the distribution is leptokuritic and if γ_2 is negative the distribution is platykuritic.

The various constants of the binomial distribution can be listed in the following table :

Constants of Binomial Distribution

Mean = np

Standard Deviation $= \sqrt{npq}$

First Moment or $\mu_1 = 0$

Second Moment or $\mu_2 = npq$

Third Moment or $\mu_3 = npq\ (q - p)$

Fourth Moment or $\mu_4 = 3n^2p^2q^2 + npq\ (1 - 6pq)$

$$\beta_1 = \frac{(q-p)^2}{npq} \qquad \beta_2 = 3 + \frac{1 - 6\,pq}{npq}$$

Importance of the Binomial Distribution

1. The outcome or results of each trial in the process are characterised as one of two types of possible outcomes. In other words, they are attributes.
2. The possibility of outcome of any trial does not change and is independent of the results of previous trials.

The following examples will illustrate the applications of binomial distribution.

SOLVED EXAMPLES

Example 1:

A coin is tossed six times. What is the probability of obtaining four or more heads ?

Solution:

When a coin is tossed the probabilities of head and tail in case of an unbiased coin are equal, i.e. $p = q = \frac{1}{2}$.

The various possibilities for all the events are the terms of the expansion $(q - p)^6$.

$$(q + p)^6 = q^6 + 6q^5p + 15q^4p^2 + 20q^3p^2 + 15q^2p^4 + 6qp^5 + p^6.$$

$\therefore$ The probability of obtaining 4 heads is

$$15q^2p^4 = 15 \times \left(\frac{1}{2}\right)^2 \left(\frac{1}{2}\right)^2 = 0.234$$

The probability of obtaining 5 heads is

$$6qp^5 = 6\left(\frac{1}{2}\right)\left(\frac{1}{2}\right)^5 = 0.094$$

The probability of obtaining 6 heads is

$$(p)^6 = \left(\frac{1}{2}\right)^6 = 0.016$$

$\therefore$ The probability of obtaining 4 more heads is

$0.234 + 0.094 + 0.016 = 0.344$

Example 2:

Assuming that half the population is vegetarian so that the chance of an individual being a vegetarian is 1/2 and assuming that 100 investigators can take sample of 10 individuals to see whether they are vegetarians, how many investigators would you expect to report that three people or less were vegetarians ?

Solution:

We are given :

n = 10, p, i.e., probability of an individual being vegetarian

$$= \frac{1}{2}\ q = 1 - p = \frac{1}{2}$$

Using binomial distribution, we have

$$P(r) = {}^nC_r\, q^{n-r}\, p^r.$$

Putting the various values, we have

$$10C_r\left(\frac{1}{2}\right)^r\left(\frac{1}{2}\right)^{10-r} = 10C_r = \left(\frac{1}{2}\right)^{1C} = \frac{1}{1024}\,10C_r$$

The probability that in a sample of 10, three or less people are vegetarian shall be given by :

$$P(0) + P(1) + P(2) + P(3)$$

$$= \frac{1}{1024}\left[{}^{10}C_0 + {}^{10}C_1 + {}^{10}C_2 + {}^{10}C_3\right]$$

$$= \frac{1}{1024}\left[1 + 10 + 45 + 120\right] = \frac{176}{1024} = \frac{11}{64}$$

Hence out of 100 investigators, the number of investigators who will report 3 or less vegetarians in a sample of 10 is $100 \times \frac{1}{64} = 17.2 = 17.$

Example 3:

The incidence of a certain disease is such that on the average 20% of workers suffer from it. If 10 workers are selected at random, find the probability that

(i) Exactly 2 workers suffer from the disease,

(ii) not more than 2 workers suffer from the disease.

Calculate the probability upto fourth decimal place.

Solution:

Probability that a worker suffers from a disease $= \frac{20}{100} = \frac{1}{5}$ i.e. $p = \frac{1}{5}$ and $q = \frac{4}{5}$.

By binomial probability law, the probability that out of 10 workers, x workers suffer from a disease is given by :

$$P(r) = {}^nC_r\, q^{n-r}\, p^r = 10C_r \left(\frac{4}{5}\right)^{10-r} \left(\frac{1}{5}\right)^r$$

$$= 10C_r \frac{4^{10-r}}{5^{10}};\ r = 0,1,2,......10.$$

(i) The required probability that exactly 2 workers will suffer from the disease is given by :

$$P_{(2)} = 10C_r \frac{4^{10-2}}{5^{10}} = \frac{45 \times 4^8}{5^{10}} = 0.302$$

(ii) The required probability that not more than 2 workers will suffer from the disease is given by :

$$P(0) + P(1) + P(2) = \frac{1}{5^{10}} [10\ C_0\ 4^{10} + 10C_1\ 4^9 + 10C_2\ 4^8] = 0.678$$

Example 4:

If the probability of defective bolts is 0.1, find (a) the mean and standard deviation for the distribution of defective bolts in a total of 500, and (b) the moment coefficient of skewness and kurtosis of the distribution.

Solution:

(a) $p = 0.1, \qquad n = 500$

Mean = np = 500 × .1 = 50.

Thus we can expect 50 bolts to be defective.

$$\sigma = \sqrt{npq}$$

$$n = 500,\ p = 0.1 \text{ and } q = 0.9$$

$$\sigma = \sqrt{500 \times 0.1 \times 0.9} = 6.71$$

(b) Moment coefficient of skewness, i.e. γ_1.

$$\gamma_1 = \sqrt{\beta_1} = \sqrt{\frac{(q-p)^2}{npq}}$$

$$= \frac{q-p}{\sqrt{npq}} = \frac{(0.9-0.1)}{6.71} = \frac{0.8}{6.71} = 0.119.$$

Since γ_1 is more than zero the distribution is positively skewed. However, skewness is very moderate.

Moment coefficient of kurtosis

$$\gamma_2 = \beta_2 - 3$$

$$\beta_2 = 3 + \frac{1-6pq}{npq} = 3 + \frac{1-6(0.1)(0.9)}{44.9}$$

$$= 3 + \frac{0.46}{44.9} = 3.01,$$

$$\gamma_1 = 3.01 - 3 = +0.01$$

Since γ_2 is positive the distribution is platykurtic

$$(q+p)^5 = q^5 + 6q^5p + 15q^4p^2 + 20q^2p^2 + 15q^2p^4 + 6pq^5 + p^6.$$

Example 5:

Eight coins are tossed at a time 256 times. Number of heads observed at each throw is recorded and the results are given below. Find the expected frequencies. What are the theoretical values of mean and standard deviation? Calculate also the mean and S.D. of the observed frequencies.

No. of heads at a throw	***Frequency***	***No. of heads at a throw***	***Frequency***
0	*2*	*5*	*56*
1	*6*	*6*	*32*
2	*30*	*7*	*10*
3	*52*	*8*	*1*
4	*67*		

Solution:

The chance of getting a head in a single throw of one coin is $\frac{1}{2}$

Hence $p = \frac{1}{2} \cdot q = \frac{1}{2}$, $n = 8$, $N = 256$.

By expanding $256\left(\frac{1}{2}+\frac{1}{2}\right)^8$ we shall get the expected frequencies of 1, 2,......, 8 heads (successes)

No. of head (X)	Frequency = N × $^nC_r\ q^{n-r}\ p^r$.
0	$256\left(\frac{1}{2}\right)^8 = 1$
1	$256 \times {}^8C_1\left(\frac{1}{2}\right)^1\left(\frac{1}{2}\right)^7 = 8$
2	$256 \times {}^8C_2\left(\frac{1}{2}\right)^2\left(\frac{1}{2}\right)^6 = 28$
3	$256 \times {}^8C_3\left(\frac{1}{2}\right)^3\left(\frac{1}{2}\right)^5 = 56$
4	$256 \times {}^8C_4\left(\frac{1}{2}\right)^4\left(\frac{1}{2}\right)^{64} = 70$
5	$256 \times {}^8C_5\left(\frac{1}{2}\right)^5\left(\frac{1}{2}\right)^3 = 56$
6	$256 \times {}^8C_6\left(\frac{1}{2}\right)^6\left(\frac{1}{2}\right)^2 = 28$
7	$256 \times {}^8C_7\left(\frac{1}{2}\right)^7\left(\frac{1}{2}\right)^1 = 8$
8	$256 \times \left(\frac{1}{2}\right)^8 = 1$
	Total = 256

The mean of the above distribution is np = 8 × 1/2 = 4.

The standard deviation is $\sqrt{npq} = \sqrt{\frac{1}{2} \times \frac{1}{2} \times 8} = \sqrt{2} = 1.414$

These are the mean and standard deviation of the expected frequency distribution. The mean and standard deviation of the observed frequency distribution shall be :

X	f	d	fd	fd^2
0	2	–4	–8	32
1	6	–3	–18	54
2	30	–2	–60	120
3	52	–1	–52	52
4	67	0	0	0
5	56	+1	+56	56
6	32	+2	+64	128
7	10	+3	+30	90
8	1	+4	+4	16
	N = 256		Σ fd = 16	$\Sigma fd^2 = 548$

$$\overline{X} = A + \frac{\Sigma fd}{N}$$

$$= 4 + \frac{16}{256} = 4.0625$$

$$\sigma = \sqrt{\frac{\Sigma fd^2}{N} - \left(\frac{\Sigma fd}{N}\right)^2}$$

$$= \sqrt{\frac{548}{256} - \left(\frac{16}{256}\right)^2}$$

$$= \sqrt{2.141 - 0.004}$$

$$= \sqrt{2.137} = 1.462$$

Example 6:

A sample of 3 lines is selected at random from a box containing 12 items of which 3 are defective. Find the possible number of defective combinations of the said 3 selected items along with probability of a defective combination.

Solution:

The usual notions we have :

$$n = 3, p = \frac{3}{12} = \frac{1}{4}, q = 1 - \frac{1}{4} = \frac{3}{4}$$

Then by the binomial probability law, the probability that there are r defective items is a given by:

No. of defective items	Probability
0	$3C_0\left(\frac{1}{4}\right)^0\left(\frac{3}{4}\right)^3 = \frac{27}{64}$
1	$3C_1\left(\frac{1}{4}\right)^1\left(\frac{3}{4}\right)^2 = \frac{27}{64}$
2	$3C_2\left(\frac{1}{4}\right)^2\left(\frac{3}{4}\right)^1 = \frac{9}{64}$
3	$3C_3\left(\frac{1}{4}\right)^3\left(\frac{3}{4}\right)^0 = \frac{1}{64}$

Example 7:

The screws product by a certain machine were checked by examining samples of 12. The following table shows the distribution of 128 samples according to the number of defective items they contained :

No. of defectives in a sample of 12 :	*0*	*1*	*2*	*3*	*4*	*5*	*6*	*7*	*Total*
No. of samples:	*7*	*6*	*19*	*35*	*30*	*23*	*7*	*1*	*128*

Fit a binomial distribution and find the expected frequencies if the chance of machine being defective is 1/2. Find the mean and variance of the fitted distribution.

Solution:

On the hypothesis that the coins are unbiased, N = 128 and n = 7. The probability of 0, 1, 2........7 defectives will be given by the expansion $\left(\frac{1}{2}+\frac{1}{2}\right)^7$

$$\left(\frac{1}{2}+\frac{1}{2}\right)^7 = {}^7C_1\left(\frac{1}{2}\right)^4\left(\frac{1}{2}\right) + {}^7C_2 + \left(\frac{1}{2}\right)^5\left(\frac{1}{2}\right)^2 + {}^7C_3\left(\frac{1}{2}\right)^4\left(\frac{1}{2}\right)^5$$

$$+ {}^7C_2\left(\frac{1}{2}\right)^3\left(\frac{1}{2}\right)^4 {}^7C_5\left(\frac{1}{2}\right)^2\left(\frac{1}{2}\right)^5 + {}^7C_6\left(\frac{1}{2}\right)\left(\frac{1}{2}\right)^6 + {}^7C_r\left(\frac{1}{2}\right)^0\left(\frac{1}{2}\right)^7$$

$$= \left(\frac{1}{2}\right)^7 = [1 + 7 + 21 + 35 + 35 + 21 + 7 + 1]$$

$$128\left(\frac{1}{2}+\frac{1}{2}\right)^7 = 128 \times \frac{1}{128} \ (1 + 7 + 21 + 35 + 35 + 21 + 7 + 1)$$

Thus the expected frequencies are :

X	0	1	2	3	4	5	6	7
fe	1	7	21	35	35	21	7	1

Mean and variance of the fitted distribution :

Mean of he binomial distribution is np and standard deviation $\sqrt{npq}$

Hence mean $= 7 \times \frac{1}{2} = 3.5$ and variance $= \frac{1}{2} \times \frac{1}{2} \times 7 = 1.75.$

Example 8:

A student obtained the following answer to a certain problem given to him. Mean = 2.4; Variance = 3.2 for a binomial distribution. Comment on the result.

Solution:

The mean of binomial distribution is np and variance npq. We are given mean = np = 2.4

Variance = npq $\qquad$ 2.4q = 3.2 or

$$q = \frac{3.2}{2.4} = 1.333$$

Since the value of q is greater than 1, the given results are inconsistent.

Example 9:

Twelve dice were thrown 4,096 times. Each 4, 5 or 6 spot appearing was considered to be a success while a 1, 2 or 3 spot was a failure. Calculate the theoretical frequencies for 0, 1, 2...., 12 successes.

Solution:

There are 4,096 trials. Since either 4, 5 or 6 is considered a success we have q = p = 1/2.

The terms of the binomial $(q + p)^n$ will give the probabilities of 0, 1, 2,.....n successes.

Here $\quad$ n = 12, q = 1/2 and p = 1/2.

$\therefore$ By expanding $4096\left(\frac{1}{2}+\frac{1}{2}\right)^{12}$

We get frequencies corresponding to 0, 1, 2,........, 12 successes.

$$4096\left(\frac{1}{4096}+\frac{12}{4096}+\frac{66}{4096}+\frac{220}{4096}+\frac{495}{4096}+\frac{792}{4096}+\frac{924}{4096}+\frac{792}{4096}+\frac{495}{4096}+\frac{220}{4096}+\frac{66}{4096}+\frac{12}{4096}+\frac{1}{4096}\right)$$

f_0 will denote observed frequencies and f_e expected frequencies. The observed frequencies cannot be in fractions but the expected frequencies may be in fractions. However, they may be approximated to the whole number.

The result can be tabulated as follows :

No. of Successes	Theoretical frequencies	No. of Successes	Theoretical frequencies
0	1	7	792
1	12	8	495
2	66	9	220
3	220	10	66
4	495	11	12
5	792	12	1
6	924		

Example 10:

The mean of the Poisson distribution is 2.25. Find the other constants of the distribution.

Solution:

We are given mean or m = 2.25

$\sigma = \sqrt{m} = \sqrt{2.25} = 1.5$

$\mu_1 = 0$

$\mu_2 = m = 2.25$

$\mu_3 = m = 2.25$

$\mu_4 = m + 3m^2 = 2.25 + 3\ (2.25)^2$

$= 2.25 + 15.1875 = 17.4375$ or 17.44 app.

$\beta_1 = \dfrac{1}{m} = \dfrac{1}{2.25} = .444.$

$\beta_2 = 3 + \dfrac{1}{m}\ 3 + .444 = 3.444.$

Example 11:

The Bombay Municipal Corporation installed 2,000 bulbs in the streets of Bombay. If these bulbs have an average life of 1,000 burning hours, with a standard deviation of 200 hours, what number of bulbs might be expected to fail in the first 700 burning hours? The table of area of the normal curve at selected values is as follows :

$\frac{X-\overline{X}}{\sigma}$	*Probability*
1.00	*0.159*
12.5	*0.106*
1.50	*0.067*

Solution:

Average lie of bulbs $\left(\overline{X}\right)$ = 1,000 hours

$$\sigma = 200 \text{ hours}$$

$$X = \text{buming hours} = 700$$

$$z = \frac{X-\overline{X}}{\sigma} = \frac{700-1{,}000}{200} = -1.5$$

Area to the left of (– 1.5) = 0.067.

∴ Number of bulbs expected to fail in the first 700 hours

$$= 0.067 \times 2{,}000 = 134.$$

Significance of the Normal Distribution

The normal distribution is mostly used for the following purposes :

1. To approximate the distribution of means and certain other quantities calculated from samples, especially large samples.
2. To approximate of "fit" a distribution of measurement under certain conditions.
3. To approximate the binomial distribution and other discrete of continuous probability distributions under suitable conditions.

The following examples shall illustrate the applications of normal distribution:

Example 12:

1,000 light bulbs with a mean life of 120 days are installed in new factory, their length of life is normally distributed with standard deviation 20 days (i) How many bulbs will expire in less than 90 days? (ii) If it is decided to replace all the bulbs together, what intervals should be allowed between replacements if not more than 10 percent should expire before replacement?

Solution:

(i) $\overline{X} = 120, \sigma = 20, X = 90$

Standard normal variate or

$$z = \frac{90 - 120}{20} = -1.5$$

Area of the curve (z = – 1.5) up to the mean ordinate – 0.4332

Area of the left of – 1.5 = 0.5 – 0.4332 = 0.0668.

Example 13:

A cigrarette company wants to promote the sales of X's cigarettes (Brand) with special advertising campaign. Fıfty out of every thousand cigarettes are rolled up in gold foil and randomly mixed with the regular (special king-sized, mentholated) cigarettes. The company offers to trade a new package of cigarettes for each gold cigarette a smoker finds in a package of Brand X. What is the probability that buyers of Brans X will find X = 0, 1, 2, 3,.. gold cigarettes in a single package of 10?

Solution:

This can be considered a Poisson experiment in which there are s = 10 trials in a pack and the probability of finding a golden cigarette (a success) is p = 50/100 = 0.05. The expected number of golden cigarettes per pack is, therefore, m = sp = 10 (0.05). Using the table of the Poisson distribution, we obtain

Number of golden cigarettes per pack	Probability
0	0.6065
1	0.3033
2	0.0758
3	0.0126
4	0.0016

We find that 60.65 per cent of the packages contain no golden cigarettes, 30.33 per cent contain one golden cigarette, 7.58 per cent contain 2 golden cigarettes, 1.26 per cent contain 3 golden cigarettes and 0.16 per cent contain 4 golden cigarettes.

Example 14:

Out of 800 families with 4 children each, what percentage would be expected to have (a) 2 boys and 2 girls, (b) at least one boy, (c) no girls, and (d) at the most 2 girls. Assume equal probabilities for boys and girls.

Solution:

(i) Probability of getting a boy $= \frac{1}{2}$ i.e., $p = \frac{1}{2}$

Probability of getting a girl $= \frac{1}{2}$ i.e., $q = \frac{1}{2}$

Probability of getting 2 boys and 2 girls

$$= {}^nC_r q^{n-r} p^r$$

$$= {}^4C_2 \left(\frac{1}{2}\right)^2 \left(\frac{1}{2}\right)^2$$

$$= \frac{4 \times 3}{2 \times 1} \times \frac{1}{16} = \frac{3}{8}.$$

Percentage of families expected to have two boys and two girls

$$= \frac{3}{8} \times 100 = 37.5\%$$

(ii) Probability of getting at least one boy means the sum of the probabilities of getting one boy and 3 girls, 2 boys and 2 girls, 3 boys and 1 girl and 4 boys and no girl.

Probability (1 boy and 3 girls) $= {}^nC_r\, q^{n-r} p^r$

$$= {}^4C_1 \left(\frac{1}{2}\right)^2 \left(\frac{1}{2}\right)^2 = \frac{4}{16} = \frac{1}{4}$$

probability (2 boys and 2 girls) $= \frac{3}{8}$

probability (3 boys and 1 girl) $= {}^4C_3 \left(\frac{1}{2}\right)^1 \left(\frac{1}{2}\right)^2 = \frac{1}{4}$

probability (4 boys and no girl) $= \left(\frac{1}{2}\right)^4 = \frac{1}{16}$

$\therefore$ Probability of getting at least one boy $= \frac{1}{4} + \frac{3}{8} + \frac{1}{16} = \frac{15}{16}$

Percentage of families expected to have at least one boy

$$= \frac{15}{16} \times 100 = 93.75\%$$

(iii) Percentage of families expected to have no girls, i.e., all boys

$$= \frac{1}{16} \times 100 = 6.25$$

(iv) Probability of getting at most 2 girls $= \; = \frac{1}{16} + \frac{1}{4} + \frac{3}{8} = \frac{11}{16}$

∴ Percentage of families expected to have at most 2 girls

$$= \frac{11}{16} \times 100 = 68.75.$$

Example 15:

You are incharge of rationing in a State affected by food shortage. The following reports were received from investigators:

Daily calories of food available per adult during current period

Area	*Mean*	*S.D.*
A	*2,000*	*350*
B	*1,750*	*100*

The estimated requirement of an adult is taken at 2,500 calories daily and the absolute minimum of 1,000. Comment on the reported figures and determines which area in poor opinion needs more urgent attention.

Solution:

Area A	**Area B**
Mean ± 3σ	Mean ± 3σ
2,000 ± (3 × 350)	1,750 ± (3 × 100)
Between 950 and 3,050 calories	Between 1,450 and 2,050 calories

Since the estimated requirement is minimum of 1,000 calories, areas A needs more urgent attention because there are people here who are getting less than 1,000 calories.

Example 16:

Certain mass-produced articles of which 0.5 per cent are defective, are packed in cartons each containing 130 articles. What proportion of cartons are free from defective articles, and what proportion contain 2 or more defectives. (Given $e^{-22} = 0.6065$).

Solution:

$$P(r) = \frac{e^{-m} m^r}{r!}, \; r = 0, 1, 2, ...$$

$P(0) = e^{-0.5} = 0.6065$ (given)

$P(1) = P(0) \times m = 0.6065 \times 0.5 = 0.30325.$

The probability of a carton with 2 or more defectives

$= 1 - 0.6065 - 0.30325 = 0.09025$

Therefore the proportion of cartons free from defective articles is 80.65% or 61% and with 2 or more defectives is 9% (approx).

Example 17:

The average monthly sales of 5000 firms are normally distributed. Its mean and standard deviation are Rs. 36000 and Rs. 10000 respectively. Find

(i) the number of firms the sales of which are over Rs. 40000.

(ii) the percentage of firms the sales of which will be between Rs. 38500 and 41000.

(iii) the number of firms the sales of which will be between Rs. 30000 and Rs. 40000. The relevant extract at the area table (under the normal curve) is given below :

Z	*0.25*	*0.40*	*0.5*	*0.6*
Area	*0.0987*	*0.1554*	*0.1915*	*0.2257*

Solution:

(i) The number of firms whose sales are over Rs. 40000.

$N = 5000, \overline{X} = 36000, \sigma = 10000, X = 40000$

$$Z = \frac{X - \overline{X}}{\sigma} = \frac{40000 - 36000}{10000} = 0.4$$

Area between Z = 0 and Z = 0.4 will be 0.1554.

Area above Z = 0.4 = 0.5 – 0.1554 = 0.3446

No of firms whose sales are over Rs. 40000

$= 5000 \times .3446 = 1723$

(ii) Percentage of firms whose sale is between Rs. 38500 and Rs. 41000

$$Z = \frac{38500 - 36000}{10000} = \frac{2500}{10000} = 0.25$$

Area = 0.987

$$\text{Again } Z = \frac{41000 - 36000}{10000} = 0.5$$

Area = 0.1915

∴ Area between 38500 & 41000 = .1915 – .0987

= 0.0928

Hence percentage = 9.28

(iii) Number of firms where sales is between Rs. 30,000 and Rs. 40,000

$$z = \frac{30,000 - 36,000}{10,000} = -0.6$$

Area = 0.2257

Again $$Z = \frac{40,000 - 36,000}{10,000} = 0.4$$

Area = 0.1554

Area between 30,000 and 40,000 = 0.2257 × 0.1554 = 0.3811

∴ No of firms = 0.3811 × 5,000 = 1905.5 or 1906.

Example 18:

The distribution of typing mistakes committed by a typist is given below: Assuming a Poisson mode, find out the expected frequencies :

No. of mistakes per page :	*0*	*1*	*2*	*3*	*4*	*5*
No. of pages :	*142*	*156*	*69*	*27*	*5*	*1*

Solution:

Calculation of Expected Frequencies by Assuming Poisson Model

No. of Mistakes X	**per page f**	**Frequency fX**
0	142	0
1	156	156
2	69	138
3	27	18
4	5	20
5	1	5
	N = 400	Σ fX = 400

$$\overline{X} = \frac{\Sigma fX}{N} = \frac{400}{400} = 1$$

This is the mean of Poisson distribution, i.e., m

$$P_{(0)} = e^{-m} = 2.7183^{-1} = \frac{1}{2.7183} = 0.3679$$

$N(P_0) = 400 \times .3679 = 147.16$

$N(P_1) = N(P_3) \times m = 147.16 \times 1 = 147.16$

$N(P_2) = N(P_1) \times \frac{m}{2} = 1.47.16 \times \frac{1}{2} = 73.58$

$N(P_3) = N(P_2) \times \frac{m}{3} = 73.58 \times \frac{1}{3} = 24.53$

$N(P_4) = N(P_3) \times \frac{m}{4} = 24.53 \times \frac{1}{4} = 6.13$

$N(P_5) = N(P_4) \times \frac{m}{5} = 6.13 \times \frac{1}{5} = 1.23.$

Thus the expected frequencies as per the Poisson model approximated to the whole number are :

No. of mistakes per Page :	0	1	2	3	4	5
No. of pages:	147	147	74	25	6	1

Example 19:

Find the probability that the value of an item drawn at random from a normal distribution with mean 20 and standard deviation 10 will be between:

(a) 10 and 15 (b) – 5 and 10, and (ii) 15 and 25.

The relevant extract of the Area Table (under the normal curve is given below) :

0.5	*1.0*	*1.5*	*2.0*	*2.5*
0.1915	*0.3413*	*0.4332*	*0.4772*	*0.4938*

Solution:

We are given $\overline{X} = 20$ and $\sigma = 10$

(i) For X = 10, $Z_1 = \frac{X-\overline{X}}{\sigma} = \frac{10-20}{10} = -1$

For X = 15, $Z_2 = \frac{X-\overline{X}}{\sigma} = \frac{15-20}{10} = -0.5$

$\therefore P = (10Z \times < 15) = 0.3413 - 0.1915 = 0.1498$

(ii) For X = – 5, $Z_1 = \frac{X-\overline{X}}{\sigma} = \frac{-5-20}{10} = -2.5$

For X = 10, $Z_2 = \frac{X-\overline{X}}{\sigma} = \frac{10-20}{10} = -1$

$\therefore P(-5 < x < 10) = 0.4938 - 0.3413 = 0.1525$

(iii) For X = 15, $Z_1 = \frac{X - \overline{X}}{\sigma} = \frac{15-20}{10} = -0.5$

For X = 25, $Z_2 = \frac{X - \overline{X}}{\sigma} = \frac{25-20}{10} = +0.5$

$\therefore$ P(15 < x < 25) = 2x – 1915 = 0.3830.

Example 20:

The income distribution of workers in a certain factory was found to be normal with mean of Rs. 500 and standard deviation equal to Rs. 50. There were 228 persons getting above Rs. 600. How many persons were there in all? (Area under the standard normal curve between heights at 0 and 2 is 0.4772).

Solution:

The probability of the persons getting above Rs. 600, i.e.,

P (X > 600) = 1 – P (X ≤ 600)

$= 1 - P\left(z \leq \frac{600 - 500}{50}\right)$

= P (Z ≤ 2)

= 1 – {P (Z ≤ 0) + P(0 ≤ Z ≤ 2)}

= 1 – {0.5000 + 0.4772}

= 1 – 0.9772 = .0228

Since there are 228 persons getting salary above Rs. 600, it means that the total number of persons is:

228/0.0228 = 10,000.

Example 21:

Among 10,000 random digits, in how many cases do we expect that the digit 3 appears at most 950 times. (The area under standard normal curve for z = 1.667 is 0.4524 approximately).

Solution:

n = 100

Probability of digit 2 appearing = 1/10

Hence p = 1/10, q = 1 – 1/10 = 9/10

np = 1000 ′ 1/10 = 1000

$$npq = 10000 \times \frac{1}{10} \times \frac{9}{19} = 900$$

Let n be the number of times the digit 3 appears among 10,000 random digits. Then n follows binomial distribution with mean 1000 and variance 900.

$$\therefore P(n \le 950) = p\left(\frac{Z \le 950 - 1000}{\sqrt{900}}\right)$$

$$= p(Z \le -1.667)$$

$$= 0.5000 - 0.4525 = 0.475$$

No. of cases = $10000 \times 0.0475 = 475$

Hence in 475 cases we may expect the digit to appear at most 950 times.

(b) Point out the fallacy if any in the following statement "The mean of a binomial distribution is 10 and its standard deviation is 4".

Solution:

The mean of binomial distribution is np and standard deviation $\sqrt{npq}$.

We are given np = 10, $\sqrt{npq} = 4$

$\sqrt{npq} = 4$ or npq 16

Putting the value of np,

$$10q = 16 \text{ or } q = \frac{16}{10} = 1.6$$

Since the value of q is exceeding one, there is some inconsistency in the given statement.

Example 22:

Is there any inconsistency in the statement, the mean of binomial distribution is 20 and its standard deviation 4? If no inconsistency is found what shall be the values of p, q and n?

Solution:

The mean of the binomial distribution is given by np and standard deviation by $\sqrt{npq}$

Here np = 20 and $\sqrt{npq} = 4$,

Since $\sqrt{npq} = 4$, npq = 16

Putting the value of np

$$20q = 16 \text{ or } q = \frac{16}{20} = 0.8$$

$$q = 8,\ p = 1 - .8 = 0.2$$

Since $\quad npq = 16$

By putting the value of p and q, we can find n,

$$n \times 2 \times .8 = 16\ n = \frac{16}{.16} = 100.$$

Example 23:

In a town 10 accidents took place in a span of 50 days. Assuming that the number of accidents per day follows the Poisson distribution, find the probability that there will be three or more accidents in a day.

Solution :

The average number of accidents per day $= \frac{10}{50} = 0.2$

P(3 or more accidents) = 1 – P(2 or less accidents)

$$= 1 - [p(0) + P(1) + P(2) + P(2)]$$

$$= 1 - \left[e^{-2} + e^{-2} \times 0.2 + \frac{e^{-2} \times 0.2 \times .2}{2}\right]$$

$$= - e^{-2} [1 + .2 + 0.02]$$

$$= - e^{-2} \times 1.22$$

$$= .8187 \times 1.22$$

$$= 1 - .999 = 0.001$$ (From table of e^{-m}).

Example 24:

Calculate the frequencies of the normal distribution which was the same mean, standard deviation and total frequency as the distribution given below or the intervals 60 – 65 – 70, etc.

X	*60 –*	*65 –*	*70 –*	*75 –*	*80 –*	*85 –*	*90 –*	*95 –*
f	*3*	*21*	*150*	*335*	*325*	*135*	*26*	*4*

Solution:

For finding frequencies of the normal distribution, first we will find out mean and standard deviation of the given distribution :

Calculation of mean and Standard Deviation

Variable	m.p	f	(m – 77.5)/5 d	fd	fd²
60 – 65	62.5	3	–3	–9	27
65 – 70	67.5	21	–2	–42	84
70 – 75	72.5	150	–1	–150	150
75 – 80	77.5	335	0	0	0
80 – 85	82.5	326	+1	+326	326
85 – 90	87.5	135	+2	+270	540
90 – 95	92.5	26	+3	+78	234
95 – 100	97.5	4	+4	+16	64
		N = 1,000		Σ fd = 489	Σ fd² = 1425

$$\overline{X} = A + \frac{\Sigma \text{ fd}}{N} \times i = 77.5 + \frac{489}{1000} \times 5 = 79.945 \text{ or } 79.95$$

$$\sigma = \sqrt{\frac{\Sigma \text{ fd}^2}{N} - \left(\frac{\Sigma \text{ fd}}{N}\right)^2} \times i = \sqrt{\frac{1425}{1000} - \left(\frac{489}{1000}\right)^2} \times 5$$

$$= \sqrt{1.425 - .239} \times 5 = 1.089 \times 5 = 5.45$$

Fitting of Normal Curve by Method of Areas

Variable	Lower Limits X	f	(X – 79.95 x	x/σ	Area Under Normal Curve from 0 to Z	Area for each class	Expected frequencies
60–65	60	3	– 19.95	–3.66	0.4999	0.0030	.0030 × 1000 = 3.0
65–70	65	21	– 14.95	–2.74	0.4969	0.0305	.0305 × 1000 = 30.5
70–75	70	150	–9.95	–1.83	0.4664	0.1480	.1480 × 1000 = 148.0
75–80	75	335	0.05	– 0.91	0.3186	0.3146	.3146 × 1000 = 314.6
80–85	80	326	+ 5.05	– 0.01	0.0040	0.3278	.3278 ′ 1000 = 327.8
85–90	85	135	+ 10.05	+ 0.93	0.3238	0.1433	.1433 × 1000 = 143.3
90 – 95	90	26	+ 15.05	+ 1.84	0.4671	0.0300	.0300 × 1000 = 30.0
95 – 100	95	4	+ 20.05	+ 2.76	0.4971	0.0028	.0028 × 1000 = 2.8
	100			+ 3.68	0.4999		
		N = 100					Total 1000

Thus, the frequencies of the normal distribution are :

X	60–65	65–70	70–75	75–80	80–85	85–90	90–95	95–100
f	3	30.5	148.5	314.5	327.8	143.3	30.0	2.8

Example 25:

The customer accounts at a certain departmental store have an average balance of Rs. 480 and a standard deviation of Rs. 160. Assuming that the account balances are normally distributed.

(i) What proportion of the accounts is over Rs. 600?

(ii) What proportion of the accounts is between Rs. 400 and Rs. 600?

(iii) What proportion of the accounts is between Rs. 240 and Rs. 360?

Solution:

Let the random X denote the balance of the customer accounts. X is normally distributed with $\overline{X} = 480$ and $\sigma = 160$. The standard normal variable Z is :

$$Z = \frac{X - \overline{X}}{\sigma} = \frac{X - 480}{100}$$

(i) when X = 600

$$Z = \frac{600 - 480}{160} = 0.75$$

$$P(X > 600) = P(Z > 0.75)$$

$$= \text{Area of the right of } Z = 0.75$$

$$= 0.5000 - 0.2734 = 0.2266$$

Hence 22.6% of the accounts have a balance in excess of Rs. 600.

(ii) Probability that the accounts lie between Rs. 400 and Rs. 600 is given by:

$P(400 \leq X \leq 600)$

when X = 400

$$Z = \frac{400 - 480}{160} = -05$$

when X = 600

$$Z = \frac{400 - 480}{160} = 0.75$$

$P(400 < x < 600)$

$= P(1 - 0.5 < Z < 0.75)$

Area between $Z = -0.5$ to $Z = 0.75$

= Area between $Z = -0.5$ to $Z = 0$) + (Area between

$= Z = 0$ to $Z = 0.75$

$= P(-0.5 \leq Z \leq 0) + P(0 \leq Z \leq 0.75)$

$= 0.1915 + 0.2734 = 0.4649$

Hence 46.69% of the accounts have an average balance between Rs. 400 and Rs. 600.

(C) We want $P(240 \leq X \leq 360)$

(Hint when $X = 240$ find Z, when $X = 360$ find Z. Calculate area between $Z = -0.75$ to $Z = -1.5$ Ans. 15.98 per cent.)

Example 26:

Fit a Poisson distribution to the following data and calculate theoretical frequencies :

Deaths	*0*	*1*	*2*	*3*	*4*
Frequency	*122*	*60*	*15*	*2*	*1*

Solution:

Fitting of Poisson Distribution

Deaths X	frequency f	fX
0	122	0
1	60	60
2	15	30
3	2	6
4	1	4
	N = 200	Σ fX = 100

$$m = \frac{\Sigma fX}{N} = \frac{100}{200} = 0.5$$

$$P_{(0)} = e^{-m} = e^{-0.5} = .6065$$

$$NP_{(0)} = N \times P_{(0)} = 200 \times .6065 = 212.30 \text{ or } 121$$

$$NP_{(1)} = NP_{(0)} \times m = 121.3 \times .5 = 65.65 \text{ or } 61$$

$$NP_{(2)} = NP_{(1)} \times \frac{m}{2} = 60.65 \times \frac{.5}{2} = 15.16 \text{ or } 15$$

$$NP_{(3)} = NP_{(2)} \times \frac{m}{3} = 15.16 \times \frac{.5}{3} = 2.53 \text{ or } 3$$

$$NP_{(4)} = NP_{(3)} \times \frac{m}{4} = 2.53 \times \frac{.5}{4} = .032 \text{ or } 0$$

Example 27:

In a manufacturing organisation, the distribution of wages was perfectly normal and the number of workers employed in the organisation was 5000. The mean wages of the workers were calculated at Rs. 800 P.M. and the standard deviation was worked out to Rs. 200. On the basis of the information estimate :

(i) the number of workers getting salary between Rs. 700 and Rs. 900.

(ii) Percentage of workers getting salary above Rs. 1000.

(iii) Percentage of workers getting salary below Rs. 600.

Solution:

(i) $$Z_1 = \frac{X-\overline{X}}{\sigma} = \frac{700-800}{200} = -0.5$$

$$Z_2 = \frac{X-\overline{X}}{\sigma} = \frac{900-800}{200} = 0.5$$

Area at $Z = -0.5 = 0.1915$

Area at $Z = +0.5 = 0.1915$

Total $= 0.3830$

Number of workers getting wages between Rs. 700 and 900 = .383 × 5000 = 1915

(ii) Percentage of workers getting salary above Rs. 1000.

$$Z = \frac{1000-800}{200} = 1$$

Area to the right of $\overline{X}$ at $Z = 1 = 0.5 - 0.3413 = 0.1587$

No. of workers = 0.1587 × 5000 = 794

∴ Percentage of workers getting salary more than 1000

$$= \frac{794}{5000} \times 100 = 15.87$$

(iii) Percentage of workers getting below Rs. 600

$$Z = \frac{X-\overline{X}}{\sigma} = \frac{600-800}{200} = -1.0$$

Area to the left of $\overline{X}$ at Z = 1 = 0.5000 – 0.3413 = 0.1587

No. of workers = 0.1587 × 5000 = 793.5

Percentage of workers getting salary below 600

$$= \frac{793.5}{5000} \times 100 = 15.87.$$

Example 28:

A aptitude test for selecting officers in a bank was conducted on 1,000 candidates, the average score is 42 and the standard deviation of scores is 24.

Assuming normal distribution for the scores, find :

(a) the number of candidates whose scores exceed 58.

(b) the number of candidates whose scores lie between 30 and 66.

Solution:

(A) Number of candidates whose score exceeds 58.

$$Z = \frac{X - \overline{X}}{\sigma} = \frac{60 - 42}{24} = 0.667$$

Area to the right of 0.667 is (0.5 – 0.2476) = 0.2524

∴ Number of candidates whose score exceeds 60 is

= 100 × 0.2524 = 252.4 or 252

(b) Number of candidates whose score exceeds 58.

$$Z = \frac{30 - 42}{24} = -0.5$$

Standard normal variate corresponding to 66

$$Z = \frac{66 - 42}{24} = 1$$

Area between Z = – 0.5 and Z = 1

= .1915 + 0.3413 = 0.5328

Number of candidates whose score lie between 30 and 66

= 1000 × 0.5328 = 532.8 or 533.

Example 29:

Eight coins are thrown simultaneously. Find the chance of obtaining.

(i) at least 6 heads,

(ii) no heads, and

(iii) all heads.

Solution:

(i) In tossing 8 coins simultaneously, the probability of getting at least six heads will be given by the sum of separate probability of getting 6 heads, no heads and 8 heads.

$$n = 8,\ r\ 6,\ 7,\ 8,\ p = \frac{1}{2}, q = \frac{1}{2}$$

$$p\ (r = 6, 7, 8) = {}^{8}C_{8}\left(\frac{1}{2}\right)^{5}\left(\frac{1}{2}\right)^{2} + {}^{2}C_{1}\left(\frac{1}{2}\right)^{7}\left(\frac{1}{2}\right) + {}^{3}C_{3}\left(\frac{1}{2}\right)^{3}\left(\frac{1}{2}\right)^{3}$$

$$= \left(28 \times \frac{1}{256}\right) + \left(8 \times \frac{1}{256}\right) + \left(1 \times \frac{1}{256}\right) = \frac{37}{256}$$

(ii) Probability of getting no head is given by

$${}^{2}C_{0}\left(\frac{1}{2}\right)^{0}\left(\frac{1}{2}\right) = 1 \times \frac{1}{256} = \frac{1}{256}$$

(iii) Probability of getting all heads is given by

$${}^{3}C_{3}\left(\frac{1}{2}\right)^{2}\left(\frac{1}{2}\right)^{3} = 1 \times \frac{1}{256} = \frac{1}{256}.$$

Example 30:

The normal rate of infection of a certain disease in animals is known to be 25%. In an experiment with 6 animals injected with a new vaccine it was observed that none of the animals caught infection. Calculate the probability of the observed result.

Solution:

Let p denote infection of the disease

$$p = \frac{25}{100} = \frac{1}{4} \text{ and } q = 1 - \frac{1}{4} = \frac{3}{4}$$

Out of 6 animals probability of 0, 1, 2, 3, 4, 5 and 6 animals catching infection will be given by 1st, 2nd, 3rd terms, etc., in the expansion of

$$\left(\frac{3}{4} + \frac{1}{4}\right)^{6}$$

$\therefore$ The probability of none of the animals being infected

$$= \left(\frac{3}{4}\right)^{6} = \frac{729}{4096}.$$

Example 31:

It is given that 3% of electric bulbs manufactured by a company are defective. Using the Poisson approximation, find the probability that a sample of 100 bulbs will contain (i) no defective, (ii) exactly one defective.

Solution:

Number of defective bulbs in a sample of 100 is 3 (since 3% bulb manufactured are defective).

Hence $m = 3$

Probability of no defective bulb in a sample of 100 is given by

$$P(0) = e^{-m} = e^{-3} = 0.05 \text{ (from the table)}$$

$$P(1) = P(0) \times m = 0.05 \times 3 = 0.15.$$

Example 32:

The income distribution of workers in a certain factory was found to be normal with mean of Rs. 500 and standard deviation equal to Rs. 50. There were 228 persons getting above Rs. 600. How many persons were there in all? (Area under the standard normal curve between heights at 0 and 2 is 0.4772).

Solution:

The probability of the persons getting above Rs. 600, i.e.,

$$P(X > 600) = 1 - P(X \le 600)$$

$$= 1 - P\left(z \le \frac{600 - 500}{50}\right)$$

$$= P(Z \le 2)$$

$$= 1 - \{P(Z \le 0) + P(0 \le Z \le 2)\}$$

$$= 1 - \{0.5000 + 0.4772\}$$

$$= 1 - 0.9772 = .0228$$

Since there are 228 persons getting salary above Rs. 600, it means that the total number of persons is:

$$228/0.0228 = 10{,}000.$$

EXERCISES

1. (a) Describe the chief characteristics of normal distribution.
   (b) What are the chief properties of normal distribution? Describe briefly the importance of normal distribution in statistical analysis.
2. (a) Explain the properties of a normal distribution.
   (b) State the conditions for a function of discrete random variable to be a distribution.

(c) Explain the salient features of binomial and normal probability distribution.

(d) Prove that the variance of a normal distribution is np q.

3. (a) What are the properties of normal curve?

(b) Prove that the mean of binomial distribution is np.

(c) Describe the chief characteristics of Poisson distribution.

(d) Find the mean and variance of Poisson distribution.

(e) What are chief characteristics of normal distribution? Under what circumstances does the binomial tend to the normal distribution?

4. (a) Suppose that life of a gas cylinder is normally distributed with a mean of 40 days and a standard deviation of 5 days. If at a time, 10,000 cylinders are issued to customers, how many will need replacement after 35 days?

(b) In a certain book, the frequency distribution for the number of words per page may be taken as approximate normal with mean 800 and standard deviation 50. If 3 pages are chosen at random, what is the probability that none of them has between 830 and 845 words each?

[0.825]

5. (a) Suppose that the life of a gas cylinder is normally distributed with a mean of 40 days and a standard deviation of 5 days. If, at a time, 10,000 cylinders are issued to customers, how many will need replacement after 35 days.

(b) A machine fills milk into milk cartons such tat the mean of the distribution of fills is 1 litre with a standard deviation of 10 ml. If the distribution of fills is normal :

(i) What percentage of the filled up cartons will contains less than 1 litre?

(ii) What percentage of the filled up cartons will contains less than 975 ml.? More than 1010 ml.?

8. A typist takes on an average 8 minutes to type a letter. It is found that she is idle for 20% of the time. (Given : $P_t(Z \leq 2{,}76) = 0.9971$)

(i) what would be the number of letters she may be receiving on an average per hour assuming Poisson arrival and exponential servicing?

(ii) What is the time (in minutes) on an average that a letter arrived will have to be on her tray before it is taken by her typing?

(iii) What is the probability that a letter sent to her would be received back fully typed within 5 minutes?

9. In a super market two girls are attending at the payment counters. If the service time for each customer is exponential with mean of the 4 minutes, and people arrive in Poisson manner at the rate of 10 per hour :

   (i) What is the probability of customer having to wait for service?

   (ii) What is the expected percentage of idle time for each girl?

10. (a) If the distribution of incomes of a group of persons be assumed to be normal with mean Rs. 500 and the standard deviation Rs. 50, estimate the proportion of individual with income (i) between Rs. 550 and Rs. 650, (ii) between Rs. 450 and Rs. 75.

    (b) The scores made by candidates in a certain test are normally distributed with mean 500 and standard deviation 100. What per cent of candidates receive (i) less than 400, (ii) between 400 and 500?

    [(a) (i) 15.735%, (ii) 14.28; (b) (i) 15.86%, (ii) 68.28%]

11. (a) Records show that the probability is 0.00002 that a car will have a flat tyre while driving over a certain bridge. Use the Poisson probability distribution to determine the probability that among 20,000 cars driven over this bridge, nor more than one will have a flat tyre.

    (b) Etona Co. Ltd. manufacturing staple pins, markets its product in packets of 1,000 and there is a small chance, 0.001 of a staple pin to be defective. The company guarantees not more than four defective pins in each packet.

    (i) What is the probability that a packet will meet the guarantee?

    (ii) If the Company sells 10,000 packets per month, what is the expected number of guarantee claims in a month?

12. (a) What are the assumptions of Poisson distribution? Explain the distribution, stating its mean and variance.

    (b) What is Poisson Distribution. Distinguish clearly the relationship between Binomial Poisson Distribution.

    (c) What is Normal distribution? Describe its properties in detail. Bring out its importance in statistics.

13. (a) The mean and standard deviation for the lifetimes of a population

of light bulbs are 1,200 and 150 hours, respectively. Assuming these lifetimes are normally distributed, what is the probability that a light bulb will last over 1,500 hours?

(b) In a certain examination of Delhi University, the mean mark was 48, and 12.7% of the candidates failed (i.e., below 40). Assuming a normal distribution of marks, find the percentage of candidates who secured first division.

14. (a) What probability model is appropriate to describe a situation where 100 misprints are distributed randomly throughout the 100 pages of a book? For this model, what is the probability that a page observed at random wil contain at least three misprint?

(b) One hundred car stereos are inspected as they came off the production line and number of defects per set is recorded below:

No. of defects	:	0	1	2	3	4
No. of sets	:	79	18	2	1	0

Fit a Poisson distribution to the above data.

15. (a) Discuss the salient features of Normal distribution. Explain the characteristics of Poisson distribution.

(b) Explain how for the normal distribution, the mean, median and mode are equal.

16. (a) How is the probability distribution of a discrete random variable defined? Define its expected value.

(b) List the chief properties of the normal distribution. Why is this distribution given a central place in statistics?

(c) Describe normal distribution and discuss its properties. Why is it so important in behavioural sciences?

7

Probability

INTRODUCTION

The word “probable” in general conversation. We need to express our ideas *quantitatively*, so as to be able to assign a numerical value to probability. One can speak of the probability that your favourite football team will win next Saturday. Such a system can be made internally consistent, and may be a valuable aid to quantitative assessment of whether or not to embark on a certain action, for example to change employment, to build a new factory, to lay a bet or to start a new experiment. Suppose the coin to be spun many times, or births to be recorded in a hospital over a long period, or a large number of seeds to be sown and the sequence of germination records studied.

MATHEMATICAL EXPECTATION

The concept of mathematical expectation is of great importance in statistical work. The mathematical expectation (also called the expected value) of a random variable is the weighted arithmetic mean of the variable, the weights used to find the mathematical expectation are all the respective probabilities of the values that the variable can possibly assume.

If X denotes a discrete random variable which can assume the values X_1, X_2, X_3, ...X_k, with respective probabilities p_1, p_2, p_3 ...., p_k where $p_1 + p_2$, $p_3 + ... + p_k = 1$ the mathematical expectation of X denoted by E(X) is defined as:

$$E(X)\ p_1 X_1 + p_2 X_2 + p_3 X_3 + ... + p_k X_k$$

Thus the expected value equals the sum of each particular value within the set (X) multiplied by the probability that X equals that particular value.

It should be noted that the concept of mathematical expectation was originally applied to games of chance and lotteries, but the notion of an expected value has become more generally applied and is now a common term in everyday parlance. Business situations frequently involve the consideration of expected values.

PROBABILITY DEFINED

The probability of a given event is an expression of likelihood or chance of occurrence of an event. A probability is a number which ranges from 0 (zero) to 1 (one) — zero for an event which cannot occur and 1 for an event certain to occur. How the number is assigned would depend on the interpretation of the term 'probability'. There is no general agreement about its interpretation and many people associate probability and chance with nebulous and mystic ideas. However, broadly speaking, there are four different schools of thought on the concept of probability.

CLASSICAL OR A PRIORI PROBABILITY

The classical approach to probability i the oldest and simple. It originated in eighteenth century in problems pertaining to games of chance, such a throwing of coins, dice or deck of cards, etc. The basic assumption underlying the classical theory is that the outcomes of a random experiment are "equally likely". The "event" whose probability is sought consists of one or more possible outcomes of the given activity such as when a die is rolled once, any one of the six possible outcomes, i.e., 1, 2, 3, 4, 5, 6, can occur. These activities are referred to in modern terminology as "experiment" which is a term that refers to processes which result in different possible outcomes or observations. The term "equally likely", though undefined, conveys the notion that each outcome of an experiment has the same chance of appearing as any other. Thus in a throw of a dice occurrence of 1, 2, 3, 4, 5, 6 are equally likely events.

The definition of probability given by French mathematician Laplace and generally adopted by disciples of the classical school runs as follows:

Example:

From a bag containing 10 black and 20 white balls, a ball is drawn at random. What is the probability that it is black?

Solution:

Total number of balls in the bag = 10 + 20 = 30

Number of black balls = 10

Probability of getting a black ball or

$$p(A) = \frac{\text{Number of favourable cases}}{\text{Total number of equally likely cases}} \text{ or } \frac{a}{n}$$

$$= \frac{10}{30} = \frac{1}{3}$$

Probability of not getting a black ball or

$$q = \frac{20}{30} = \frac{2}{3}$$

Thus, $p+q = \frac{1}{3} + \frac{2}{3} = 1.$

Shortcomings of the Classical Approach : The classical definition of probability given above suffers from certain limitations. First, the definition cannot be applied whenever it is not possible to make a simple enumeration of cases which can be considered equally likely. For example, how does it apply to probability of rain? What are the possible cases ? We might think that there are two possibilities 'rain' or 'no rain'. But at any given time it will not usually be agreed that they are equally likely. Similarly, if a person jumps from the top of Qutab Minar the probability of his survival will not be 50 percent since survival and death, i.e., the two mutually exclusive and exhaustive outcomes, are not equally likely.

The classical approach also fails to answer questions like "what is the probability that a male will die before the age of 60?" "What is the probability that a bulb will burn less than 2,000 hours?", etc. All these are legitimate questions which we want to bring into the realm of probability theory. Real life situations unlikely and disorderly as they often are make it difficult and at time impossible to apply classical probability concept.

RELATIVE FREQUENCY THEORY OF PROBABILITY

In the 1800s, British statisticians, interested in a theoretical foundation for calculating risk of losses in life insurance and commercial insurance, began defining probabilities from statistical data collected on births and deaths. Today this approach is called relative frequency of occurrence.

This classical definition is difficult or impossible to apply as soon as we deviate from the fields of coins, dice, cards and other simple games of chance. Secondly, the classical approach may not explain actual results in certain cases. For example, if a coin is tossed 10 times we may get 6 heads and 4 tails. The probability of a head is thus 0.6 and that of a tail 0.4. However, if the experiment is carried out a large number of times we should expect approximately equal number of heads and tails. As n increases, i.e., approaches ∞ (infinity), we find that the probability of getting a head or tail approaches 0.5. The probability of an event can thus be defined as the relative frequency with which it occurs in an indefinitely large number of trials. If an event occurs a times out of n, its relative frequency is $\frac{a}{n}$ when n becomes infinity is called the limit of the relative frequency.

Symbolically, $P(A) = \lim_{n \to \infty} \frac{a}{n}$

Theoretically, we can never obtain the probability off an event a given by the above limit. However, in practice we can only try to have a close estimate convenience, the estimate of P(A) can be written as if it were actually P(A) and the relative frequency definition of probability may be expressed as :

$$P(A) = \frac{a}{n}.$$

In the relative frequency definition the fact that he probability is the value which is approached by $\frac{a}{n}$ when n becomes infinity, emphasises a very important point, i.e., probability involves a long-term concept. This means that if we toss a coin only 10 times, we may not get exactly 5 heads and 5 tails. However, as the experiment is carried out larger and larger number of times say, coin is thrown 10,000 times, we can expect heads and tails very closer to 50 percent.

The two approaches, classical and empirical, though seemingly same, differ widely. In the former, P(A) and $\frac{a}{n}$ were practically equal when n was large whereas in the latter we say that P(A) is the limit $\frac{a}{n}$ as n tends to infinity. In the second approach, thus, the probability itself is the limit of the relative frequency as the number of observations increases indefinitely.

The probability obtained by following relative frequency definition is called a posteriori or empirical probability as distinguished from a priori probability obtained by following the classical approach.

A clear distinction between a priori and empirical probability is quite important for proper understanding of the concept of probability. First a priori probability is normally encountered in problems dealing with games of chance—for example, dice and card. Normally, we think a of a priori probability as being deductive in nature, i.e., from cause to effect and based on theory instead of the evidence of experience or experimentation. On the other hand, probability derived from past experience is called empirical probability and is used in many practical problems, the class example being the preparation of insurance mortality tables, which, obviously, are based on past experience. In the analysis of most practical business problems, these two kinds of probability concepts are widely used.

SUBJECTIVE APPROACH TO PROBABILITY

The subjective approach to assigning probabilities was introduced in the year 1926 by Frank Ramsey in his book. The Foundation of Mathematics and other Logical Essays. The concept was further developed by Bernard Koopman, Richard Good and Leonard Savage. The subjective probability is defined as the probability assigned to an event by an individual based on whatever evidence is available. Hence such probabilities are based the beliefs of the person making the probability statement. For example, if a teacher wants to find out the probability of Mr. X topping in M. Com. examination in Delhi University this year, he may assign a value between zero and one according to his degree of belief for possible occurrence. He may take into account such factors as the past academic performance, the views of his other colleagues, the attendance record, performance in periodic test, etc., and arrive at a probability figure.

The application of personalistic concept to statistical problems has occurred virtually entirely in the post-World War II period, particularly in connection with statistical decision theory. This concept emphasises the fact that since probability of an event is the degree of belief or degree of confidence placed in the occurrence of an event by a particular individual based on the evidence available to him, different individuals may different in their degrees of confidence even when offered the same evidence. This evidence may on sit of relative frequency of occurrence data and any other quantitative or non-quantitative information. Persons might arrive at different probability assignments because of differences in values, experience and attitudes, etc. If an individual believes that it is unlikely that an event will occur, he will assign a probability close to zero to its occurrence. On the other hand, if he believes that it is very likely that the event will occur, he will assign a probability close to one.

The personalistic approach is very broad and highly flexible. It permits probability assignment to events for which there may be no objective data, or for which there may be a combination of subjective and objective data. However, one has to be very careful and consistent in the assignment of these probabilities otherwise the decisions made may be misleading. Used with care the concept is extremely useful in the context of situations in business decision-making.

AXIOMATIC APPROACH TO PROBABILITY

The axiomatic approach to probability was introduced by the Russian mathematician A. N. Kolmogorov in the year 1933. Kolmogorov axiomised the theory of probability and his book Foundations of Probability, published in 1933, introduces probability as a set function and is considered as a classic. When this approach is followed, no precise definition of probability is given, rather we give certain axioms or postulates on which probability calculations are based. The whole field of probability theory for calculations are based. The whole field of probability theory for finite sample spaces is based upon the following three axioms :

(1) The probability of an event ranges from zero to one. In the event cannot take place its probability shall be zero and if it is certain, i.e., bound to occur. Its probability shall be one.
(2) The probability of the entire sample space is 1, i.e., P(S) = 1.
(3) If A and B are mutually exclusive (or disjoint) events then the probability of occurrence of either A or B denied by P (A ∪ B) shall be given by :

$$P(A \cup B) = P(A) + P(B)$$

It may be pointed out that out of the four interpretations of the concept of probability, each has it own merits and one may use whichever approach is convenient and appropriate for the problem under consideration. Experts disagree about which approach is the proper one to use.

CONDITIONAL PROBABILITY

The multiplication theorem explained above is not applicable in case of dependent events. Two events A and B are said to be dependent when B can occur only when A is known to have occurred (or vice versa). The probability attached to such an event is called the conditional probability and is denoted by P (A/B) or, in other word, probability of A given that B has occurred.

If two events A and B are dependent, then the conditional probability of B given A is :

$$P(B/A) = \frac{P(AB)}{P(A)}$$

Proof. Suppose a_1 is the number of cases for the simultaneous happening of A and B out of $a_1 + a_2$ cases in which A can happen with or without happening of B.

$$\therefore \qquad P(B/A) = \frac{a_1}{a_1 + a_2} = \frac{a_1 / n}{(a_1 + a_2)/n} = \frac{P(AB)}{P(A)}$$

Similarly it can be shown that :

$$P(A/B) = \frac{P(AB)}{P(A)}.$$

The general rule of multiplication in its modified form in terms of conditional probability becomes :

$$P(A \text{ and } B) = P(B) \times P(A/B)$$

or

$$P(A \text{ and } B) = P(A) \times P(B/A)$$

For three events A, B and C, we have

$$P(ABC) = P(A) \times P(B/A) \times P(C/AB)$$

i.e., the probability of occurrence of A, B and C is equal to the probability of A times the probability of B given that A has occurred, time the probability of C given that both A and B have occurred.

SOLVED EXAMPLES

Example 1:

Find the probability of drawing a queen, a king and a knave in that order from a pack of cards in three consecutive draws, the cards drawn not being replaced.

Solution:

The probability of drawing a queen $= \frac{4}{52}$.

The probability of drawing a king after a queen has been drawn $= \frac{4}{51}$

The probability of drawing a knave given that a queen and king have been drawn $= \frac{4}{50}$

Since they are dependent events, the required probability of the compound event is :

$$\frac{4}{52} \times \frac{4}{51} \times \frac{4}{50} = \frac{64}{1{,}32{,}600} = 0.00048.$$

Example 2:

A company uses a 'selling aptitude test' in the selection of salesmen. Past experience has shown that only 70% of all persons applying for a sales position achieved a classification "dissatisfactory" in actual selling, whereas the remainder were classified as "satisfactory", 85% had scored a passing grade on the aptitude test. Only 25% of those classified unsatisfactory, had passed the test on the basis of this information. What is the probability that a candidate would be a satisfactory salesman given that he passed the aptitude test?

Solution:

If stands for a 'satisfactory' classification as a salesman and P stands for 'passing the test', then the probability that a candidate would be "satisfactory" salesman given that he passed the aptitude test is :

$$P(S/P) = \frac{(0.07)\ (0.85)}{(0.70)\ (0.85) + (0.30)\ (0.25)}$$

$$= \frac{0.595}{0.595 + 0.75} = 0.888$$

The result indicates that the tests are of value in screening candidates. Assuming no change in the type of candidates applying for the selling positions, the probability that a random applicant would be satisfactory is 70%. On the other hand, if the company only accepts an applicant if he passed the test, the probability increases to 0.888.

Example 3:

There are three alternative proposals before a businessman to start a new project:

Proposal A : Profit of Rs. 5 lakhs with a probability of 0.6 or a loss of Rs. 80,000 with a probability of 0.4.

Proposal B : Profit of Rs. 10 lakhs with a probability of 0.4 or a loss of Rs. 2 lakhs a probability of 0.6.

Proposal C : Profit of Rs. 4.5 lakhs with a probability of 0.8 or a loss of Rs. 50,000 with a probability of 0.2.

If he wants to maximise the profits and minimise the loss, which proposal should he prefer.

Solution:

We would calculate the mathematical expectation of each of the proposal.

$E(X) = p_1 X_1 + p_2 X_2 + ... + p_k X_k$

Proposal A : Expected Value = 5,00,000 (0.6) – 80,000 (0.4)

= 3,00,000 – 32,000 = 2,68,000.

Proposal B : Expected Value = 4,50,000 (0.8) – 50,000 (0.2)

= 3,60,000 – 10,000 = 3,50,000.

Since expected value is highest in case of proposal C, hence he should prefer proposal C.

Example 4:

An industrial salesman wants to know the average of units he sells per sales call. He checks his past sales records and comes up with the following probabilities:

Sales in Units	*0*	*1*	*2*	*3*	*4*	*5*
Probability	*0.15*	*0.20*	*0.10*	*0.05*	*0.30*	*0.20*

What is the average number of units he sells per sales call ?

Solution:

The salesman wants to know the average number of units he sells per sales call. This is the same thing as saying that he wants to know the expected value of each sales call, where a sales call is the random variable x. The expected value is calculated by the formula :

$E(x) = p_1x_1 + p_2x_2 + p_3x_3 + ...$

$= 0.15(0) + 0.20(1) + 0.10(2) + 0.05(3) + 0.30(4) + 0.20(5)$

= 0 + 0.2 + 0.2 + 0.15 + 1.2 + 1.0 = 2.75

Thus he would expect to sell 2.75 or 3 units on each sales call.

Example 5:

A firm plans to bid Rs. 300 per tonne for a contract to supply 1,000 tonnes of a metal. It has two competitors A and B and it assumes that the probability that A will bid less than Rs. 300 per tonne is 0.3 and that B will bid less than Rs. 300 per tonne is 0.7. If the lowest bidder gets all the business and the firms bid independently, what is the expected value of the contract to the firm?

Solution:

There are two competitors A and B and the lowest bidder gets the contract. Value of Plan = 300 × 1,000 = Rs. 3,00,000

Contractor A : Probability that bid is less than Rs. 300 per tonne = 0.3
Probability that bid is Rs. 300 or more = 0.7

Contractor B : Probability that bid is less than Rs. 300 per tonne = 0.7
Probability that bid is Rs. 300 or more per tonne = 0.3

(1) If both bids less than Rs. 300
Probability is 0.3 × 0.7 = 0.21
Therefore plan value is 3,00,000 × 0.21 = 63,000

(2) If only A bids less than 300 and B bids more than 300,
Probability is 0.3 × 0.3 = 0.09
Therefore plan value is 3,00,000 × 0.09 = 27,000

(3) B bids less than 300 while A bids more than 300
Probability is 0.7 × 0.7 = 0.49
Therefore plan value is 3,00,000 × 0.49 = 1,47,000
Therefore value of plan is Rs. = 63,000 + 27,000 + 1,47,000
= 2,37,000

Hence expected value of plan is Rs. 2,37,000.

Example 6:

Under an employment promotion programme it is proposed to allow sale of newspapers on the buses during off-peak hours. The vendor can purchase the newspapers at a special concessional rate of 25 paise per copy against the selling price of 40 paise. Any unsold copies are however, a dead loss. A vendor has estimated the following probability distribution for the number of copies demanded.

No. of copies	*15*	*16*	*17*	*18*	*19*	*20*
Probability	*.04*	*.19*	*.33*	*.26*	*.11*	*.07*

How many copies should be ordered so that his expected profit will be maximum ?

Solution:

Profit per copy = Selling Price – Purchasing Price.
= 40 Paise – 25 Paise = 15 Paise.

Expected Profit

Expected Profit = No. of copies × probability × Profit per copy.

Computation of Expected Profit.

No. of copies	*Probability*	*Profit per copy*	*Expected Profit (paise)*
15	0.04	15	9
16	0.19	15	46
17	0.33	15	84
18	0.26	15	70
19	0.11	15	31
20	0.07	15	21

17 copies will give the maximum expected profit of 84 paise.

Example 7:

A company has two plants to manufacture scooters. Plant I manufactures 70% of the scooters and Plant II manufactures 30%. At plant I, 80% of scooters are rated standard quality and at plant II, 90& of scooters are rated standard quality. A scooter is picked up at random and is found to be of standard quality. What is the chance that it has come from plant I, or plant II ?

Solution:

Let A_1 = The event of drawing a scooter produced by plant I

A_2 = The event of drawing a scooter produced by plant II

E = Event of drawing a standard quality scooter produced either by plant I or plant II.

From the first information :

$P(A_1) = 70\% = 0.7, P(A_2) = 30\% = 0.3$

From the additional information :

$$P(E/A_1) = \frac{80}{100} = 0.80; P(E/A_2) = \frac{9}{100} = 0.90$$

$$P(E \cap A_1) = P(E/A_1) \times P(A_1) = 0.8 \times 0.7 = 0.56$$

$P(E \cap A_2) = P(E/A_2) = P(E/A_2) \times P(A_2) = 0.9 \times 0.3 = 0.27$

The probability that a standard quality scooter chosen at random is manufactured by plant I is given by the Bayes' theorem as :

$$P(A_1/E) = \frac{P(E/A_{1)}) \times P(A_1)}{P(E/A_1) \times P(A_1) + P(E/A_2) \times P(A_2)} = \frac{0.56}{0.56 + 0.27} = \frac{56}{83}$$

Similarly we can obtain $P(A_2/E)$

Example 8:

In an examination 30% students have failed in mathematics 20% of the students have failed in chemistry and 10% have failed in both mathematics and chemistry. A student is selected at random. What is the probability that

(i) The student has failed in mathematics if it is known that he has failed in chemistry.

(ii) What is the probability that the student has failed in mathematics to chemistry ?

Solution:

Let A and B denote the event that the student has failed in mathematics and chemistry respectively. Then we have:

$$P(A) = 0.30,\ P(B) = 0.20,\ P(AB) = 0.10$$

(i) Probability that the student selected at random has failed in mathematics if it is known that he has failed in chemistry.

$$= P(A/B) = \frac{P(AB)}{P(B)} = \frac{0.10}{0.20} = 0.50$$

(ii) Probability that the student selected at random has failed either in mathematics or chemistry $= P(A \cup B) = P(A) + P(B) - P(AB)$

$$= 0.30 + 0.20 - 1.0 = 0.40$$

Example 9:

Set up a sample space for the single toss of a pair of fair dice. From the sample space, determine the probability that the sum in tossing a pair of dice is either 7 or 11.

Solution:

Sample space for the single toss of a pair of dice.

(1, 6)	(2, 6)	(3, 6)	(4, 6)	(5, 6)	(6, 6)
(1, 5)	(2, 5)	(3, 5)	(4, 5)	(5, 5)	(6, 5)
(1, 4)	(2, 4)	(3, 4)	(4, 4)	(5, 4)	(6, 4)
(1, 3)	(2, 3)	(3, 3)	(4, 3)	(5, 3)	(6, 3)
(1, 2)	(2, 2)	(3, 2)	(4, 2)	(5, 2)	(6, 2)

(1, 1) (2, 1) (3, 1) (4, 1) (5, 1) (6, 1)

P(A) = Sum of probabilities associated with each point in A = 6/36.

P(B) = Sum of probabilities associated with each point in B = 2/36.

P(A+B) = Sum of probabilities of points in A, in B or in both

= (6 + 2)/36 = 8/36 = 2/9.

Example 10:

You note that your officer is happy in 60% cases of your calls. You have also noticed that if he is happy, he accedes to your requests with a probability of 0.4, whereas if he is not happy, he accedes to your requests with a probability of 0.1. You call on him one day and he accedes to your request. What is the probability of his being happy ?

Solution:

The probability that he is happy and accedes to requests

$= .6 \times .4 = .24$

The probability that he is unhappy and accedes to requests

$= .4 \times .1 = .04$

Total probability of acceding to requests

$= .24 + .04 = 0.28$

The probability of his being happy if he accedes to requests

$$= \frac{0.24}{0.28} = 0.857.$$

Example 11:

The data for the promotion status and academic qualification regarding 100 employees of a company is an follows :

	MBA	***Academic qualifications Non - MBA***	***Total***
Promotional status			
Promoted	*12*	*48*	*60*
Not promoted	*18*	*22*	*40*
Total	*30*	*70*	*100*

At random one employee is picked up. What is the probability that

(i) he is an MBA,

(ii) he is promoted,

(iii) he is promoted given that he is an MBA, and

(iv) he is an MBA given that he is promoted ?

Solution:

Let the various events be defined as follows :

A : an employee is an MBA

B : an employee is promoted

C : an MBA employee is promoted

D : a promoted employee is an MBA

(a) $P(A)=\frac{30}{100}=0.3$

(b) $P(B)=\frac{60}{100}=0.6$

(c) $P(c)=\frac{12}{30}=0.4$

(d) $P(d)=\frac{12}{60}=0.2$

Example 12:

A consignment is offered to two firms X and Y for Rs. 1,00,000. The following table shows the probability at which the firms will be able to sell it at different prices :

Probability		***Selling Price (Rs.)***		
	80,000	***90,000***	***1,05,000***	***1,10,000***
X	*0.2*	*0.3*	*0.4*	*0.1*
Y	*0.25*	*0.2*	*0.5*	*0.05*

Which firm X or Y will be more inclined towards the offer ?

Solution:

Firm X

$$E(X) = (80{,}000 \times .2) + (90{,}000 \times .3) + (1{,}05{,}00 \times 0.4) + (1{,}10{,}00 \times .1)$$
$$= 16{,}000 + 27{,}000 + 42{,}000 + 11{,}000 = 96{,}000$$

Firm Y

$$E(Y) = (80{,}000 \times .25) + (90{,}000 \times .2) + (1{,}05{,}000 \times .5) + (1{,}10{,}000 \times .05)$$
$$= 20{,}000 + 18{,}000 + 52{,}500 + 5{,}500 = 96{,}000$$

Since the expectation is the same, both the firms would be equally inclined towards the offer.

Example 13:

A committee of 4 people is to be appointed from 3 officers of the production department, 4 officers of the purchase department, two officers of the sales department and 1 chartered accoutant. Find the probability of forming the committee in the following manner :

(b) It should have at least one from the purchase department.

(c) The chartered accountant must be on the committee.

Solution:

Total no. of officers = 3 + 4 + 2 + 1 = 10.

4 persons can be chosen out of 10 in $^{10}C_4$ ways

$$= \frac{10 \times 9 \times 8 \times 7}{4 \times 3 \times 2 \times 1} = 210 \text{ ways.}$$

(i) 1 Person may be chosen from each category in :

$$^3C_1 = 3,\ ^4C_1 = 4,\ ^2C_1 = 2,\ ^1C_1 = 1 \text{ ways}$$

∴ Probability of forming a committee having one from each category

$$= \frac{3 \times 4 \times 2 \times 1}{210} = \frac{24}{210} = 0.114$$

(ii) Let P(A) represent the probability of at least one purchase officer being chosen. Then P(B), is the probability of none being purchase officer, and in this case, all the four persons are to be taken from the other six persons not belonging to the purchase department, which may be done in

$$^6C_4 = {}^6C_2 = \frac{6 \times 5}{2 \times 1} = 15 \text{ ways}$$

$$\therefore P(B) = \frac{15}{210} = \frac{1}{14}$$

and $\quad P(A) = 1 - P(B) = 1 - \frac{1}{14} = \frac{13}{14}$

(iii) When the only chartered accountant is always included in the committee, we have to choose any three persons out of a total of 9 persons, which may be done in $^9C_2$ se

$$= \frac{9 \times 8 \times 7}{3 \times 2 \times 1} = 84 \text{ ways}$$

Hence the required probability $= \frac{84}{210} = \frac{2}{5}$

Example 14:

The daily production of a machine producing a very complicated item gives the following probabilities for the number of items produced:

$$P(A) = 0.20,\ P(2) = 0.35,\ P(3) = 0.45$$

Furthermore, the probability of defective items being produced is 0.02. Defectives are assumed to occur independently. Determine the probability of no defectives during a day's reduction.

Solution:

Let A be the event that no defective item is produced during a day, then by total probability theorem

$$P(A) = P(1).\ P(A/1) + P(2).\ P(A/2) + P(3).\ P(A/3)$$

The probability that a defective item is produced is given to be 0.02, therefore, the probability that a non - defective item is produced will be 1– 0.02 = 0.98. Also defectives are assumed to occur independently, therefore

$P(A/1) = 0.98,\ P(A/2) = (0.98)\ (0.98).$

$P(A/3) = (0.98)\ (0.98)\ (0.98)$

$$P(A) = (0.20)\ (0.98) + (0.35)\ (0.98)^2 + (0.45)\ (0.98)^2$$

$$= 0.1960 + 0.3361 + 0.4322 = 0.9643.$$

Hence the probability of no defectives during a day's production is 0.9643.

Example 15:

A problem in statistics is given to two students A and B. The odds in favour of A solving the problem are 6 to 9 and against B solving the problem 12 to 10. If A and B attempt, find the probability of the problem being solved.

Solution:

$$P(\text{A's solving the problem}) = \frac{6}{6+9} = \frac{6}{15}$$

$$P(\text{B's solving the problem}) = 1 - \frac{12}{12+10} = \frac{5}{11}$$

$$P(\text{A's not solving the problem}) = 1 - \frac{6}{15} = \frac{9}{15}$$

P(Both A and B are unable to solve the problem)

$$= \frac{9}{15} \times = \frac{6}{11} = \frac{18}{55}$$

$$P(\text{the problem being solved}) = 1 - \frac{18}{55} = \frac{37}{55} = 0.673$$

Example 16:

(i) A class consists of 80 students, 25 of them are girls and 55 boys, 10 of them are rich and remaining poor, 20 of them are fair complexioned. What is the probability of selecting a fair complexioned rich girl?

(ii) Explain why there must be a mistake in the following statement:

Solution:

(i) Probability of selecting

$$\text{a fair complexioned person} = \frac{20}{80} = \frac{1}{4}$$

$$\text{Probability of selecting a rich person} = \frac{10}{80} = \frac{1}{8}$$

$$\text{The probability of selecting a girl} = \frac{25}{80} = \frac{5}{16}$$

The probability of selecting a fair complexioned rich girl

$$= \frac{1}{4} \times \frac{1}{8} \times \frac{5}{16} = \frac{5}{512} = 0.0098$$

(ii) We know that probability of happening of an event can never exceed one. In the question given.

P(A) + P(B) + P(C)..........

= 0.11 + 0.23 + 0.37 + 0.16 + 0.09 + 0.05 = 1.01

Since the total probability is greater than one, there is some mistake in the statement given.

Example 17:

Three groups of workers contain 3 men and one woman, 2 men and 2 women, and 1 man and 3 women respectively. One worker is selected at random from each group. What is the probability that the group selected consists of 1 man and 2 women ?

Solution:

There are three possibilities in this case :

(i) Man is selected from the first group and women from 2nd and 3rd groups; or

(ii) Man is selected from the 2nd group and women from the 1st and 3rd groups; or

(iii) Man is selected from the 3rd group and women from 1st and 2nd groups.

∴ The probability of selecting a group of 1 man and 2 women

$$= \left(\frac{3}{4} \times \frac{2}{4} \times \frac{3}{4}\right) + \left(\frac{2}{4} \times \frac{1}{4} \times \frac{3}{4}\right) + \left(\frac{1}{4} \times \frac{1}{4} \times \frac{2}{4}\right)$$

$$= \frac{9}{32} + \frac{3}{32} + \frac{1}{32} = \frac{13}{32}$$

Example 18:

An artical manufactured by a company consists of two parts A and B. In the process of manufacture of part A, 9 out of 100 are likely to be defective. Similarly, 5 out of 100 are likely to be defective in the manufacture of part B. Calculate the probability that the assembled part will not be defective.

Solution:

The assembled part will be defective if any of the part is defective .

∴ The probability of the assembled part being defective.

$$= P[\text{that any of the part is defective}]$$

$$= P[A \cup B]$$

$$= P(A) + P(B) - P(AB)$$

$$= \frac{9}{100} + \frac{5}{100} - \frac{9}{100} \times \frac{5}{100}$$

$$= 0.09 + 0.05 - 0.0045 = 0.1355$$

∴ The probability that assembled part is not defective $= 1 - 0.1355$

$$= 0.8645.$$

Example 19:

A bag contains 6 white, 4 red and 10 blacks balls. Two balls are drawn at random. Find the probability that they will both be black.

Solution:

Total number of balls in the bag

$$\therefore = 6 + 4 + 10 = 20$$

Two balls can be drawn from 20 in ${}^{20}C_2$ ways

$$= \frac{20 \times 19}{2 \times 1} = 190 \text{ ways}$$

And 2 balls can be drawn from 10 blacks balls in ${}^{10}C_2$ or

$$= \frac{10 \times 9}{2 \times 1} 45 \text{ ways.}$$

∴ The probability that the two balls drawn at random are black

$$= \frac{45}{90} = 0.237$$

Example 20:

Three horses A,B and C are in a race. A is twice as likely to win as B and B is twice as likely to win as C. What are the respective probability of winning ?

Solution:

We have A : B : C = 4 : 2 : 1

Probability of winning for horse A = 4/7

Probability of winning for horse B = 2/7

Probability of winning for horse C = 1/7

Example 21:

An investment consultant predicts that the odds against the price of a certain stock will go up during the next week are 2 : 1 and odds in favour of the price remaining the same are 1 : 3. What is the probability that the price of the stock will go down during the next week?

Solution:

We have:

P (price of a certain stock not going up) = 2/3

P (price of a certain stock remaining same) = 1/4

$\therefore$ The probability that the price of the stock will go down during the next week

= p (price of the stock not going up and not remaining same)

= P (price of the stock not going up × P (price of the stock not remaining same)

$$= \frac{2}{3} \times \left(1 - \frac{1}{4}\right) = \frac{2}{3} \times \frac{3}{4} = \frac{1}{2} = 0.5$$

Example 22:

A bag contains 2 white and 3 black balls. Four persons A, B, C, and D in the order named each draw one ball and do not replace it The person to draw a white ball receives Rs. 200. Determine their expectations.

Solution:

Since only 3 black balls are contained in the bag, one person must win in the first attempt.

Probability that A wins = 2/5

$\therefore$ A's expectations = $\frac{2}{5} \times 200$ = Rs. 80.

Probability that A loses and B wins

$$= \frac{3}{5} \times \frac{2}{4} = \frac{3}{10}$$

$\therefore$ B's expectation = $\frac{3}{10} \times 200$ = Rs. 60.

Probability that A and B lose but C wins

$$= \left(\frac{3}{5}\right)\left(\frac{2}{4}\right)\left(\frac{2}{3}\right) = \frac{1}{5}$$

$$\text{C's expectation} = \frac{1}{5}\times 200 = \text{Rs.}40$$

Probability that A, B and C lose and D wins

$$= \left(\frac{3}{5}\right)\left(\frac{2}{4}\right)\left(\frac{1}{3}\right)\left(\frac{2}{2}\right) = \frac{1}{10}$$

$$\therefore \text{D's expectation} = \frac{1}{10}\times 200 = \text{Rs.}20.$$

Thus the expectation of A, B, C and D respectively are Rs. 80, Rs. 60, Rs. 40 and Rs. 20.

Example 23:

A and B play for a prize of Rs. 1,000. A is to throw a dice first and is to win if he throws 6. If he fails B is to throw and is to win if he throws 6 or 5. If he fails, A is to throw again and to win if he throw 6, 5, or 4, and so on. Find their respective expectations.

Solution:

Probability of A's winning in the 1st throw

= 1/6

Probability of B's winning in the 2nd throw

$$= \frac{5}{6}\times\frac{2}{6} = \frac{5}{18}$$

Probability of A's winning in the 3rd throw

$$= \frac{5}{6}\times\frac{4}{6}\times\frac{3}{6} = \frac{5}{18}$$

Probability of B's winning in the 4th throw

$$= \frac{5}{6}\times\frac{4}{6}\times\frac{3}{6}\times\frac{4}{6} = \frac{5}{27}$$

Probability of A's winning in the 5th throw

$$= \frac{5}{6}\times\frac{4}{6}\times\frac{3}{6}\times\frac{2}{6}\times\frac{5}{6} = \frac{25}{324}$$

Probability of B's winning in the 6th throw

$$= \frac{5}{6}\times\frac{4}{6}\times\frac{3}{6}\times\frac{2}{6}\times\frac{1}{6}\times\frac{6}{6} = \frac{5}{324}$$

Total of A's chances of success

$$= \frac{1}{6}+\frac{5}{18}+\frac{25}{324}=\frac{169}{324}$$

Total of B's chances of success

$$= \frac{5}{18}+\frac{5}{27}+\frac{5}{324}=\frac{155}{324}$$

Their respective expectations are

A $= \frac{169}{324}\times 1{,}000 = \text{Rs.}521.6$

B $= \frac{155}{324}\times 1{,}000 = \text{Rs.}478.4$

Example 24:

A Bag contains 8 white and 4 red balls are drawn at random. What is the probability that 2 of them are red and 3 white ?

Solution:

Total number of balls in the bag = 8 + 4 = 12

Number of balls drawn = 5

5 balls can be drawn from 12 in $^{12}C_5$ ways, 2 red balls can be drawn from 4 red in $^4C_2$ ways and 3 white balls can be drawn from 8 white balls in $^8C_3$ ways.

∴ The number of favourable cases $^4C_2 \times {}^3C_3$ and the required probability is

$$P = \frac{^4C_2 \times {}^8C_3}{^{12}C_5}$$

$$= \frac{4\times 3}{2\times 1}\times\frac{8\times 7\times 6}{3\times 2\times 1}\times\frac{5\times 4\times 3\times 2\times 1}{12\times 11\times 10\times 9\times 8}=\frac{14}{33}=0.424.$$

Example 25:

One bag contains 4 white and 2 black balls. Another contains 3 white and 5 black balls. If one ball is drawn from each bag, find the probability that (a) both are white, (b) are black, and (c) one is white and one is black.

Solution:

Probability of drawing a white ball from the first bag = 4/6

Probability of drawing a white ball from the second bag = 3/8

(a) Since the events are independent the probability that both the balls are white $= \frac{48}{6}\times\frac{3}{8}=\frac{1}{4}$

(b) Probability of drawing a black ball from the first bag = 2/6

Probability of drawing a black ball from the second bag = 5/8

Probability that both are black = $\frac{2}{6} \times \frac{5}{8} = \frac{5}{24}$

(c) That event "one is white one is black" is the same as event "either the first is white and the second is black or the first is black and the second is white."

$\therefore$ The Probability that one is white and one is black

$$= \left(\frac{4}{6}\right)\left(\frac{5}{8}\right) + \left(\frac{2}{6}\right)\left(\frac{3}{8}\right)$$

$$= \frac{20}{48} + \frac{6}{48} = \frac{13}{24}$$

Example 26:

A bag contains 5 white and 8 red balls. Two drawing of 3 balls are made such that

(a) the balls are replaced before the second trial, and

(b) the balls are not replaced before the second trial.

Find the probability that the first drawing will give 3 white and the second 3 red balls in each case.

Solution (a):

When balls are replaced:

Total number of balls in the bag = 5 + 8 = 13.

3 balls can be drawn from 13 in $^{13}C_3$ ways.

3 white balls can be drawn from 5 in $^{5}C_3$ ways.

3 red balls can be drawn from 8 in $^{8}C_3$ ways.

$\therefore$ The probability of drawing 3 white balls in the first trial is

$$= \frac{^{5}C_3}{^{13}C_3} = \frac{5}{143}$$

and the probability of 3 red balls at the second trial is

$$= \frac{^{5}C_2}{^{12}C_2} = \frac{28}{143}$$

$\therefore$ The probability of the compound event is

$$= \frac{5}{143} \times \frac{28}{143} = \frac{140}{20449} = 0.007$$

(b) When balls are not replaced:

At the first trial 3 white balls can be drawn in $^5C_3$ ways.

$\therefore$ The probability of drawing 3 balls at the first trial is

$$= \frac{^5C_3}{^{13}C_3} = \frac{5}{143}$$

When the white balls have been drawn and not replaced, the bag contains 2 white and 8 red balls. Therefore, at the second trial 3 balls can be drawn from 10 in $^{10}C_3$ ways and 3 red balls can be drawn from 8 in $^5C_3$ ways.

$\therefore$ The probability of 3 red balls in the second trial $= \frac{^8C_3}{^{10}C_3} = \frac{7}{15}$

$\therefore$ The probability of the compound event $= \frac{5}{143} \times \frac{7}{15} = \frac{7}{429} = 0.016.$

Example 27:

Five men in a group of 20 are graduates. If 3 men are picked out of 20 at random

(i) what is the probability that all are graduates, and

(ii) what is the probability of at least one being graduate ?

Solution:

Probability of finding one graduate is

$$= \frac{5}{20} = \frac{1}{4}$$

Probability of 1 graduate out of 3 selected

$$= {}^3C_1 \left(\frac{1}{4}\right)^1 \left(\frac{3}{4}\right)^2 = \frac{27}{64}$$

Probability of 2 graduates out of 3 selected

$$= {}^2C_2 \left(\frac{1}{4}\right)^2 \left(\frac{3}{4}\right)^1 = \frac{9}{128}$$

Probability of 3 graduates out of 3 selected

$$= {}^3C_3 \left(\frac{1}{4}\right)^3 \left(\frac{3}{4}\right)^0 = \frac{1}{64}$$

Therefore, the prob. of at least one graduate out of 3 selected

$$= \frac{27}{64} + \frac{9}{128} + \frac{1}{64} = \frac{65}{128}$$

Example 28:

Prove that the sum of the probabilities of all possibilities in two independent events amount to certainty.

Solution:

Let the probability of success and failure in the first event be p_1 and q_1 and in the second event p_2 and q_2. Then,

The chance of success in the first event and success in the second event is

$p_1 \times p_2$

the chance of success in the first and failure in the second event is

$p_1 \times q_2$

the chance of failure in the first event and success in the second event is

$q_1 \times p_2$

the chance of failure in the first event and failure in the second event is

$q_1 \times q_2$

These are all the possibilities, and the sum of these possibilities is:

$= (p_1 \times p_2) + (p_1 \times q_2) + (q_1 \times p_2) + (q_1 \times q_2)$

$= p_1 (p_2 + q_2) + q_1 (p_2 + q_2)$

$= (p_1 + q_1) (p_2 + q_2)$

$= 1 \times 1 = 1 \; (\therefore \; p_1 + q_1 = p_2 + q_2 = 1).$

Example 29:

What is the probability that a leap year, selected at random, will contain 53 Sunday ?

Solution:

A leap year consists of 366 days and, therefore, contains 52 complete weeks and 2 days extra. These 2 days may make the following 7 combinations:

1. Monday and Tuesday
2. Tuesday and Wednesday
3. Wednesday and Thursday
4. Thursday and Friday
5. Friday and Saturday
6. Saturday and Sunday
7. Sunday and Monday

Of these seven equally likely cases only the last two are favourable. Hence the required probability = 2/7.

Example 30:

A University has to select an examiner from a list of 50 persons. 20 of them women and 30 men, 10 of them knowing Hindi and 40 not, 15 of them being teachers and the remaining 35 not. What is the probability of the University selecting a Hindi-knowing woman teacher ?

Solution:

Probability of selecting a woman = 20/50

Probability of selecting a teacher = 15/50

Probability of selecting a Hindi-Knowing candidate = 10/50

Since the events are independent the probability of the University selecting a Hindi-knowing woman teacher

$$= \frac{20}{50} \times \frac{15}{50} \times \frac{10}{50} = \frac{3}{125} = 0.024$$

Example 31:

The probability that a boy will get a scholarship is 0.9 and that a girl will get is 0.8. What is the probability that at least one of them will get the scholarship ?

Solution:

The probability that a boy will get a scholarship = 0.9

The probability that a girl will get a scholarship = 0.8

The probability that at least one of them will get the scholarship is

P(A) + P(B) – P(AB)

= .9 + .8 – (.9 × .8)

= 1.7 – .72 = 0.98.

Example 32:

A box contains 3 red and 7 white balls. One ball is drawn at random and in its place a ball of the other colour is put in the box. Now one ball is drawn at random from the box. Find the probability that it is red.

Solution:

At the second draw red ball can be drawn in two ways if at the first draw ball drawn is red or at the first draw the ball drawn is white.

When the first ball drawn is red:

The probability of drawing a red ball = 3/10

Now in the box a white ball is put in place of the red ball drawn so the box contains 2 red and 8 white balls.

Hence, the probability of drawing a red ball = 2/10

When the first ball drawn is white:

The probability of drawing a white ball = 7/10

Now in the box a red ball is put in place of the white ball drawn. There are thus 4 red and 6 white balls in the box.

∴ The probability of drawing a red ball = 4/10

Hence the probability of drawing a red ball

$$= \left(\frac{3}{10}\right) \times \left(\frac{2}{10}\right) + \left(\frac{7}{10}\right) \times \left(\frac{4}{10}\right)$$

$$= \frac{6}{100} + \frac{28}{100} = \frac{34}{100} = 0.34$$

Example 33:

(a) A can solve 90 percent of the problems given in a book and B can solve 70 percent. What is the probability that at least one of them will solve a problem selected at random ?

(b) In a single throw of two dice, what is the probability of obtaining a total of at least 10 ?

Solution:

(a) Probability that A will not be able to solve the problem

$$= 1 - \frac{9}{10} = \frac{1}{10}$$

Probability that B will not be able to solve the problem

$$= 1 - \frac{7}{10} = \frac{3}{10}$$

Probability that none of them will be able to solve the problem

$$= \frac{1}{10} \times \frac{3}{10} = \frac{3}{100}$$

Hence the probability that at least one of them will solve the problem

$$= 1 - \frac{3}{100} = \frac{97}{100}$$

(b) Two dice can be thrown together in $6 \times 6 = 36$ ways.

We have to find the probability of getting a sum of at least ten, i.e., either ten, eleven or twelve.

The probability of getting a total 10 in a single throw of two dice (6,4) (4,6) (5,5)

= 3/36

The probability of getting a total 11 in a single throw of two dice (6,5) (5,6)

= 2/36

Probability of getting a total 12 in a single throw of two dice (6,6)

= 1/36

Since the events are mutually exclusive the probability of obtaining a sum of at least 10 in a single throw of two dice $= \frac{3}{36} + \frac{2}{36} + \frac{1}{36} = \frac{6}{36} = \frac{1}{6}$

Example 34:

In a product testing procedure, each radio on an assembly line must pass two inspection points before being packaged for shipment. The probability is p1 = 0.7 that a defective radio is detected at the first inspection point and p2 = 0.8 that a defective radio is detected at the second inspection point. What is the probability that a defective radio will be packaged for shipment ?

Solution:

A defective radio has to pass two inspection points before it is packaged. The probability that such a radio will pass the first inspection point is P(A) = (1 – P1) = 0.03. The probability of passing the second inspection point given that the defective radio passed the first point, is P(B/A) = (1 – P2) = 0.20. The probability that a defective radio will pass both inspection points is, therefore,

P(A and B) = (1 – P1) = (1 – P2) = (0.30) (0.20) = .06

We conclude that about 6 percent of the defective radios will be packaged for shipment under the present inspection plan.

Example 49:

A manufacturing firm produces steel pipes in three plants with daily production volumes of 500, 1,000 and 2,000 units respectively. According to past experience, it is known that the fraction of defective output produced by the three plants are respectively .005, .008 and 0.10. If a pipe is selected from a day's total production and found to be defective, find out

(i) From which plant the pipe comes ?

(ii) What is the probability that it came from the first plant ?

Solution (i):

According to Bayes' theorem we have from the problem the following events:

A1: Production volume of first plant = 500 units
A2: Production volume of second plant = 1,000 units
A3: Production volume of third plant = 2.000 units
E: a defective item.

From these events, we see that P(A/E) is the probability that the item is produced by the plant, given that the item is defective. Also P(A ∩ E) is the probability that the items are produced by the plant and are defective. Information in the problem gives the following probabilities in connection with the random selection of a pipe from a day's total production:

(a) Prior probabilities:

$$P(A1) = \frac{500}{500 + 1000 + 2000} = \frac{1}{7}$$

$$P(A2) = \frac{1000}{3500} = \frac{2}{7};\ P(A3) = \frac{2000}{3500} = \frac{4}{7}$$

(b) Likelihoods:

P(E/A1) = 0.005; P(E/A2) = 0.008; P(E/A3) = 0.10

(c) Joint Probabilities:

$$P(A1 \cap E) = P(A1)\ P(E/A1) = \left(\frac{1}{7}\right)(0.005) = 0.005/7$$

$$P(A2 \cap E) = P(A2)\ P(E/A2) = \left(\frac{2}{7}\right)(0.008) = 0.016/7$$

$$P(A3 \cap E) = P(A3)\ P(E/A3) = \left(\frac{4}{7}\right)(0.10) = 0.040/7$$

$$P(E) = \Sigma\ P(A1)\ P(E/A1) = 0.061/7$$

(d) Posterior Probabilities:

$$P(A1/E) = \frac{P(A1 \cap E)_1}{P(E)} = \frac{0.005/7}{0.061/7} = \frac{5}{61}$$

$$P(A2/E) = \frac{P(A2 \cap E)}{P(E)} = \frac{0.016/7}{0.061/7} = \frac{16}{61}$$

$$P(A3/E) = \frac{P(A3 \cap E)}{P(E)} = \frac{0.040/7}{0.061/7} = \frac{40}{61}$$

Since P(A2/E) is by far the greatest posterior probability, it is then most probable that the defective item has been drawn from the output of the third plant As a check on the above calculations, the sum of all the posterior probabilities must be unity.

(ii) Probability that the pipe came from the first plant

$$= \frac{\frac{1}{7} \times 0.005}{\left(\frac{1}{7} \times 0.005\right) + \left(\frac{2}{7} \times 0.008\right) + \left(\frac{4}{7} \times .010\right)}$$

$$= \frac{.0007}{.0007 + .0023 + .0057} = \frac{.0007}{.0087} = 0.0805.$$

Example 35:

The odds against A speaking the truth are 4 : 6 while the odds in favour of B speaking the truth are 7 : 3.

(i) What is the probability that A and B contradict each other in stating the same fact ?

(ii) If A and B agree on a statement, what is the probability that this statement is true ?

Solution:

Let us define the events

E1 = A speaks the truth

E2 = B speaks the truth

Then $\overline{E1}$ and $\overline{E2}$ represent the complementary events that A and B tell lie respectively. We are given:

$$P(E1) = \frac{6}{10}\frac{3}{5};\ P(\overline{E1}) = 1 - \frac{3}{5} = \frac{2}{5}$$

$$P(E2) = \frac{7}{10};\ P(\overline{E2}) = 1 - \frac{7}{10} = \frac{3}{10}$$

(i) The event E that A and B contradict each other on an identical point can happen in the following mutually exclusive ways:

(a) A speaks the truth and B tells a lie, i.e., the event $E1 \cap \overline{E2}$ happens.

(b) A tells a lie and B speaks the truth, i.e., the event $\overline{E1} \cap E2$ happens.

By applying addition theorem, we get:

$$P(E) = P(E1 \cap \overline{E2}) + P(\overline{E1} \cap E2)$$

$= P(E1), P(\overline{E2}) + P(\overline{E1}). P(E2)$

$$= \left(\frac{3}{5}\times\frac{3}{10}\right) + \left(\frac{2}{5}\times\frac{7}{10}\right) = \frac{23}{50}$$

(ii) If a statement is true, the probability that both A and B agree in asserting that it is true is given by multiplication theorem of probability as:

$$= \frac{3}{5} \times \frac{7}{10} = \frac{21}{50}$$

The probability that both A and B agree in asserting that the statement is false is:

$$\frac{2}{5} \times \frac{3}{10} = \frac{6}{50}$$

Hence the required probability that the statement is true:

$$\frac{21/50}{(21/50) + (6/50)} = \frac{21}{27}$$

Example 36:

A Product is assembled from three components X, Y and Z, the probability of these components being defective is respectively 0.01, 0.02 and 0.05. What is the probability that the assembled product will not be defective?

Solution:

Let A, B and c denote the respective probabilities of components X, Y and Z being defective. We are given

P(A) = 0.01, P(B) = 0.02, P(C) = 0.05

P(A or B or C) = P(A) + P(B) + P(C) – P(AB) – P(BC) – P(AC) + P(ABC)

= 0.01 + 0.02 + 0.05 – 0.0002 – 0.0010 – 0.0005 + 0.00001

= 0.0784.

Hence the probability that the assembled product will not be defective = 1 – .07831 = 0.92169 or 0.922.

Example37:

An artical manufactured by a company consists of two parts A and B. In the process of manufacture of part A9 out of 100 are likely to be defective. Similarly A out of 100 are likely to be defective in the manufacture of part B. Calculate the probability that the assembled parts will not be defective.

Solution:

P(Part A is defective) = 9/100 = 0.09

P(Part A is not defective) = 1 – 0.09 = 0.91

P(Part B is not defective) = 5/100 = 0.05

P(Part B is not defective) = 1 – .05 = .95

P(Assembled parts will not be defective)

= P(Part A is not defective) × P(Part B is not defective)

= 0.91 × 0.95 = 0.864.

Example 38:

Items produced by a certain process each may have one or both of two types of defects, A and B. It is known that 20 percent of the items have type A defects and 10 percent have type B defects. Furthermore, 6 percent are known to have both types of defects. What is the probability that a randomly selected item will be defective ?

Solution:

We are given that

$P(A) = 0.20,\ P(B) = 0.10,\ P(AB) = 0.06$

∴ Required probability will be given by

$P(A \text{ or } B) = P(A) + P(B) - P(AB) = 0.20 + 0.10 - 0.06 = 0.24$

Hence the probability that a randomly selected item will be defective is 0.24.

Example 39:

A bag contains 30 balls numbered from 1 to 30. One ball is drawn at random. Find the probability that the number of the ball drawn will be a multiple of (a) 5 or 7, and (b) 3 or 7.

Solution:

The probability of the number being multiple of 5 is

$$P\,(5, 10, 5, 20, 25, 30) = \frac{6}{30}$$

The probability of the number being multiple of 7 is

$$P\,(7, 14, 21, 28) = \frac{4}{30}.$$

Since the events are mutually exclusive the probability of the number being a multiple of 5 or 7b will be

$$\frac{6}{30} + \frac{4}{30} = \frac{10}{30} = \frac{1}{3}$$

The probability of the number being multiple of 3 is

$$P\,(3, 6, 9, 12, 15, 18, 21, 24, 27, 30) = \frac{10}{30}$$

The probability of the number being multiple of 7 is

$$P\ (7, 14, 21, 28) = \frac{4}{30}$$

Since 21 is a multiple of 3 as well as 7, the drawing of the ball numbered 21 entails the occurrence of both the events and hence the probability of getting a number which is multiple of 3 or 7 is

$$= \frac{10}{30} + \frac{4}{30} - \frac{1}{30} = \frac{13}{30}.$$

Example 40:

The Managing Committee of Vaishalli Welfare Association formed a sub-committee of 5 persons to look into electricity problem. Profiles of the 5 persons are :

1. male age 40
2. male age 43
3. female age 38
4. female age 27
5. male age 65

If a chairperson has to be selected from this, what is the probability that he would be either female or over 30 years ?

Solution:

P (female or over 30) = P (female) + P (over 30) – P (female and over 30)

$$= \frac{2}{5} + \frac{4}{5} - \frac{1}{5} = \frac{5}{5} = 1.$$

Multiplication Theorem

This theorem states that if two events A and B are independent, the probability that they both will occur is equal to the product of their individual probability. Symbolically, if A and B are independent, then

$$P\ (A \text{ and } B) = P\ (A) \times P\ (B)$$

The theorem can be extended to three or more independent events.

Thus $P\ (A, B \text{ and } C) = P\ (A) \times P\ (B) \times P\ (C)$

Proof of the Theorem. If an event A can happen in n_1 ways of which a_1 are successful and the event B can happen in n_2 ways of which a_2 are successful, we can combine each successful event in the first with each successful event in the second case. Thus, the total number of successful happenings in both cases is $a_1 \times a_2$. Similarly, the total number of possible cases is $n_1 \times n_2$.

Then by definition the probability of the occurrence of both events is

$$\frac{a_1 \times a_2}{n_1 \times n_2} = \frac{a_1}{n_1} \times \frac{a_2}{n_2}$$

But $$\frac{a_1}{n_1} = P\ (A)$$

and $$\frac{a_2}{n_2} = P\ (B).$$

$\therefore$ P (A and B) = P (A) × P (B).

In a similar way the theorem can be extended to three or more events.

Example 41:

A problem in statistics is given to five students A, B, C, D and E. Their chances of solving it are $\frac{1}{2}, \frac{1}{3}, \frac{1}{4}, \frac{1}{5}$ and $\frac{1}{6}$. *What is the probability that the problem will be solved ?*

Solution:

Probability that A fails to solve the problem is $1 - \frac{1}{2} = \frac{1}{2}$

Probability that B fails to solve the problem is $1 - \frac{1}{3} = \frac{2}{3}$

Probability that C fails to solve the problem is $1 - \frac{1}{4} = \frac{3}{4}$

Probability that D fails to solve the problem is $1 - \frac{1}{5} = \frac{4}{5}$

Probability that E fails to solve the problem is $1 - \frac{1}{6} = \frac{5}{6}$

Since the events are independent the probability that all the five students fail to solve the problem is :

$$\frac{1}{2} \times \frac{2}{3} \times \frac{3}{4} \times \frac{4}{5} \times \frac{5}{6} = \frac{1}{6}$$

The problem will be solved if anyone of them is able to solve it.

$\therefore$ The probability that the problem will be solved $= 1 - \frac{1}{6} = \frac{5}{6}$

Example 42:

A man wants to marry a girl having qualities; white complexion—the probability of getting such a girl is one in twenty; handsome dowry—the probability of getting this is one in fifty; westernised manners and etiquettes—the probability here is one in hundred. Find out the probability of his getting married to such a girl when the possession of these three attributes is independent.

Solution:

Probability of a girl with white complexion

$$= \frac{1}{20} = 0.05$$

Probability of a girl with handsome dowry

$$= \frac{1}{50} = 0.02$$

Probability of a girl with westernised manners

$$= \frac{1}{100} = 0.01$$

Since the events are independent, the probability of simultaneous occurrence of all these qualities

$$= \frac{1}{20} \times \frac{1}{50} \times \frac{1}{100} = 0.05 \times 0.02 \times 0.01 = 0.0001.$$

If we are given n independent events $A_1, A_2, A_3 ..., A_n$ with respective probability of occurrence as $p_1, p_2, p_3, ..., p_n$, then the probability of occurrence of at least one of the n events $A_1, A_2, A_3, ..., A_n$ are can be determined as follows:

p (happening of at least one of the events) = 1

– p (happening of none of the events).

The following example hall illustrate the application of the above principle.

Example 43:

A bag contains 5 white and 3 black balls. Two balls are drawn at random one after the other without replacement. Find the probability that both balls drawn are black.

Solution:

Probability of drawing a black ball in the first attempt is

$$P(A) = \frac{3}{5+3} = \frac{3}{8}$$

Probability of drawing the second black ball given that the first ball drawn is black

$$P(B/A) = \frac{2}{5+2} = \frac{2}{7}$$

∴ The probability that both balls drawn are black is given by

$$P(AB) = P(A) \times P(B/A) = \frac{3}{8} \times \frac{2}{7} = \frac{3}{28}.$$

EXERCISES

1. Explain the various approaches to probability. Are these approaches contradictory?

   What is conditional probability? Explain with the help of an example.
2. Explain with examples the various school of thought on probability. Discuss the importance of probability in business decision-making.
3. (a) Explain the concept of probability and point out its role in business decision-making.

   (b) What is conditional probability? Explain with the help of an example. Discuss the importance of probability in statistics.
4. (a) Define the concepts of conditional probability and independent events.

   (b) Explain what do you understand by the term probability. Discuss its importance in decision-making.
5. (a) Explain the concepts of simple, joint, conditional and marginal probabilities. Give examples to illustrate these probabilities.

   (b) State and explain Baye's theorem and bring out its importance in probability, theory.

   (c) State the relationship between ${}^nP_r$ and ${}^nC_r$.

   (d) In how many ways can the word 'management' be arranged?
6. (a) Distinguish between conditional probability and joint probability under conditions of statistical dependence with help of an example.

   (b) What do you mean by empirical approach to probability? Discuss the importance of probability in statistics.
7. (a) State and prove the multiplicative theorem of probability. How is the result modified when the events are independent?

   (b) State and prove the addition and multiplication theorems of probability for two mutually exclusive events.
8. (a) State the addition and multiplication theorems of probability and give two different examples illustrating the application of these theorems.

   (b) Define independent and mutually exclusive events. Can two events be mutually exclusive and dependent simultaneously? Support your answer with an example.
9. (a) Explain with examples the difference between 'independent' and 'mutually exclusive' events in the theorem of probability. State the theorem of total and compound probabilities.

(b) State the addition theorem of probability for two events : (a) when they are mutually exclusive, and (b) when they are not mutually exclusive.

10. (a) Define probability and explain the importance of this concept in Statisues.

(b) Explain the various approaches to probability. Are these approaches contradictory?

(c) Define probability. Briefly explain the different schools of thoughts on probability.

(d) When are two events said to be independent in the probability sense? Give examples of dependent and independent events.

11. (a) (i) Explain the theorems of probability.

(ii) What is joint probability?

(iii) What is inverse probability?

12. (a) Define mutually exclusive events. Stage and prove addition theorem of probability for two events.

(b) Define probability. Does probability always relate to one event only ? Give examples.

(c) Examine critically the different schools of thought on probability.

13. (a) A player tosses four fair coins. He wins Rs. 16 if four tails occur; Rs. 8 if three tails occur; Rs. 4 if two tails occur and Rs. 2 if one tail occurs. If the game is to be fair how much should he win or lose in case no tail occurs?

(b) There are 100 students in a college class of which 36 are boys studying Statistics and 13 are girls not studying Statistics. If there are 55 girls in all, find the probability that a boy picked up at random is not studying Statistics.

14. (a) The odds again student X solving Business Statistics problem are 8:6 and odds in favour of student Y solving the sample problem are 14:16:

(i) What is the chance that the problem will be solved if they both try?

(ii) What is the probability that they both working independently of each other, solve the problem?

(iii) What is the probability that neither solves the problem?

(b) In a colony, 5,000 persons are residing, out of which 1,200 are above 30 years of age and 3,000 are females. Out of 1,200 who are above 30 years, 200 are females. Suppose after a person is chosen you are told that the person is a female. What is the probability that 'she is above 30 years of age?

15. (a) Five coins whose faces are marked 2 and 3 are thrown. What is the probability of obtaining a total of 12?

(b) If it rains, a taxi driver can earn Rs. 100 per day. If it is fair, he can lose Rs. 10 per day. If the probability of rain is 0.4, what is his expectation?

16. (a) A husband and wife appear for an interview for two vacancies in the same post. The probability of husbands selection is 1/3 and that of wife's selection is 1/5. Find the probability that :

(i) both of them will be selected.

(ii) only one of them will be selected.

[Refer to similar question in illustrations.]

(b) A company has 300 employees of whom 20 are men. When questioned, 180 people agreed that they were happy in their work. What is the probability that a man employee in the company is unhappy?

(c) A bag contains 6 white, 4 red and 10 black balls. Two balls are drawn at random. Find the probability that they will both be black.

(d) In a class of 50 students, 20 play football and 16 play basket ball. If 10 play both the games, find out how many play neither.

17. (a) What are mutually exclusive events? State which of the following are mutually exclusive :

(i) A : a red card, B : an ace in a draw of a card from a pack.

(ii) A : a total of 7, B: an odd number on each die in a simultaneous throw of 2 dice. Give reasons for you answer.

(b) A manufacturing firm produces T.V. sets in three plants with daily production value of 250, 500 and 1,000 units respectively. According to past experience, it is known that the fractions of defective output produced by the three plants are respectively, .005, .008 and .010. If a T.V. set is selected from a day's total production and found to be defective, find out (i) from which plant the T.V. comes, (ii) what is the probability that it comes from the second plant?

18. (a) A bag contains 8 red and 7 white balls. Two balls are drawn from the bag. Find the probability that (i) they are both red (ii) they are both white (iii) one red and the other white.

(b) In the play of two dice, the throw loses if the first throw is 2, 4 or 12. He wins if the first throw is a 5 or 11. Find the ratio between his probability of losing and probability of winning in the first throw.

19. (a) A company is planning to market a new product A. If its competitor does not market a similar product, the probability that product A will do well is 0.8. If the competitor also markets similar product, the probability that product A will do well is 0.4. It is believed that the probability that the competitor will market a similar product is 0.6. What is the probability that product A will do well if it is marketed?
[0.56]

    (b) The probabilities of the students A, B, C solving a problem are respectively 1/2, 1/3, 1/4. What is the probability that the problem is solved if they attempt to solve it independently?
[3/4]

20. (a) Each of the three identical jewellery boxes has 2 drawers. In each drawer of the first box there is a gold watch. In each drawer of the second box there is a silver watch. In one drawer of the third box there is a gold watch while in the other drawer there is a silver watch. If we select a box at random, open one of the dawers and find it to contain a silver watch, what is the probability that the other drawer has the gold watch?

    (b) A and B won dice for a prize of Rs. 11 which is to be won by the player who first throws 6. If A has the first throw what are their respective expectations?

    (c) A player tosses three fair coins. He wins Rs. 12 if three tails occur, Rs. 7 if two tails occur and Rs. 2 if only one tail occurs. If the game is to be fair, how much should he win or lose in case no tail occurs?

21. (a) Suppose it is 9 : 7 against a person A, who is now 35 years of age living till he is 95, and 3:2 against a person B, now 45 years of age living till he is 75. Find the chance that at least one of these persons will be alive 30 years hence.
[53/80]

    (b) A bag contains 10 red, 5 white and 4 blue balls. If 4 balls are drawn at random, determine the probability that - (i) all 4 are blue balls, and (ii) 1 red and 2 are white.
[1/3167, 100/969]

22. (a) If the probability that the value of certain stock will remain the same is 0.46, the probabilities that its value will increase by 50 paise or Re. 1.00 per share are respectively 0.17 and 0.23 and

the probability that its value will decrease by Re. 0.25 per share is 0.14, what is the expected gain per share?

(b) From a sales force of 150 people, one will be chosen to attend a special sales meeting. If 52 are single and 72 are college graduates, and 3/4 of 52 that are single are college graduates, what is the probability that a sales person selected at random will be neither single nor a college graduate?
[13, 130]

23. (a) A speaks truth in 60 percent cases and B speaks truth in 70 percent cases. In what percentage of cases are they likely to contradict each other in stating the same fact? [46%]

(b) A speaks truth is 80% cases and B in 90% of the cases. In what percentage of cases they are likely to contradict each other in stating the same fact?

24. (a) A can solve a problem of statistics in 4 out of 5 chances and P can do it in 2 out of 3 chances. If both A and B try to solve the problem, find the probability that it will be solved.

(b) (i) In how many ways can be alphabets of the word LOOPHOLE be arranged?

(ii) In throw of 2 rectangular dice containing numbers from 1 to 6 on their 6 sides, what are the probabilities of getting (a) a total of 12. and (b) a total of 10.

(iii) The probability of a cricket team winning match at Kanpur is 2/5 and losing match at Delhi is 1/7. What is the probability of the team winning at least one match?

(iv) The probability of selling a tin of ghee of four different trademark in a shop are given below:

Trade Mark :	A	BC		D
Probability of a tin being sold :	0.24	0.30	0.50	0.40

A buyer purchased a tin of ghee from that shop. What is the probability that it was of 'A' trade mark?

(v) A and B enter into a bet according to which A will get Rs. 200 if it rains on that day and will lose Rs. 100 if it does not rain. The probability of raining on that day is 0.7. What is the mathematical expectation of A?

25. (a) Two cards are drawn from a well-shuffled pack of 52 cards. Find the probability that they are both aces if the first is (i) replaced, and (ii) not replaced.
[(i) 1/16; (ii) 1/221]

(b) Urn A contains 5 red balls and 5 black bals, urn B contains 4 red and 8 black balls and urn C contains 3 red and 6 black balls. A ball is drawn from A, colour unknown, and put into B. Then a ball is drawn from B, colour unknown, and put into C. What is the probability that ball now drawn from C will be red? [87/260]

26. (a) A problem of Statistics is given to three students for solution. Their probabilities of solving it are 1/2, 1/3 and 1/4 respectively, What is the probability that the problem will be solved? [3/4]

(b) A group consists of 100 men and 80 women, 40 among men and 50 among women are graduates. If one person is selected at random from the group, find the probability the person is either a woman or a graduate.

27. (a) Two computers A and B are to be marketed. A salesman, who is assigned the job of finding customers for them, has 60% and 40% chances respectively of succeeding in case of computers A and B. The two computers can be sold independently. Given that he was able to sell at least one computer, what is the probability that computer A has been sold?

(b) State and explain Bayes' Theorem. From a pack of cards, two cards are drawn, the first being replaced before the second is drawn, what is the probability that first is a heart and second is not a king?

28. (a) The probability that India wins a cricket Test match against England is given to be 1/3. If India and England play three Test matches, what is the probability that:

(i) India will lose all the three Test matches?

(ii) India will win at least one Test match?

[(i) 8/27; (ii) 19/27]

(b) The probabilities of X, Y and Z becoming managers are 4/9, 2/9 and 1/3 respectively. The probabilities that the Bonus scheme will be introduced if X, Y and Z become managers, are 3/10, 1/2 and 4/5 respectively, (i) What is the probability that the Bonus scheme will be introduced, (ii) what is the probability that the manager appointed was X?

[(i) 23/45; (ii)6/23]

29. (a) Three contractors A, B and C are bidding for the construction of a new cinema hall. An expert in industry believes that A has

exactly half the chance that B has; B, in turn, is 4/5 as likely as C wins the contract. What is the probability for each to win the contract if the expert's estimates are accurate?

[A2/ 11, B4/11, C5/11]

(b) Radio values are tested by subjecting them to a number of electric shocks. Each shock has an independent chance 0.8 of destroying the valves. How many shocks must be given to the valve in order that the probability of the valve being destroyed is 0.99?

[3 shocks]

30. (a) An investment firm purchases three stocks for one week trading purposes. It assesses the probability that the stocks will increase in value over the week as 0.8, 0.75 and 0.69 respectively. What is the probability that all the three stocks will increase, assuming that the movements of these stocks are independent? Is this a reasonable assumption?

(b) Two events A and B are mutually exclusive:

P (A) = 1/5 and P (B) = 1/3, find the probability that

(i) either A or B will occur, (ii) both A and B will occur, (iii) neither A nor B will occur.

31. (a) The probability that a man will be believe for 25 years is 3/5 and the probability that his wife will be alive for 25 years is r/ 3. Find the probability that (i) both will be alive (ii) only the man will be alive (iii) only the wife will be alive (iv) at lest one will be alive.

[(i) 2/5, (ii) 1/5, (iii) 4/15, (iv) 3/4]

(b) A manager has two assistants and he bases his decision on information supplied independently by each of them. The probability that he makes a mistake in his thinking is 0.005. The probability that an assistant gives wrong information is 0.3. Assuming that the mistakes made by the manager are independent of the information give by the assistants, find the probability that he reaches a wrong decision.

[0.513]

32. (a) Out of 5 vowels and 7 consonants, how many words can be made taking 3 vowels and 3 consonants?

(b) A problem in Statistics is given to the three students P, Q and R whose chances of solving it are 1/3, 1/4 and 2/5 respectively. What is the probability that the problem will be solved?

33. (a) A card is drawn at random from a full pack of 52 playing cards. What is the probability that the drawn card will be either a black card or an Ace?

    (b) Find the probability of drawing a queen, a king and jockey in that order from a pack of cards in three consecutive draws, the cards drawn not being replaced.

    (c) A person is known to hit the larger in 3 out of 4 shots, whereas another person is known to his the larger in 2 out of 3 shots. Find the probability of the target being his when both of them try.

34. (a) If a player plays a game of chance where he can win Rs. 1,000 with probability 0.5, win Rs, 500 with probability 0.3 and lose Rs. 3,000 with probability 0.2, what is his expected gain in one play of the game?

    (b) Explain whether or not each of the following claims could be correct:

    (i) A businessman claims the probability that he will get contract A is 0.5 and that he will get contract B is 0.20. Furthermore, he claims that the probability of getting A or B is 0.50

    (ii) A market analyst claims that the probability of selling ten million pounds of plastic A or five million pounds of plastic B is 0.60. He also claims that the probability of selling ten million pounds of A and five million pounds of B is 0.45.

35. Two computers, say, A and B, are to be marketed, A salesman, who is assigned the job of finding customers for them, has 60% and 40% chances respectively of succeeding in case of computers A and B. The computers can be sold independently. Given that he was able to sell at least one computer, what is the probability that A has been sold?

    [0.79]

36. Out of 800 families with 4 children each, what percentage would be expected to have (a) 2 boys and 2 girls, (b) at least one boy, (c) no girls, and (d) at most 2 girls. Assume equal probability for boys and girls?

    [(a) 37% (b) 93.75% (c) 6.15% (d) 68.75%]

8

Correlation Analysis

INTRODUCTION

According to *Craxton and Cowden "When the relationship is of quantitative nature, the appropriate statistical tool for discovering and measuring the relattonship, and expressing it in a brief formula is known as correlation"*. Although the correlations analysis deals with the association between two or more variables and determines the extent of relationship, it does not give any idea about cause and effect relationship. Historically, the study of the prediction of one variable from the knowledge of another preceded the development of measures of correlation. In 1885 *Francis Gallon* published a paper called *"Regression towards Mediocrity in Hereditary Stature."* Galton was interested in predicting the physical characteristics of offspring from a knowledge of the physical characteristics of the parents. He observed, for example, that the offsprings of tall parents tended on the average to be shorter than their parents, whereas the offsprings of short parents tended on the average to be taller than their parents. In parties we come across a large number of problems involving the use of two or more than two variables. If two quantities vary in such a way that movements in one are accompanied by movements in the other, these quantities are correlated.

For example, there exists some relationship between age of husband and age of wife, price of commodity and amount demanded, increase in rainfall up to a point and production of rice, an increase in the number of television licences and number of cinemagoers, etc. The degree of relationship between the variables under consideration is measured through the Correlation.

For second aspect a reference may be made to chapter on Tests of Significance. The third aspect in the analysis, that of establishing the cause-effect relation, is difficult to be treated statistically. An extremely high and significant correlation between the increase in smoking and increase in lung cancer would not prove that smoking causes lung cancer. The proof of a cause and effect relation can be developed only by means of an exhaustive study of the operative elements themselves.

PROPERTIES OF THE COEFFICIENT OF CORRELATION

The following are the important properties of the correlation coefficient. *r*:

1. The coefficient of correlation lies between – 1 and + 1. Symbolically.
   $- \leq r \leq +1$ or $|r| \leq 1$

Proof:

Let x and y be deviations of X and Y series from their mean and σ_x and σ_y be their standard deviations. Expand the function :

$$\Sigma = \left(\frac{x}{\sigma_x} + \frac{y}{\sigma_y}\right)^2 = \Sigma\left[\frac{x^2}{\sigma_x^2} + \frac{y^2}{\sigma_y^2} + \frac{2xy}{\sigma_x \sigma_y}\right]$$

$$= \frac{\Sigma x^2}{\sigma_x^2} + \frac{\Sigma y^2}{\sigma_y^2} + \frac{2\Sigma xy}{\sigma_x \sigma_y}$$

But $$\frac{\Sigma x^2}{\sigma_x^2} = N \left[\because \sigma_x^3 = \frac{\Sigma x^2}{N} \text{ or } \frac{\Sigma x^2}{\sigma_x^2} = \frac{\Sigma x^2}{\Sigma x^2} \times N = N\right]$$

Similarly, $$\frac{y^2}{\sigma_y^2} = N$$

Also, $$= \frac{2xy}{\sigma_x \sigma_y} = 2Nr \qquad \left[\because r = \frac{\Sigma xy}{N\sigma_x \sigma_y}\right]$$

Hence

$$\Sigma\left(\frac{x}{\sigma_x} + \frac{y}{\sigma_y}\right)^2 = N + N + 2Nr = 2N + 2Nr = 2N(1 + r).$$

But $\Sigma\left(\frac{x}{\sigma_x} + \frac{y}{\sigma_y}\right)$ is the sum of squares of real quantities and as such it cannot b negative; at the most it can be zero.

$\therefore \qquad 2N(1 + r \geq 0)$

Hence r cannot be less than – 1; at the most it can be – 1.

Similarly, by expanding $\left(\frac{x}{\sigma_x} + \frac{y}{\sigma_y}\right)^2$ it can be shown that this is equal to 2 N (r – 1).

The again cannot be negative : at the most it can be zero.

$\because$ r cannot be grater than + 1 : at the most it can be + 1.

Hence $\quad -1 \leq r \leq +1.$

2. The coefficient of correlation is independent of change of scale and origin of the variable X and Y.

Proof:

By change of origin we mean subtracting some constant from every given value of X and Y and by change of scale we mean dividing or multiplying every value of X and Y by some constant.

We know that $r_{xy} = \dfrac{\Sigma(X-\overline{X})(Y-\overline{Y})}{\sqrt{\Sigma(X-\overline{X})^2 (Y-\overline{Y})^2}}$

where $\overline{X}$ and $\overline{Y}$ refer to actual means of X and Y series.

Let us now change the scale and origin. Deduct a fixed quantity a from X and b from Y. Also divide X and Y series by a fixed value i and c. After these changes are introduced, new values of x and y obtained from original X and Y shall be

$$x = \frac{X-a}{i} \text{ and } y = \frac{Y-b}{c}$$

$$\text{Mean of } x = \frac{\Sigma\left(\frac{(X-a)}{i}\right)}{n} = \frac{\frac{\Sigma X - Na}{i}}{N} = \frac{\Sigma X - Na}{Ni}$$

But
$$\frac{\Sigma X - Na}{Ni} = \frac{X-a}{i}$$

Thus, $\quad \text{mean of } x = \dfrac{\overline{X}-a}{i}$

Similarly, it can be shown that mean of $y = \dfrac{\overline{Y}-b}{c}$

The value of coefficient of correlation r, for new set of values will be

$$r_{xy} = \frac{\Sigma\left(\frac{X-a}{i} - \frac{\overline{X}-a}{i}\right)\Sigma\left(\frac{Y-b}{c} - \frac{\overline{Y}-b}{c}\right)}{\sqrt{\Sigma\left(\frac{X-a}{i} - \frac{\overline{X}-a}{i}\right)^2 \Sigma\left(\frac{Y-b}{c} - \frac{\overline{Y}-b}{c}\right)^2}}$$

$$= \frac{\Sigma\left(\frac{X-a-\overline{X}-a}{i}\right)\Sigma\left(\frac{Y-b-\overline{Y}-b}{c}\right)}{\sqrt{\Sigma\left(\frac{X-a-\overline{X}-a}{i}\right)^2 \Sigma\left(\frac{Y-b-\overline{Y}-b}{c}\right)^2}}$$

$$= \frac{\dfrac{\Sigma(X-\overline{X})(Y-\overline{Y})}{ic}}{\sqrt{\dfrac{\Sigma(X-\overline{X})^2}{i^2} \times \dfrac{\Sigma(Y-\overline{Y})^2}{c^2}}}$$

$$= \frac{\Sigma(X-\overline{X})(Y-\overline{Y})/ic}{\sqrt{\Sigma(X-\overline{X})^2 (Y-\overline{Y})^2/i^2c^2}} = \frac{\Sigma(X-\overline{X})(Y-\overline{Y})}{\sqrt{\Sigma(X-\overline{X})^2\,\Sigma(Y-\overline{Y})^2}}$$

Thus the coefficient of correlation is independent of change of scale and origin.

3. The coefficient of correlation is the geometric mean of two regression coefficient.

   Symbolically, $r = \sqrt{b_{xy} \times b_{yx}}$

4. The degree of relationship between the two variables is symmetric as shown below :

$$r_{xy} = r_{yx}$$

$$r_{xy} = \frac{\Sigma xy}{N\sigma_x \sigma_y} = \frac{\Sigma yx}{N\sigma_y \sigma_x} = r_{yx}.$$

COEFFICIENT OF DETERMINATION

One very convenient and useful way of interpreting the value of coefficient of correlation between two variables is to use square of coefficient of correlation, which is called coefficient of determination. The coefficient of determination thus equals r^2. If the value of $r = 0.9$, r^2 will be 0.81 and this would mean that 81 percent of the variation in the dependent variable has been explained by the independent variable. The maximum value of r^2 is unity because it is possible to explain all of the variation in Y, but it is not possible to explain more than all of it.

The coefficient of determination (r^2) is defined as the ratio of the explained variance to the total variance.

$$\text{Coefficient of determination} = \frac{\text{Explained Variation}}{\text{Total Variance}}$$

The ratio of explained variance to total variance is frequently called coefficient of non-determination. The coefficient of non-determination is denoted by K^2 and its square root is called the coefficient of alienation, or K. The K and K^2 value may also be used as the measure of the degree or relationship between two variables. For example, the higher the unexplained variance with respect to total variance, the higher will be the value of K^2 and the value of K. However, r^2 and r are more convenient in interpreting the result of Correlation.

It is much easier to understand the meaning of r^2 than r and, therefore, the coefficient of determination should be preferred in presenting the result of Correlation. Tuttle has beautifully pointed out that "the coefficient of correlation has been grossly overrated and is used entirely too much. Its square, the coefficient of determination, is a much more useful measure of the linear covariation of two variables. The reader should develop the habit of squaring every correlation coefficient he finds cited or stated before coming to any conclusion about the extent of the linear relationship between the two correlated variables."

The relationship between r and r^2 may be noted—as the value of r decreases from its maximum value of 1, the value of r^2 decreases much more rapidly, r will of course always be larger than r^2, unless $r^2 = 0$ or 1, 0 when $r = r^2$.

r	r^2	r	r^2
0.90	0.81	0.60	0.36
0.80	0.64	0.50	0.25
0.70	0.49	0.40	0.16

Thus the coefficient of correlation is 0.707 when just half the variance in Y is due to X. It should be clearly noted that the fact that a correlation between two variables has a value of $r = 0.60$ and the correlation between two other variables has a value of $r = 0.30$ does not demonstrate that the first correlation is twice as strong as the second. The relationship between the two given values of r can better be understood by computing the value of r^2. When $r = 0.06$, $r = 0.036$ and when $r = 0.30$, $r^2 = 0.90$. This implies that in the first case 36% of the total variation is explained whereas in the second case 90% of the total variation is explained.

The coefficient of determination is a highly useful measure. However, it is often misinterpreted. The term itself may be misleading in that it implies that the variable X stands in a determining or causal relationship to the variable Y. The statistical evidence itself never establishes the existence of such causality. All that the statistical evidence can do is to define covariation, that term being used in a perfectly neutral sense. Whether causality is present or not and which way it runs if it is present, must be determined on the basis of evidence other than the quantitative observations. However, r^2 is always a positive number. It cannot tell whether the relationship between the two variables is positive or negative. Thus the square root of r^2 i.e., $\sqrt{r^2} = \pm r$ is frequently computed to indicate the direction of the relationship, in addition to indicating the degree of relationship. Since the range of r^2 is from

0 to 1, the coefficient of correlation r will vary within the range of $\sqrt{0}$ to $\sqrt{1}$, or from ± 1. The + (plus) sign of *r* will indicate positive correlation,

whereas the – (minus) sign will mean a negative correlation.

CORRELATION OF GROUPED DATA

When the number of observations is large, the data are often classified into two-way frequency distribution called a correlation table.

The class intervals for Y are listed in the captions or column headings, and those for X are listed in the stubs at the left of the table, the order can also be reversed. The frequencies for each cell of the table are determined by either tallying or card sorting just as in the case of frequency distribution of a single variable.

The formula for calculating the coefficient of correlation is :

$$r = \frac{N\Sigma f d_x d_y - \Sigma f d_x \Sigma f d_y}{\sqrt{N\Sigma f d_x^2 - (\Sigma f d_x)^2}\sqrt{N\Sigma f d_y^2 - (\Sigma f d_y)^2}}$$

ASSUMPTION OF THE PEARSONIAN COEFFICIENT

Karl Pearson's coefficient of correlation is based on the following assumptions :

1. The two variables under study are affected by a larger number of independent causes so as to form a normal distribution. Variables like height, weight, price, demand, supply, etc., are affected by such forces that a normal distribution is formed.
2. There is linear relationship between the variables, i.e., when the two variables are plotted on a scatter diagram a straight ling will be formed by the points so plotted.
3. There is a cause and effect relationship between the forces affecting the distribution of the items in the two series. If such a relationship is not formed between the variables, i.e., if the variables are independent, there cannot be any correlation. For example, there is no relationship between income and height because the forces that affect these variables are not common.

MERITS AND LIMITATIONS OF THE PEARSONIAN COEFFICIENT

Amongst the mathematical methods used for measuring the degree of relationship. Karl Pearson's method is most popular. The correlation coefficient summarizes in one figure not only the degree of correlation but also the direction, i.e., whether correlation is positive or negative.

However, the utility of this coefficient depends in part on a wide knowledge of the meaning of this 'yardstick' together with its limitations. The chief limitations of the method are :

1. As compared with other methods this method takes more time to compute the value of correlation coefficient.
2. The value of the coefficient is unduly affected by the extreme items.
3. Great care must be exercised in interpreting the value of this coefficient as very often the coefficient is misinterpreted.
4. The correlation coefficient always assumes linear relationship regardless of the fact whether that assumption is correct or not.

INTERPRETING COEFFICIENT OF CORRELATION

The coefficient of correlation measures the degree of relationship between two sets of figures. As the reliability of estimates depends upon the closeness of the relationship it is imperative that utmost care be taken while interpreting the value of coefficient of correlation, otherwise fallacious conclusions can be drawn.

Unfortunately, the interpretation of the coefficient of correlation depends very much on experience. The full significance of r will only be grasped after working out a number of correlation problems and seeing the kinds of data that give rise to various values of *r*. The investigator must know his data thoroughly in order to avoid errors of interpretation and emphasis. He must be familiar, or become familiar, with all the relationships and theory which bear upon the data and should reach a conclusion based on logical reasoning and intelligent investigation on significantly related matters. However, the following general rules are given which would help in interpreting the value of *r* :

1. When $r = -1$, it means there is perfect negative relationship between the variables.
2. When $r = +1$, it means there is perfect positive relationship between the variables.
3. When $r = 0$, it means that there is no relationship between the variables, i.e., the variables are uncorrelated.
4. The closer r is to $+1$ or -1, the closer the relationship between the variables and the closer r is to 0, the less close the relationship. Beyound this it is not safe to go. The full interpretation of r depends upon circumstances one of which is the size of the sample. All that can really be said that when estimating the value of one variable from

the value of another, the higher the value of r the better the estimates.

5. The closeness of the relationship is not proportional to r. If the value of r is 0.8 it does not indicate a relationship twice as close as one of 0.4. It is, in fact, very much closer.

Conditions for the Use of Probable Error

The measure of probable error can be properly used only when the following three conditions exist :

1. The statistical measure for which the P.E. is computed must have been calculated from a sample.
2. The data must approximate a normal frequency curve (bell-shaped curve).
3. The sample must have been selected in an unblased manner and the individual items must be independent.

However, these conditions are generally satisfied and as such the reliability of the correlation coefficient is determined largely on the basis of exterior tests of reasonableness which are often of a statistical character.

RANK CORRELATION COEFFICIENT

The Karl Pearson's method is based on the assumption that the population being studied is normally distributed. When it is known that the population is not normal or when the shape of the distribution is not known, there is need for a measure of correlation that involves no assumption about the parameter of the population.

It is possible to avoid making any assumptions about the populations being studied by ranking the observations according to size and basing the calculations on the ranks rather than upon the original observations. It does not matter which way the items are ranked, item number one may be the largest or it may be the smallest. Using ranks rather than actual observations gives the coefficient of rank correlation.

This method of finding out covariablity or the lack of it between two variables was developed by the British Psychologist Charles Edward Spearman in 1904. This measure is especially useful when quantitative measures for certain factors (such as in the evaluation of leadership ability or the judgment of female beauty) cannot be fixed, but the individual in the group can be arranged in order thereby obtaining for each individual a number indicating his (her) rank in the group. Spearman's rank correlation coefficient is defined as :

$$R = 1 - \frac{6\Sigma D^2}{N(N^2-1)} \text{ or } 1 - \frac{6\Sigma D^2}{N^3 - N}$$

where R denotes rank coefficient of correlation and D refers to the difference of rank between paired items in two series.

Consider a set of n individuals, ranked according to two characters X and Y.

Individual	$A_1\ A_2\ ...\ A_1\ ...\ A_n$
Ranking (X)	$x_1\ x_2\ ...\ x_1\ ...\ x_n$
Ranking (Y)	$y_1\ y_2\ ...\ y_1\ ...\ y_n$

We know that,

$$\bar{x} = \frac{1}{N}\Sigma x_1 = \frac{1}{X}[1 + 2 + 3 ... + N] = \frac{N+1}{2}$$

$$\bar{y} = \frac{1}{N}\Sigma y_1 = \frac{1}{Y}[1 + 2 + 3 ... + N] = \frac{N+1}{2}$$

So, $\bar{x} = \bar{y}$

$$\sigma_x^2 = \frac{N^2 - 1}{12} = \sigma_y^2 \text{ or } r = \frac{\mu_{11}}{\sigma_x \sigma_y}$$

Now if d_i stands for the difference in ranks of ith individual, we have

$$d_i = x_i = y_i = (x_i - x) - (y_i - y) = x_i' - y_i'$$

where x_i' and y_i' are deviations of x_i and y_i for the mean $\bar{x}$ and $\bar{y}$

$$\Sigma\, d_i^2 = \Sigma\, (x_i' - y_i')^2$$

$$= \Sigma\, x_i'^2 + \Sigma\, y_i'^2 - 2\,\Sigma\, x_i'\, y_i'$$

$$\Sigma d_i^2 = \frac{1}{N}\,\Sigma x_i'^2 + \frac{1}{N}\,\Sigma y_i'^2 - \frac{2\Sigma s_i'\, y_i'}{N}$$

$$= \sigma_x^2 + \sigma_y^2 + 2\mu_{11}$$

$$= \sigma_x^2 + \sigma_y^2 - 2r\sigma_x\sigma_y$$

$$= 2\sigma_x^2 - 2r\sigma_x^2$$

$$= 2\sigma_x^2\,(1 - r)$$

$$\frac{1}{N} = \Sigma d_i^2 = 2\sigma_x^2\,(1 - r)$$

$$(1 - r) = \frac{1}{N}\,\frac{\Sigma d_i^2}{2\sigma_x^2}$$

$$(1 - r) = \frac{1}{N}\,\frac{\Sigma d_i^2}{\frac{2(N^2 - 1)}{12}} = \frac{1}{N}\,\frac{6\,\Sigma d_i^2}{(N^2 - 1)}$$

$$r = 1 - \frac{6\,\Sigma d_i^2}{N(N^2 - 1)} \text{ or } 1 - \frac{6\,\Sigma d_i^2}{N^3 - N}$$

The value of this coefficient interpreted in the same way as Karl Pearson's correlation coefficient, ranges between + 1 and – 1. When r_2 is + 1 there is complete agreement in the order of the ranks and the ranks are in the same direction. When r_2 is – 1 there is complete agreement in the order of the ranks and they are in opposite directions. This shall be clear from the following:

R_1	R_2	D $(R_1 - R_2)$	D^2	R_1	R_2	D $(R_1 - R_2)$	D^2
1	1	0	0	1	3	– 2	4
2	2	0	0	2	2	0	0
3	3	0	0	3	1	2	4
			$\Sigma D^2 = 0$				$\Sigma D^2 = 8$

$$R = 1 - \frac{6\,\Sigma D^2}{N^3 - N}$$

$$R = 1 - \frac{6\,\Sigma D^2}{N^3 - N}$$

$$= 1 - \frac{6 \times 0}{3^3 - 2} = 1 - 0 = 1$$

$$= 1 - \frac{6 \times 8}{3^3 - 2} = 1 - 2 = -1$$

Features of Spearman's Correlation Coefficient

1. The Spearman's correlation coefficient is nothing but Karl Pearson's correlation coefficient between the ranks. Hence, it can be interpreted in the same manner as Pearsonian correlation coefficient.
2. Spearman's correlation coefficient is distribution-free or non-parametric because no strict assumptions are made about the form of population from which sample observations are drawn.
3. The sum of the difference of ranks between two variables shall be zero. Symbolically. $\Sigma\, d = 0$.

In rank correlation we may have two types of problems :

a. Where ranks are given.

b. Where ranks are not given.

Where Ranks are Given. Where actual ranks are given to us the steps required for computing rank correlation are :

(i) Take the differences of the two ranks, i.e., $(R_1 - R_3)$ and denote these differences by D.

(ii) Square these differences and obtain the total ΣD^2.

(iii) Apply the formula $R = 1 - \frac{6 \Sigma D^2}{N^3 - N}$.

SIGNIFICANCE OF THE STUDY OF CORRELATION

The study of correlation is of immense use in practical life because of the following reasons :

a. Most of the variables show some kind of relationship. For example, there is relationship between price and supply, income and expenditure, etc. With the help of Correlation we can measure in one figure the degree of relationship existing between the variables.

b. Once we know that two variables are closely related, we can estimate the value of one variable given the value of another. This is known with the help of regression analysis discussed in the next chapter.

c. Correlation contributes to the understanding of economic behaviour, aids in locating the critically important variables on which other depend, may reveal to the economist the connection by which disturbances spread and suggest to him the paths through which stabilizing forces may become effective.

In business, Correlation enables the executive to estimate costs, sales, prices and other variables on the basis of some other series with which these costs, sales, or prices may be functionally related. Some of the guesswork can be removed from decisions when the relationship between a variable to b estimated and the one or more other variables on which it depends are close and reasonably invariant.

However, it should be noted that coefficient of correlation is one of the most widely used and also one of the most widely abused statistical measure. It is abused in the sense that one sometimes overlooks the fact that correlation measures are nothing but the strength of linear relationship and that it does not necessarily imply a cause-effect relationship.

a. The effect of correlation is to reduce the range of uncertainly. The prediction based on Correlation is likely to be more valuable and near to reality.

b. Progressive development in the methods of science and philosophy has been characterized by increase in the knowledge of relationship or correlations. In nature also one finds multiplicity of interrelated forces.

CORRELATION AND CAUSATION

Correlation helps us in determining the degree of relationship between two or more variables—it does not tell us anything about cause and effect relationship. Even a high degree of correlation does not necessarily mean that a relationship of cause and effect exists between the variables or, simply stated, correlation does not necessarily simply causation or functional relationship though the existence of causation always implies correlation. By itself it establishes only covariation. The explanation of a significant degree of correlation may be any one. Combination of the following reasons :

(i) The correlation may be due to pure chance, especially in a small sample. We may get a high degree of correlation between two variables in a sample but in the universe there may not be any relationship between the variables at all. This is especially so in case of small samples. Such a correlation may arise either because of pure random sampling variation or because of the bias of the investigator in selecting the sample. The following example shall illustrate the point :

Income (Rs.) :	3500	3600	3700	3800	3900
Weight (Ibs) :	120	140	160	180	200

The above data show a perfect positive relationship between income and weight, i.e., as the income is increasing the weight is increasing and the rate of change between two variables is the same.

(ii) Both the correlated variables may be influenced by one or more other variables. It is just possible that a high degree of correlation between the variables may be due to some causes affecting each variable or different causes affecting each with the same effect. For example, a high degree of correlation between the yield per acre of rice and tea may be due to the fact that both are related to the amount of rainfall. But none of the two variables is the cause of the other. To take another example : suppose the correlation of teachers salaries and the consumption of liquor over a period of year comes out to be 0.9, this does not prove that teachers drink : nor does it prove that liquor sale increases teachers salaries. Instead, both variables move together because both are influenced by a third variable—Long-run growth in national income and population.

(iii) Both the variables may be mutually influencing each other so that neither can be designated as the cause and the other the effect. There may be a high degree of correlation between the variables but it may be difficult to pinpoint as to which is the cause and which is the effect. This is especially likely to be so in case of economic variables. For example, such variables as demand and supply, price and production, etc. mutually interact. To take a specific case, it is a well-known principle of economics that as the price of a commodity increases its demand goes down and so price is the cause and demand the effect. But it is also possible that increased demand of a commodity due to growth of population or other reasons may exercise an upward pressure on price. Now, the cause is the increased demand, the effect the price. Thus, at times it may become difficult to explain from the two correlated variables which is the cause and which is the effect because both may be resting on each other.

The above points clearly bring out the fact that a mathematical relationship implies nothing in itself about cause and effect. In general, if factors A and B are correlated. It may be that (1) A causes to be sure but it might also be that (2) B causes A, (3) A and B influence each other continuously or intermittently (4) A and B are influenced by C or (5) the correlation is due to chance. In many instances extremely high degree of correlation between two variables may be obtained when no meaning can be attached to the answer. There is, for example, extremely high correlation between some series representing the production of pigs and the production of pig iron, yet no one has ever believed that this correlation has any meaning or that it indicates the existence of a cause-effect relation. By itself, it establishes only covariation. Correlation observed between variables that cannot conceivably be casually related is called spurious or nonsense correlation. More appropriately, we should remember that it is interpretation of the degree of correlation that is spurious, not the degree of correlation itself. The high degree of correlation indicates only the mathematical result. We should reach a conclusion based on logical reasoning and intelligent investigation of significantly related matters. It may also be pointed out that errors in Correlation include not only reading causation into spurious correlation but also interpreting spuriously a perfectly valid relationship.

KINDS OF CORRELATION

Correlation is described or classified in several different ways. Three of the most important ways of classifying correlation are :

1. Positive or negative.
2. Simple, partial and multiple.
3. Linear and non-linear.

Positive and Negative Correlation : Whether correlation is positive (direct) or negative (inverse) would depend upon the direction is change of the variables. If both the variables are varying in the same direction, i.e. if as one variable is increasing the other, on an average, is also increasing or, if as one variable is decreasing the other, on an average, is also variables are varying in opposite directions, i.e., as one variable is increasing, the other is decreasing or vice versa, correlation is said to be negative. The following examples would illustrate the difference between positive and negative correlation.

I. POSITIVE CORRELATION

X :	10	12	15	18	20	X :	80	70	60	40	30
Y :	15	20	22	25	37	Y :	50	44	30	20	10

II. NEGATIVE CORRELATION

X :	20	30	40	60	80	X :	100	90	60	40	30
Y :	40	30	22	15	10	Y :	10	20	30	40	50

Simple, Partial and Multiple Correlation : The distinction between simple, partial and multiple correlation is based upon the number of variables studied. When only two variables are studied it is a problem of simple correlation. When three or more variables are studied it is a problem of either multiple or partial correlation. In multiple correlation three or more variables are studied simultaneously. For example, when we study the relationship between the yield of rice per acre and both the amount of rainfall and the amount of fertilizers used, it is a problem of multiple correlation. On the other hand, in partial correlation we recognize more than two variables, but consider only two variables to be influencing each other the effect of other influencing variables being kept constant. For example, in the rice problem taken above if we limit our Correlation of yield and rainfall to periods when a certain average daily temperature existed it becomes a problem relating to simple correlation only.

Linear and Non-Linear (Curvilinear) Correlation : The distinction between linear and non-linear correlation is based upon the constancy of the ratio of change between the variables. If the amount of change in one variable tends to bear constant ratio to the amount of change in the other variable then the correlation is said to be linear. For example, observe the following two variables X and Y :

X :	10	20	30	40	50
Y :	70	140	210	280	350

It is clear that the ratio of change between the two variables is the same. If such variables are plotted on a graph paper all the plotted points would fall on a straight line.

Correlation would be called not-linear or curvilinear if the amount of change in one variable does not bear a constant ratio to the amount of change in the other variable. For example, if we double the amount of rainfall the production of rice or wheat, etc., would not necessarily be doubled. It may be pointed out that in most of the practical situations, we find a non-linear relationship between the variables. However, since techniques of analysis for measuring non-linear correlation are far more complicated than those for linear correlation, we generally make an assumption that the relationship between the variables is of the liner type.

Merits and Limitations of the Method

Merits : Following are the merits of scatter diagram method :

a. It is not influenced by the size of extreme items whereas most of the mathematical methods of finding correlation are influenced by extreme items.
b. Making a scatter diagram usually is the first step in investigating the relationship between two variables.
c. It is simple and non-mathematical method of studying correlation between the variables. As such it can be easily understood and a rough idea can very quickly be formed as to whether or not the variables are related.

Limitations : By applying this method we can get an idea about the direction of correlation and also whether it is high or low. But we cannot establish the exact degree of correlation between the variables as is possible by applying the mathematical methods.

Graphic Method

When this method is used the individual values of the two variables are plotted on the graph paper. We thus obtain two curves, one for X variable and another for Y variable. By examining the direction and closeness of the two curves so drawn we can infer whether or not the variables are related. If both the curves drawn on the graph are moving in the same direction (either upward or downward) correlation is said to be positive. On the other hand, if the curves are moving in the opposite directions correlation is said to be negative. The following example shall illustrate the method.

KARL PEARSON'S COEFFICIENT OF CORRELATION

Of the several mathematical methods of measuring correlation, the karl Pearson's method, popularly known as Pearson's coefficient of correlation, is most widely used in practice. The Pearson coefficient of correlation is denoted by the symbol r. It is one of the very few symbols that are used universally for describing the degree of correlation between two series. The formula for computing Pearsonian r is :

$$r = \frac{\Sigma xy}{N\sigma_x \sigma_y} \quad \text{...(i)}$$

Here $x = (X - \overline{X}); y = (Y - \overline{Y})$

σ_x = Standard deviation of series X

σ_y = Standard deviation of series Y

N = Number of pairs of observations

r = the (product moment) correlation coefficient.

This method is to be applied only where deviations of items are taken from actual mean and not from assumed mean.

The value of the coefficient of correlation as obtained by the above formula shall always lie between ± 1. When r = + 1, it means there is perfect positive correlation between the variables. When r = – 1, it means there is perfect negative correlation between the variables. When r = 0, it means there is no relationship berween the two variables. However, in practice such values of r as + 1, – 1, and 0 are rare. We normally get values which lie between + 1 and – 1 such as + 0.8, – 0.26, etc. The coefficient of correlation describes not only the magnitude of correlation but also its direction. Thus, + 0.8 would mean that correlation is positive because the sign of r is + and the magnitude of correlation is 0.8. Similarly – 0.26 means low degree of negative correlation.

The above formula for computing Pearson's coefficient of correlation can be transformed to the following from which is easier to apply.

$$r^* = \frac{\Sigma xy}{\sqrt{\Sigma x^2 \times \Sigma y^2}} \quad \text{...(ii)}$$

where $x = (X - \overline{X})$ and $y = (Y - \overline{Y})$

It is obvious that while applying this formula we have not to calculate separately the standard deviation of X and Y series as is required by formula (i). This simplifies greatly the task of calculating correlation coefficient.

Steps.

a. Take the deviations of Y series from the mean of Y and denote these deviations by y.
b. Square these deviations and obtain the total, i.e., Σx^2.
c. Multiply the deviations of X and Y series and obtain the total, i.e., $\Sigma x y$.
d. Substitute the values of Σxy, Σx^2 and Σy^2 in the above formula.
e. Take the deviations of X series from the mean of X and denote these deviations by x.

f. Square these deviations and obtain the total, i.e., Σx^2.

The following examples will the procedure :

SOLVED EXAMPLES

Example 1:

Making use of the data summarized below, calculate the coefficient of correlation, r_{12} :

Case	X_1	X_2	*Case*	X_1	X_2
A	*10*	*9*	*E*	*12*	*11*
B	*6*	*4*	*F*	*13*	*13*
C	*9*	*6*	*G*	*11*	*8*
D	*10*	*9*	*H*	*9*	*4*

Solution:

Calculation of Coefficient of Correlation.

Case	X_1	$(X_1 - \overline{X}_1)$ x_1	x_1^2	X_2	$(X_2 - \overline{X}_2)$ x_2	x_2^2	$x_1 x_2$
A	10	0	0	9	+ 1	1	0
B	6	– 4	16	4	– 4	16	16
C	9	– 1	1	6	– 2	4	2
D	10	0	0	9	+ 1	1	0
E	12	+ 2	4	11	+ 3	9	6
F	13	+ 3	9	13	+ 5	25	15
G	11	+ 1	1	8	0	0	0
H	9	– 1	1	4	– 4	16	4
N = 8	$\Sigma X_1 = 80$	$\Sigma x_1 = 0$	$\Sigma x_1^2 = 32$	$\Sigma X_2 = 64$	$\Sigma x_2 = 0$	$\Sigma x_2^2 = 72$	$\Sigma x_1 x_2 = 43$

$$\overline{X}_1 = \frac{\Sigma X_1}{N} = \frac{80}{8} = 10$$

$$\overline{X}_2 = \frac{\Sigma X_2}{N} = \frac{64}{8} = 8$$

$$r_{12} = \frac{\Sigma x_1 x_2}{\sqrt{\Sigma x_1^2 \times \Sigma x_2^2}}$$

$\Sigma x_1 x_2 = 43, \Sigma x_1^2 = 32, \Sigma x_2^2 = 72$

Substituting the value

$$r_{12} = \frac{43}{\sqrt{32 \times 72}} = \frac{43}{\sqrt{2304}} = \frac{43}{48} = +0.896$$

Note : It may be noted that the above formula is the same as given earlier,

i.e.,
$$r = \frac{\Sigma xy}{\sqrt{\Sigma x^2 \times \Sigma y^2}}$$

The only difference is that of the symbols. Since in the question we are given series X_1 and X_2 we have changed the symbols in the formula accordingly.

Direct Method of Finding out Correlation Coefficient

Correlation coefficient can also be calculated without taking deviations of items either from actual mean or assumed mean, i.e., actual X and Y values. The formula in such a case is :

$$r = \frac{N\Sigma XY - (\Sigma X)(\Sigma Y)}{\sqrt{N\Sigma X^2 - (\Sigma X)^2}\sqrt{N\Sigma Y^2 - (\Sigma Y)^2}}$$

This formula would give the same answer as we get when deviations of items are taken from actual mean or assumed mean. The following example shall illustrate the point.

Example 2:

Calculate coefficient of correlation from the following data :

X :	*100*	*200*	*300*	*400*	*500*	*600*	*700*
Y :	*30*	*50*	*60*	*80*	*100*	*110*	*130*

Solution:

To simplify calculation let every value of X be divided by 100 and every value of Y by 10 and denote these series by X' and Y'.

Computation of Correlation Coefficient.

X	*X/100*	$(X' - \overline{X})$		*Y*	*Y/10*	$(Y' - \overline{Y})$		
	X'	*x*	x^2	x^2	*Y'*	*y*	y^2	*xy*
100	1	− 3	9	30	3	− 5	25	15
200	2	− 2	4	50	5	− 3	9	6
300	3	− 1	1	60	6	− 2	4	2
400	4	0	0	80	8	0	0	0
500	5	1	1	100	10	+ 2	4	2
600	6	+ 2	4	110	11	+ 3	9	6
700	7	+ 3	9	130	13	+ 5	25	15
		$\Sigma x = 0$	$\Sigma x^2 = 28$			$\Sigma y = 0$	$\Sigma y^2 = 76$	$\Sigma xy = 46$

$$r = \frac{\Sigma xy}{\sqrt{\Sigma x^2 . \Sigma y^2}}$$

$$x = (X - \overline{X}); \ y = (Y - \overline{Y})$$

$$\Sigma \, xy = 46, \ \Sigma \, x^2 = 28, \ \Sigma \, y^2 = 76$$

$$r = \frac{46}{\sqrt{28 \times 76}} = +0.997$$

When Deviations are taken from an Assumed Mean

When actual means are in fractions, say the actual means of X and Y series are 20.167 and 29.23, the calculation of correlation by the method discussed above would involve too marry calculations and would take a lot of time. In such a case we make use of the assumed mean method for finding out correlation. When deviations are taken from an assumed mean the following formula is applicablc :

$$r = \frac{N \Sigma d_x d_y - \Sigma d_x \times \Sigma d_y}{\sqrt{N \Sigma d_x^2 - (\Sigma d_x)^2} \ \sqrt{N \Sigma d_y^2 - (\Sigma d_y)^2}}$$

where d_x refers to deviations of X series from an assumed mean, i.e., $(X - \overline{X})$.

Similarly, d_y refers to deviations of Y series from an assumed mean i.e., $(Y - \overline{Y})$.

$\Sigma \, d_x \, d_y$ = sum of the product of the deviations of X and Y series from their assumed means.

$\Sigma \, d_x^2$ = sum of the squares of the deviations of X series from an assumed means.

$\Sigma \, d_y^2$ = sum of the squares of the deviations of Y series from an assumed means.

$\Sigma \, d_x$ = sum of the deviations of X series from an assumed means.

$\Sigma \, d_y$ = sum of the deviations of Y series from an assumed means.

It may be pointed out that there are many variations of the above formula. For example, the above formula may be written as :

$$r = \frac{N \Sigma d_x d_y - \{(\Sigma d_x) \times (\Sigma d_y)\}}{\sqrt{N \Sigma d_x^2 - (\Sigma d_x)^2} \ \sqrt{N \Sigma d_y^2 - (\Sigma d_y)^2}}$$

But the formula given above is the easiest to apply.

Steps :

a. Substitute the value of $\Sigma \, d_x \, d_y$, $\Sigma \, d_x \, . \, \Sigma \, d_y$, $\Sigma \, d_x^2$ and $\Sigma \, d_y^2$ in the formula given above.

b. Multiply d_x and d_y and obtain the total $\Sigma\, d_x\, d_y$.

c. Square d_y and obtain the total $\Sigma\, d_y^2$.

d. Square d_x and obtain the total $\Sigma\, d_x^2$.

e. Take the deviations of Y series from an assumed mean and denote these deviations by d_y and obtain the total i.e. $\Sigma\, d_y$.

f. Take the deviations of X series from an assumed mean and denote these deviations by d_x and obtain the total, i.e., $\Sigma\, d_x$.

It may be pointed out that this form is simplest to apply. The following examples shall illustrate the procedure.

Example 3:

Given :

Total of the product of deviations of X and Y series = 3,044

Number of pairs of observations = 10

Total of the deviations of X series = – 170

Total of the deviations of Y series = – 20

Total of the squares of deviations of X series = 8,288

Total of the squares of deviations of Y series = 2,264.

Find out the coefficient of correlation when the assumed means of X series and Y series are 82 and 68 respectively.

Solution:

We are given :

$\Sigma\, d_x\, d_y = 3{,}044$, $\Sigma\, d_x = -170$, $\Sigma\, d_y = -20$,

$\Sigma\, d_x^2 = 8{,}288$, $\Sigma\, d_y^2 = 2{,}264$, $N = 10$.

Applying the formula $r = \dfrac{N\Sigma d_x d_y - \Sigma d_x \Sigma d_y}{\sqrt{N\Sigma d_x^2 - (\Sigma d_x)^2}\,\sqrt{N\Sigma d_y^2 - (\Sigma d_y)^2}}$

$$= \frac{10 \times 3044 - (-170)(-20)}{\sqrt{10 \times 8288 - (-170)^2}\,\sqrt{10 \times 2264 - (20)^2}}$$

$$= \frac{30444 - 3400}{\sqrt{82880 - 28900}\,\sqrt{22640 - 400}}$$

$$= \frac{27040}{\sqrt{53980}\,\sqrt{22240}}$$

$$= \frac{27040}{232.336 \times 149.131} = \frac{27040}{34648.5} = +0.7804$$

Another form of the formula when deviations from the assumed average.

$$r = \frac{\Sigma\, d_x\, d_y - N(\overline{X} - A_x)(\overline{Y} - A_y)}{N \sigma_x\, \sigma_y}$$

Where $\Sigma\, d_x\, d_y$ = sum of deviations from the assumed average.

$\overline{X}$ = actual mean of the X series,

$\overline{Y}$ = actual mean of Y series,

A_x = assumed mean of X series,

A_y = assumed mean of Y series.

The following example shall illustrate the application of the above formula.

Example 4:

The following table gives the distribution of items of production and also the relatively defective items among them, according to size groups. Find the correlation coefficient between size and defect in quality and its probable error.

Size-group	:	*15–16*	*16–17*	*17–18*	*18–19*	*19–20*	*20–21*
No. of items	:	*200*	*270*	*340*	*360*	*400*	*300*
No. of defective items	:	*150*	*162*	*170*	*180*	*180*	*114*

Solution:

Let mid-point of size be denoted by X and % of defective items by Y.

Calculation of Correlation Coefficient.

Size	*m.p.* *X*	*(X – 17.5)* d_x	d_x^2	*% of Defective items* *Y*	d_y	d_y^2	$d_x d_y$
15–16	15.5	– 2	4	75	+ 25	625	– 50
16–17	16.5	– 1	1	60	+ 10	100	– 10
17–18	17.5	0	0	50	0	0	0
18–19	18.5	+ 1	1	50	0	0	0
19–20	19.5	+ 2	4	45	– 5	25	– 10
20–21	20.5	+ 3	9	38	– 12	144	– 36
		$\Sigma d_x = 3$	$\Sigma d_x^2 = 19$		$\Sigma d_y = 18$	$\Sigma d_y^2 = 894$	$\Sigma d_x d_y = -106$

$$r = \frac{N\Sigma d_x d_y - \Sigma d_x \Sigma d_y}{\sqrt{N\Sigma d_x^2 - (\Sigma d_x)^2}\sqrt{N\Sigma d_y^2 - (\Sigma d_y)^2}}$$

$$= \frac{6 \times -106 - 3 \times 18}{\sqrt{6 \times 19 - (3)^2}\sqrt{6 \times 894 - (18)^2}}$$

$$= \frac{-636 - 54}{\sqrt{105}\sqrt{5040}} = -\frac{690}{727.46} = -0.949$$

$$\text{P.E.}_r = 0.6745\frac{1 - r^2}{\sqrt{N}} = 0.6745\frac{1 - (8)^2}{\sqrt{6}}$$

$$= 0.6745 \times \frac{.36}{2.449} = 0.099$$

Example 5:

Calculate correlation coefficient from data of illustration 3 by the direct method, i.e., without taking the deviation of items from actual or assumed mean.

Solution:

Calculation of correlation coefficient by the direct method

$$r = \frac{N\Sigma XY - (\Sigma X)(\Sigma Y)}{\sqrt{N\Sigma X^2 - (\Sigma X)^2}\sqrt{N\Sigma Y^2 - (\Sigma Y)^2}}$$

Calculation of Correlation Coefficient (Direct Method).

X	X^2	*Y*	Y^2	*XY*
9	81	15	225	135
8	64	16	256	128
7	49	14	196	98
6	36	13	169	78
5	25	11	121	55
4	16	12	144	48
3	9	10	100	30
2	4	8	64	16
1	1	9	81	9
$\Sigma X = 45$	$\Sigma X^2 = 285$	$\Sigma Y = 108$	$\Sigma Y^2 = 1,356$	$\Sigma XY = 597$

$N = 9, \Sigma XY = 597,$

$\Sigma X = 45, \Sigma Y = 108,$

$\Sigma X^2 = 285, \Sigma Y^2 = 1,356$

$$r = \frac{9 \times 597 - 45 \times 108}{\sqrt{9 \times 285 - (45)^2}\ \sqrt{9 \times 1356 - (108)^2}}$$

$$r = \frac{5373 - 4860}{\sqrt{2565 - 2025}\ \sqrt{12{,}204 - 11{,}664}}$$

$$r = \frac{513}{\sqrt{540 \times 540}} = \frac{513}{540} = +0.95.$$

Calculation of Correlation Coefficient when Change of Scale and Origin is made

Since r is a pure number, shifting the origin and changing the scale of series does not affect its values.

Example 6:

If r = 0.6 and N = 64, find out the probable error of the coefficient of correlation and determine the limits for population r.

Solution:

$$P.E._r = 0.6745\ \frac{1 - r^2}{\sqrt{N}}$$

$$r = 0.6 \text{ and } N = 64$$

$$P.E._r = 0.64745\ \frac{1 - (0.6)^2}{\sqrt{(64)}}$$

$$= \frac{0.6745 \times 0.64}{8} = 0.054.$$

Limits of population correlation = 0.6 ± 0.054

= 0.546 to 0.654.

Example 7:

Calculate the coefficient of correlation from the following data of the Spearman's. Rank difference method :

Price of Tea (*Rs.*)	*Price of Coffee* (*Rs.*)	*Price of Tea* (*Rs.*)	*Price of Coffee* (*Rs.*)
75	120	60	110
88	134	80	140
95	150	81	142
70	115	50	100

Solution:

Calculation of Spearman's Correlation Coefficient.

Price of Tea (Rs.)	R_1	*Price of Coffee (Rs.)*	R_2	$(R_1 - R_2)^2$ D^2
75	4	120	4	0
88	7	134	5	4
95	8	150	8	0
70	3	115	3	0
60	2	110	2	0
80	5	140	6	1
81	6	142	7	1
50	1	100	1	0
				$\Sigma D^2 = 6$

$$R = 1 - \frac{6\,\Sigma D^2}{N^3 - N} = 1 - \frac{6 \times 6}{8^3 - 8}$$

$$= 1 - \frac{36}{512 - 8}$$

$$= 1 - 0.071 = +\,0.329$$

Example 8:

Ten competitors is a beauty contest are ranked by three judges in the following order :

1st judge	*1*	*6*	*5*	*10*	*3*	*2*	*4*	*9*	*7*	*8*
2nd judge	*3*	*5*	*8*	*4*	*7*	*10*	*2*	*1*	*6*	*9*
3rd judge	*6*	*4*	*9*	*8*	*1*	*2*	*3*	*10*	*5*	*7*

Use the rank correlation coefficient to determine which pair of judges has the nearest approach to common tastes in beauty.

Solution:

In order to find out which pair of judges has the nearest approach to common tastes in beauty, we compare Rank Correlation between the judgments of :

(i) 1st judge and 2nd judge,

(ii) 2nd judge and 3rd judge, and

(iii) 1st judge and 3rd judge.

Rank by 1st judge R_1	*Rank by 2nd judge* R_2	*Rank by 3rd judge* R_3	$(R_1 - R_2)^2$ D^2	$(R_2 - R_3)^2$ D^2	$(R_1 - R_3)^2$ D^2
1	3	6	4	9	25
6	5	4	1	1	4
5	8	9	9	1	16
10	4	8	36	16	4
3	7	1	16	36	4
2	10	2	64	64	0
4	2	3	4	1	1
9	1	10	64	81	1
7	6	5	1	1	4
8	9	7	1	4	1
N = 10	N = 10	N = 10	$\Sigma D^2 = 200$	$\Sigma D^2 = 214$	$\Sigma D^2 = 60$

Rank correlation between the judgment of 1st and 2nd judges :

$$R = 1 - \frac{6\,\Sigma D^2}{N^3 - N}$$

Here we have directly calculated D^2 because D's are not required in applying the formula.

$$\therefore \quad R_{(I \text{ and } II)} = 1 - \frac{6 \times 200}{10^3 - 10} = 1 - \frac{1200}{990} = 1 - 1.212 = -0.212$$

Rank correlation between the judgments of 2nd and 3rd judges :

$$R = 1 - \frac{6\,\Sigma D^2}{N^3 - N}$$

$$\therefore \quad R_{(II \text{ and } III)} = 1 - \frac{6 \times 214}{10^3 - 10} = 1 - 1.297 = -0.297$$

Rank correlation between judgments of 1st and 3rd judge :

$$R_{(I \text{ and } III)} = 1 - \frac{6\,\Sigma D^2}{N^3 - N} = 1 - \frac{6 \times 60}{10^3 - 10} = 1 - \frac{360}{990} = 0.636.$$

Since coefficient is maximum in the judgments of the first and third judges we conclude that they have the nearest approach to common tastes in beauty.

Where Ranks are Not Given : When we are given the actual data and not the ranks. It will be necessary to assign the ranks. Ranks can be assigned by taking either highest value as I or the lowest value of 1. But whether we start with the lowest value or the highest value we must follow the same method in case of both the variables.

Example 9:

The ranking of 10 students in two subjects A and B are as follows :

A	*B*	*A*	*B*
6	3	4	6
5	8	9	10
3	4	7	7
10	9	8	5
2	1	1	2

Calculate rank correlation coefficient.

Solution:

Calculation of Rank Correlation Coefficient.

R_1	R_2	$(R_1 - R_2)^2$ D^2
6	3	9
5	8	9
3	4	1
10	9	1
2	1	1
4	6	4
9	10	1
7	7	0
8	5	9
1	2	1
		$\Sigma D^2 = 36$

$$R = 1 - \frac{6\,\Sigma D^2}{N^3 - N} = 1 - \frac{6 \times 36}{10^3 - 10}$$

$$= 1 - \frac{216}{990} = 0.782$$

Example 10:

Two ladies were asked to rank 7 different types of lipsticks. The ranks give by them are as follows :

Lipsticks	*A*	*B*	*C*	*D*	*E*	*F*	*G*
Neelu	2	1	4	3	5	7	6
Neera	1	3	2	4	5	6	7

Calculate Spearman's rank correlation coefficient.

Solution:

Calculation of Spearman's Rank Correlation Coefficient.

$X\ (R_1)$	$Y\ (R_2)$	$(R_1 - R_2)\ D$	D^2
2	1	+ 1	1
1	3	– 2	4
4	2	+ 2	4
3	4	– 1	1
5	5	0	0
7	6	+ 1	1
6	7	– 1	1
			$\Sigma D^2 = 12$

$$R = 1 - \frac{6\ \Sigma D^2}{N^3 - N} = 1 - \frac{6 \times 12}{7^3 - 7} = 1 - 0.214 = 0.786.$$

Example 11:

Calculate Speaman's coefficient of correlation between marks assigned to ten students by judges X and Y in a certain competitive test as shown below:

S. No. :	*1*	*2*	*3*	*4*	*5*	*6*	*7*	*8*	*9*	*10*
Marks by judge X :	*52*	*53*	*42*	*60*	*45*	*41*	*37*	*38*	*25*	*27*
Marks by judge Y :	*65*	*68*	*43*	*38*	*77*	*48*	*35*	*30*	*25*	*50*

Solution:

First assign ranks and then calculate rank correlation coefficient

Computation of Spearman's Coefficient of Correlation.

Marks by judge X	R_x	**Marks by judge Y**	R_y	$(R_x - R_y)$ D^2
52	8	65	8	0
53	9	68	9	0
42	6	43	5	1
60	10	38	4	36
45	7	77	10	9
41	5	48	6	1
37	3	30	3	0
38	4	32	2	4
25	1	25	1	0
27	2	50	7	25
				$\Sigma D^2 = 76$

$$R = 1 - \frac{6\,\Sigma D^2}{N^3 - N} = 1 - \frac{6 \times 76}{10^3 - 10} = 1 - 0.461 = -\,0.539.$$

Example 12:

Quotations of Index Numbers of security prices of a certain joint stock company are given below :

Year	*Debenture price*	*Share price*
1	*97.8*	*73.2*
2	*99.2*	*85.8*
3	*98.8*	*78.9*
4	*98.3*	*75.8*
5	*98.4*	*77.2*
6	*96.7*	*87.2*
7	*97.1*	*83.8*

Using rank correlation method, determine the relationship between debenture prices and share prices.

Solution:

Calculation of Rank Correlation Coefficient.

X	R_x	Y	R_y	$(R_x - R_y)^2$ D^2
97.8	3	93.2	1	4
99.2	7	85.8	6	1
98.8	6	78.9	4	4
98.3	4	75.8	2	4
98.4	5	77.2	3	4
96.7	1	87.2	7	36
97.1	2	83.8	5	9
				$\Sigma D^2 = 62$

$$R = 1 - \frac{6\,\Sigma D^2}{N^3 - N} = 1 - \frac{6 \times 62}{7^3 - 7}$$

$$= 1 - \frac{372}{336} = 1 - 1.107 = -\,0.107$$

Equal Ranks : In some cases it may be found necessary to rank two or more individuals or entries as equal. In such a case it is customary to give each individual an average rank. Thus, if two individuals are ranked equal at fifth place, they are each given the rank $\frac{5+6}{2}$, that is 5.5 while. It three are ranked equal at fifth place, they are given the rank $\frac{5+6+7}{3} = 6$. In other words, where two or more items are to be ranked equal the rank assigned for purposes of calculating coefficient of correlation is the average of the ranks which these individuals would have gat had they differed slightly from each other. Where equal ranks are assigned to some entries an adjustment in the above formula for calculating the rank coefficient of correlation is made.

The adjustment consists of adding $\frac{1}{12}(m^3 - m)$ to the value of ΣD^2, where M stands for the numbers of items whose ranks are common. If there are more than one such group of items with common rank, this value is added as many times the number of such groups. The formula can thus be written

$$R = 1 - \frac{6\left\{\Sigma D^2 + \frac{1}{12}(m^3 \quad m) + \frac{1}{12}(m^3 - m) + \ldots\right\}}{N^3 - N}$$

Example 13:

Obtain the rank correlation coefficient between the variables X and Y from the following pairs of observed values.

X :	*50*	*55*	*5*	*50*	*55*	*60*	*50*	*65*	*70*	*75*
Y :	*110*	*110*	*15*	*125*	*140*	*115*	*130*	*120*	*115*	*160*

Solution:

For finding ranks correlation coefficient first rank two various values. Taking lowest as 1 and next higher as 2 etc.

X	*Rank X* R_1	*Y*	*Rank Y* R_2	$(R_1 - R_2)$ D^2
50	2	110	1.5	0.25
55	4.5	110	1.5	9.00
65	7.5	115	4	12.25
50	2	125	7	25.00
55	4.5	140	9	20.25
60	6	115	4	4.00
50	2	130	8	36.00
65	7.5	120	6	2.25
70	9	115	4	25.00
75	10	160	10	00.00
				$\Sigma D^2 = 134$

It may be noted that in series X, 50 has repeated thrice (m = 3), 55 has been repeated twice (m = 2), 65 has been repeated twice (m = 2). In series Y, 110 has been repeated twice (m = 2) and 115 twice (m = 3).

$$R = 1 - \frac{6\left\{\begin{aligned}&\Sigma D^2 + \frac{1}{12}(m^3 - m) + \frac{1}{12}(m^3 - m) + \frac{1}{12}(m^3 - m)\\ &\qquad + \frac{1}{12}(m^3 - m) + \frac{1}{12}(m^3 - m)\end{aligned}\right\}}{N^3 - N}$$

$$R = 1 - \frac{6\left\{\begin{aligned}&134 + \frac{1}{12}(3^3 - 3) + \frac{1}{12}(2^3 - 2) + \frac{1}{12}(2^3 - 2)\\ &\qquad + \frac{1}{12}(2^3 - 2) + \frac{1}{12}(3^3 - 3)\end{aligned}\right\}}{10^3 - 10}$$

$$= 1 - \frac{6[134 + 2 + .5 + .5 + .5 + 2]}{990}$$

$$= 1 - 6\,\frac{[139.5]}{990} = 1 - \frac{837}{990} = 1 - 845 = 0.155$$

Merits and Limitations of the Rank Method

Merits : The merits of the Rank Method can be discussed here :

1. This is the only method that can be used where we are given the ranks and not the actual data.
2. Where the data are of a qualitative nature like honesty, efficiency, intelligence, etc., this method can be used with great advantage. For example, the workers of two factories can be ranked in order of efficiency and the degree of correlation established by applying this method.
3. Even where actual data are given, rank method can be applied for ascertaining correlation.
4. This method is simpler to understand and eastern to apply compared to the Karl Pearson's method. The answers obtained by this method and the Karl Pearson's method will be the same provided no value is repeated, i.e., all the items are different.

Limitations The method is however associated with a few limitations also :

a. Where the number of items exceeds 30 the calculations become quite tedious and require a lot of time. Therefore, this method should not be applied where N exceeds 30 unless we are given the ranks and not the actual values of the variables.

b. This method cannot be used for finding out correlation in a grouped frequency distribution.

When to use Rank Correlation Coefficient? The rank method has principal uses :

a. The initial data are in the form of ranks.

b. If N is fairly small (say, not more than 25 or 30) rank method is sometimes applied to interval data as an approximation to the more time consuming r. This requires that the interval data be transferred to rank orders for both variables. If N is much in excess of 30, the labour required in ranking the scores becomes greater the is justified by the anticipated saving of time through the rank formula.

Concurrent Deviation Method

This method of studying correlation is the simplest of all the methods. The only thing that is required under this method is to find out the direction of change of X variable and Y variable. The formula applicable is :

$$r_c = \pm \sqrt{\pm\left(\frac{2C - n}{n}\right)}$$

where r_c stands for coefficient of correlation by the concurrent method; C stands for the number of concurrent deviations or the number of positive signs obtained after multiplying D_x with D_y.

n = Number of pairs of observation compared.

Steps.

a. Find out the direction of change of X variable, i.e., as compared with the first value, whether the second value is increasing or decreasing or is constant. If it is increasing put a + sign; If it is decreasing put a – sign (minus) if it is constant put zero. Similarly, as compared to second value, find out whether the third value is increasing, decreasing or constant. Repeat the same process for other values. Denote this column by D_x.

b. In the same manner as discussed above, find out the direction of change of Y variable and denote this column by D_y.

c. Multiply D_x with D_y and determine the value of C, i.e., the number of positive sign.

d. Apply the above formula, i.e.,

$$r_c = \pm \sqrt{\pm\left(\frac{2C - n}{n}\right)}$$

Note. The significance of ± signs, both (inside the under-root and outside the under-root, is that we cannot take the under-root of minus sign. Therefore, if $\frac{2C - n}{n}$ is negative, this negative value multiplied with the minus sign inside would make it positive and we can take the under-root. But the ultimate result would be negative. If $\frac{2C - n}{n}$ is positive then, of course, we get a positive value of the coefficient of correlation.

Example 14:

Calculate coefficient of concurrent deviation from the following data :

Price	*Imports*	*Price*	*Imports*
368	*22*	*384*	*26*
384	*21*	*395*	*24*
385	*24*	*403*	*29*
361	*20*	*400*	*28*
347	*22*	*385*	*27*

Solution:

Calculation of Coefficient of Concurrent Deviation.

Price *X*	*Direction of change of variable X* D_x	*Imports* *Y*	*Direction of change of variable Y* D_y	$D_x D_y$
368		22		
384	+	21	–	–
385	+	24	+	+
361	–	20	–	+
347	–	22	+	–
384	+	26	+	+
395	+	24	–	–
403	+	29	+	+
400	–	28	–	+
384	–	27	–	+
				C = 6

$$r_c = \pm\sqrt{\pm\left(\frac{2C - n}{n}\right)}; C = 6, n = 9$$

$$r_c = \pm\sqrt{\pm\left(\frac{2\times 6 - n}{9}\right)} = \pm\sqrt{\pm .333} = +0.577.$$

Example 15:

Calculate Karl Pearson's Coefficient of Correlation between age and playing habits from the data given below. Also calculate probable error and comment on the value :

Age	:	*20*	*21*	*22*	*23*	*24*	*25*
No. of Students	:	*500*	*400*	*300*	*240*	*200*	*160*
Regular Players	:	*400*	*300*	*180*	*96*	*60*	*24*

Solution:

Let us first find the percentage of regular players and then calculate correlation between age and percentage.

Age X	*(X – 22)* d_x	d_x^2	*No. of Students*	*Regular Players*	*% or Regular Players (Y)*	*(Y – 50)* d_y	d_y^2	$d_x d_y$
20	– 2	4	500	400	80	+ 30	900	– 60
21	– 1	1	400	300	75	+ 25	625	– 25
22	0	0	300	180	60	+ 10	100	0
23	+ 1	1	240	96	40	– 10	100	– 10
24	+ 2	4	200	60	30	– 20	400	– 40
25	+ 3	9	160	24	15	– 35	1225	– 105
$\Sigma X =$ 135	Σd_x = 3	Σd_x^2 = 19			$\Sigma Y =$ 300	Σd_y = 0	$\Sigma d_y^2 =$ 3350	$\Sigma d_x d_y$ = – 240

$$r = \frac{N\Sigma d_x d_y - \Sigma d_x d_y}{\sqrt{N\Sigma d_x^2 - (\Sigma dx)^2}\sqrt{N\Sigma d_y^2 - (\Sigma d_y)^2}}$$

$$= \frac{(6\times -240) - (3\times 0)}{\sqrt{6\times 19 - (3)^2}\sqrt{6\times 3350}}$$

$$= \frac{-1440}{\sqrt{105\times 20100}} = \frac{-1440}{1452.756} = -0.991$$

$$\text{P.E.r} = 0.6745\,\frac{1-r^2}{\sqrt{N}}$$

$$= 0.6745\,\frac{1-(.991)^2}{\sqrt{6}}$$

$$= \frac{0.6745 \times .018}{2.449}$$

$$= 0.005.$$

Example 16:

The ranks of the same 15 students in two subjects A and B are given below. The two numbers within brackes denote the ranks of the same student in A and B respectively.

(1, 10), (2, 7), (3, 2), (4, 6), (5, 4), (6, 8), (7, 3), (8, 1), (9, 1), (10, 15), (11, 19), (12, 5), (13, 14), (14, 12), (15, 13).

Find the Spearman's Rank Correlation Coefficient.

Solution:

Calculation of Spearman's Rank Correlation Coefficient.

R_A	R_B	$(R_A - R_B)^2$ D^2
1	10	81
2	7	25
3	2	1
4	6	4
5	4	1
6	8	4
7	3	16
8	1	81
9	11	4
10	15	25
11	9	4
12	5	49
13	14	1
14	12	4
15	13	4
		$\Sigma D^2 = 304$

$$R = 1 - \frac{6 \Sigma D^2}{N^3 - N}$$

$$= 1 - \frac{6 \times 304}{15^3 - 15}$$

$$= 1 - \frac{1824}{3360} = 1 - 0.543 = 0.457.$$

Example 17:

Find the coefficient of the correlation for the following :

Cost :	*39*	*65*	*62*	*90*	*82*	*75*	*25*	*98*	*36*	*78*
Sales :	*47*	*53*	*58*	*86*	*62*	*68*	*60*	*91*	*51*	*84*

Solution:

Calculation of Coefficient of Correlation By Karl Pearson's Method.

X	$(X - \overline{X})$ $\overline{X} = 65$ x	x^2	Y	$(Y - \overline{Y})$ $\overline{Y} = 66$ y	y^2	xy
39	26	676	47	– 19	361	+ 494
65	0	0	53	– 13	169	0
62	– 3	9	58	– 8	64	+ 24
90	+ 25	625	86	+ 20	400	+ 500
82	+ 17	289	62	– 4	16	– 68
75	+ 10	100	68	+ 2	4	+ 20
25	– 40	1600	60	– 6	36	+ 240
98	+ 33	1089	91	+ 25	625	+ 825
36	– 29	841	51	– 15	225	+ 435
78	+ 13	169	84	+ 18	324	+ 234
$\Sigma X =$ 650	$\Sigma x = 0$	$\Sigma x^2 =$ 5398	$\Sigma Y =$ 660	$\Sigma y = 0$	$\Sigma y^2 =$ 2224	$\Sigma xy =$ 2704

$$r = \frac{\Sigma xy}{\sqrt{\Sigma x^2 \times \Sigma y^2}}$$

$$= \frac{2704}{\sqrt{5398 \times 2224}}$$

$$= \frac{2704}{3464.85} = +0.78.$$

Example 18:

With the following data in 6 cities calculate the coefficient of correlation by Pearson's method between the density of population and death rate :

City	*Area in kilometres*	*Population in 1000*	*No. of deaths*
A	*150*	*30*	*300*
B	*180*	*90*	*1440*
C	*100*	*40*	*560*
D	*60*	*42*	*840*
E	*120*	*72*	*1224*
F	*80*	*24*	*312*

Solution:

First we will calculate density of population and death rate and denote them by X and Y.

$$\text{Density} = \frac{\text{Population}}{\text{Area}}; \ \text{Death rate} = \frac{\text{No. of Deaths}}{\text{Population}} \times 1000$$

City	*Density* X	*(X – 450)/100* x	x^2	*Death* Y	*(Y – 15)* y	y^2	xy
A	200	– 2.5	6.25	10	– 5	25	+ 12.5
B	500	+ 0.5	0.25	16	+ 1	1	+0.5
C	400	–0.5	0.25	14	– 1	1	+0.5
D	700	+ 2.5	6.25	20	+ 5	25	+ 12.5
E	600	+ 1.5	2.25	17	+ 2	4	+ 3.0
F	300	– 1.5	2.25	13	– 2	4	+ 3.0
		$\Sigma x = 0$	$\Sigma x^2 = 17.5$	$\Sigma Y = 90$	$\Sigma y = 0$	$\Sigma y^2 = 60$	$\Sigma xy = 32$

$$r = \frac{\Sigma\, xy}{\sqrt{\Sigma\, x^2 \times \Sigma y^2}} = \frac{32}{\sqrt{17.5 \times 60}} = \frac{32}{32.404} = +0.988.$$

EXERCISES

1. (a) Explain what is meant by correlation between two variables. What are the methods of finding the existence of correlation? How can it be measured?

   (b) Define Karl Pearson's coefficient of correlation. What does it intend to measure? Show that it is not affected by the change of origin and scale.

2. (a) Explain what is meant by correlation between two variables. What are the methods of finding the existence of correlation? How can it be measured?

3. (a) What is Spearman's rank correlation coefficient? How does it difference from Karl Pearson's coefficient of correlation.

   (b) Differentiate between the coefficient of correlation and coefficient of determination.

4. (a) Define Rank Correlation. Write down Spearman's formula for rank correlation coefficient. State the advantages of Spearman's rank correlation over Karl Pearson's correlation coefficient.

   (b) Define coefficient of correlation between two variables. What is the range of the coefficient? Interpret the results: $r=0$. $r=\pm 1$.

5. (a) What is 'Spurious' or 'non-sensical correlation'? Explain with an example.

   (b) What is coefficient of Rank Correlation? Bring out its usefulness. How does it differ from coefficient of correlation?

6. (a) Explain what is meant by coefficient of correlation between two variables. What are the different methods of finding correlation? Distinguish between Positive and Negative correlation.

   (b) Define coefficient of correlation and mention its important properties.

7. (a) In a bivariate sample, the sum of squares of difference between the ranks of observed values of two variables is 231 and the correlation coefficient between them is 0.4. Find the number of pairs.

   (b) Explain the concepts of (i) Coefficient of determination, and (ii) Probable error. How are they helpful in interpreting correlation?

   (c) Interpret the value of 0 and +1 for the product moment correlation coefficient.

   (b) Define rank correlation coefficient. How is it determined and when it preferred to Karl Pearson's coefficient?

8. (a) What is scatter diagram? How does it help in studying the correlation between two variables in respect of both of its direction and degree?

   (b) What are the methods of calculating coefficient of correlation?

9. (a) Explain the meaning and significance of the concept of correlation. Does it always signify cause and effect relationship between two variables? How is the coefficient of correlation interpreted?

   (b) Explain the meaning of correlation between two variables. State the methods available for finding the correlation.

10. (a) What is correlation? Give the properties of Karl Pearson's coefficient of correlation.
    (b) What is correlation? Clearly explain, with suitable illustrations, its role in dealing with business problems.
11. (a) Prove that the correlation coefficient is unaffected by the change of origin or scale.
    (b) Define rank correlation coefficient. When is it preferred to Karl Pearsons's coefficient of correlation?
12. Explain the meaning of correlation. State the extreme values of the coefficient of correlation and interpret them.
13. State the properties of Pearson's coefficient of correlation. How do you interpret a calculated value of r? Explain the term "probable error of r."
14. (a) What is correlation? Explain the significance of positive, negative or zero coefficient of correlation.
    (b) Explain the term correlation and give methods to measure the same.
15. (a) Define Karl Pearson's coefficient of correlation. Show that it is independent of change of scale and origin.
    (b) Prove that two independent variables are uncorrelated. By giving an example, show that the converse is not true. Explain the reason?